Food Marketing and Labelling

This book provides in-depth coverage of the interconnection among food marketing, labelling, and technology to enhance consumers', practitioners', and scholars' understanding of the modern food market. In 14 chapters, it explores the marketing, labelling, and technological aspects of the food industry. It briefly discusses crucial aspects of food marketing, from consumer preferences and branding strategies to regulation in food labelling, technological advancement, and sustainable practices. It allows readers to understand the holistic view of food marketing, labelling, and technology and their interrelationship. Throughout, it includes several case studies and practical examples.

Key Features

- Analyses consumer psychology behind food choice and marketing strategy
- Covers food labelling regulations and compliance and nutritional requirements extensively
- Includes sustainable and ethical dimensions of food marketing, labelling, and technology for guiding readers on responsible practices

Food Marketing and Labelling

A Practical Guide

Debasish Biswas
Debarun Chakraborty
Atanu Manna

CRC Press
Taylor & Francis Group
Boca Raton London New York

CRC Press is an imprint of the
Taylor & Francis Group, an **informa** business

First edition published 2026
by CRC Press
4 Park Square, Milton Park, Abingdon, Oxon, OX14 4RN

and by CRC Press
2385 NW Executive Center Drive, Suite 320, Boca Raton FL 33431

CRC Press is an imprint of Informa UK Limited

British Library Cataloguing-in-Publication Data
A catalogue record for this book is available from the British Library

ISBN: 978-1-032-75300-3 (hbk)
ISBN: 978-1-032-73676-1 (pbk)
ISBN: 978-1-003-47330-5 (ebk)

DOI: 10.1201/9781003473305

Typeset in Times
by SPi Technologies India Pvt Ltd (Straive)

Contents

Acknowledgment

The journey of creating *Food Marketing and Labelling: A Practical Guide* has been both inspiring and enriching. This book is the result of collaborative efforts, unwavering support, and valuable contributions from numerous individuals and organizations. We are deeply grateful to everyone who played a part in bringing this work to fruition.

First and foremost, we extend our deepest gratitude to the Director of the Indian Institute of Management Nagpur, Prof. Bhimaraya Metri, whose visionary leadership and steadfast support have been a source of inspiration throughout the development of this book.

We express our heartfelt gratitude to the Indian Institute of Management Nagpur and Vidyasagar University for providing the intellectual and academic environment that nurtured the development of this book. Their unwavering encouragement and resources have been instrumental in shaping the ideas and insights presented here.

We would like to extend our special thanks to the industry professionals and subject matter experts who shared their experiences and knowledge, enriching the content with real-world relevance. The inclusion of case studies and practical examples would not have been possible without their generous contributions.

To our families and loved ones, your unwavering patience, understanding, and support throughout this journey have been our greatest source of strength. Your encouragement has motivated us to persevere, and we dedicate this book to you with immense gratitude.

Finally, we acknowledge the tireless efforts of our publisher and editorial team, who ensured that this book met the highest standards of quality and clarity. Their meticulous attention to detail and guidance have been invaluable in presenting our ideas in the most effective manner.

We hope this book serves as a valuable resource for readers from all walks of life, sparking meaningful discussions and fostering innovations that shape the future of the food industry. Thank you all for being a part of this journey.

Sincerely
Prof. Debasish Biswas
Prof. Debarun Chakraborty
Mr. Atanu Manna

Authors

Dr. Debasish Biswas is presently serving as an Associate Professor & Head in the Department of Business Administration at Vidyasagar University, West Bengal. He joined Vidyasagar University on 2 April 2012. He was awarded M.Phil. in Commerce from Periyar University in 2008 and a PhD degree in Commerce by the University of North Bengal on 7 September 2011. His area of research was "Productivity & Industrial Relations." Dr. Biswas obtained an M.Com. in 2000 and a MBA (HR) in 2002 from Burdwan University. He also obtained a Diploma in Labour Laws in 2003 and a Post Graduate Diploma in Hospital Management in 2006 from Annamalai University. He has also obtained a Postgraduate Diploma in Marketing Management, a Postgraduate Diploma in Entrepreneurship, and a Postgraduate Diploma in Financial Management from Annamalai University in 2015, 2016, and 2017, respectively. Dr. Biswas also qualified for NET in Human Resource Management in 2008. He has more than 12 years of postgraduate teaching experience. His areas of specialization are Accounting & Finance, Human Resources, and Marketing. Dr. Biswas is a prolific researcher. He has presented 110 research papers in different management areas at national and international seminars and conferences. He has authored 86 research papers in national and international journals of repute and has acted as a resource person at different national and international seminars in different parts of India. He is a life member of the Indian Commerce Association and the All India Accounting Association. He has authored 27 books and contributed 23 chapters in various edited books. He received the Best Business Academic of the Year Award -2016 (Gold Medal) at the 69th All India Commerce Conference organized by Lucknow University, Lucknow. Dr. Biswas has developed study materials for the MBA course for the Distance Education wing of Tejpur Central University, Assam, study materials for the M.Com. course for the Distance Education wing of Vidyasagar University, West Bengal, and study materials for the BBA course for the Distance Education wing of Odisha State Open University, Sambalpur. Ten research scholars have been awarded PhD under his supervision. He is a reviewer of several renowned journals indexed in Scopus, Web of Science and has been invited to contribute resources to a variety of corporate training programmes.

Prof. Debarun Chakraborty has 17 years of work experience (15 years in academic institutions and 2 years in industry). At present, he is working as an Associate Professor in Marketing & Chairperson Online MBA at the Indian Institute of Management Nagpur. Prof. Chakraborty has completed a BE in Chemical and an MBA in Marketing and he has qualified for UGC NET in Management (twice), in June 2012 and December 2012. In 2017, he completed his PhD in Management (Marketing) in the field of Consumer Behaviour. Prof. Chakraborty has written six books on management, which have been published by a number reputed publishers. He has also written 38 research articles in ABDC A & 2 articles in ABDC A* category journals on consumer behaviour, technology adoption, sustainability and alike. He has also participated in more than 50 national and international seminars/

conferences. He has guided two Research Scholars (PhD) under the auspices of the Faculty of Management, Symbiosis International (Deemed University), Pune.

Prof. Chakraborty recently appeared in Elsevier's top 2% scientists list. A few papers are also currently under review at the ABS 4 FT 50 level. He has received many awards, including the "Amity Excellence Award for Best Faculty in the area of Marketing" during the International Business Summit and Research Conference Inbush Era World Summit 2020, organized by Amity University, Noida which took place between the 18th and the 20th February 2020; and the "INDAM Best Paper of the Track Award" (Track: Consumer Behaviour & Retail Management) at the 23rd International Conference in Sixth Biennial Conference of the Indian Academy of Management on Architecting Indian Management Scholarship in the Era of Disruption on 2–4 January 2020 organized by the Indian Institute of Management, Tiruchirappalli (IIM, Trichy); and the "Conference Best Paper Award – Faculty Category" (2nd Prize with US$750) in the ISDSI Global Conference on "Assessing the Determinants of Patient Experience: A Mixed-method approach," held on 27–29 December 2022, organized by the Jagdish Sheth School of Management (JAGSoM), Bengaluru; and many others.

His recent article on healthcare apps was covered and published by the *Times of India* (Nagpur Edition) on 18 March 2022 under the heading "City Scholar Researches Role of Healthcare Apps." One of his papers, "Marching Towards 'Heart Work': Connecting in New Ways to Thrive Amidst COVID-19 Crisis," which was published in *Conflict Resolution Quarterly*, received the award for the most downloaded paper published in that journal in the year 2022 from the publishers, Wiley. He has written four articles on COVID-19, all of which have been listed on the World Health Organization (WHO) database. His interview, "Heart Work: New Exigency, Will Have Strongest Effect on Staff," was published in the *Times of India* (Nagpur Edition) on 25 July 2021. Prof. Chakraborty is also on the editorial boards of various top journals published by Taylor & Francis, Emerald, Sage, Wiley & IGI Global.

Atanu Manna is an accomplished researcher, academic, and author with expertise in geography and human resource management. He holds an MSc in Geography and an MBA with a specialization in Human Resource Management, and he is currently pursuing a PhD at Vidyasagar University. His interdisciplinary research bridges environmental studies and management. He has published 12 research papers and contributed to one book and two book chapters in renowned academic platforms. He was recently honored with the Best Paper Award at the ISDSI 2024 Conference for his pioneering research on app-related studies. Additionally, he provides consultation to several organizations and individual clients, helping them develop strategies based on a data-driven approach. His expertise includes designing predictive models, generating insightful visualizations, and offering strategic guidance that enhances decision-making processes. He has also achieved significant academic distinctions, including qualifying for the UGC NET and JRF in both Geography and Human Resource Management. A life member of the Indian Commerce Association, he regularly presents his research at prestigious conferences, contributing valuable insights on financial inclusion, entrepreneurial activities, and environmental factors. Beyond his professional achievements, he enjoys reading and writing, continually exploring new ideas that enrich both academic discourse and societal progress.

Foreword

The food industry is currently undergoing an unprecedented transformation driven by shifting consumer preferences, technological advancements, and evolving regulatory frameworks. In this rapidly changing landscape, the interplay of food marketing, labelling, and technology is more critical than ever before in shaping the choices of consumers and defining the success of food businesses. This book, *Food Marketing and Labelling: A Practical Guide*, emerges as a timely and comprehensive exploration of these interwoven dimensions, offering invaluable insights to a wide spectrum of readers – from industry professionals to academic scholars and students.

Over the last few decades, the food industry has witnessed profound changes. What was once a straightforward process of producing and selling food has become a multifaceted domain involving innovative marketing strategies, complex labelling requirements, and cutting-edge technological advancements. Today's consumers demand transparency, sustainability, and quality, compelling businesses to adopt approaches that are not only profitable but also socially and environmentally responsible. This book delves into these trends, shedding light on the intricate mechanisms that govern modern food systems.

The chapters are meticulously structured to guide the reader to a holistic understanding of food marketing, labelling, and technology. The book begins with an overview of the food marketing landscape, tracing its historical evolution and highlighting the factors that have shaped it over time. It then transitions into an in-depth examination of consumer behaviour, offering valuable insights into how people make decisions about food choices, and how these decisions are influenced by marketing strategies and branding.

Subsequent chapters explore critical aspects such as food product development, advertising, and promotion techniques. Readers will gain a thorough understanding of how companies design and position products to appeal to target markets. The role of packaging design and its impact on consumer perception is another focus area, emphasizing the psychological and aesthetic considerations that drive successful branding. Furthermore, the book addresses pricing strategies and distribution channels, providing a detailed look at how food businesses manage logistics and reach their customers effectively.

An essential component of this book is its emphasis on food labelling and regulatory compliance. Modern consumers rely heavily on labels to make informed decisions, and the book unpacks the complexities of nutritional labelling, allergen declarations, and health claims. It also delves into the stringent regulations that govern these practices, offering guidance to industry professionals on navigating compliance while maintaining consumer trust.

Technology, a transformative force across industries, plays a pivotal role in food production, processing, and marketing. This book highlights the ways in which technological advancements have revolutionized food safety, quality assurance, and supply chain management. From smart labelling systems to innovative processing techniques, the integration of technology into the food sector is thoroughly examined, showcasing its potential to enhance efficiency and consumer satisfaction.

Sustainability and ethical considerations are central to the discourse in this book. As the world grapples with the challenges of climate change and resource scarcity, the food industry faces growing pressure to adopt sustainable practices. This book provides a roadmap for integrating sustainability into marketing and production processes, addressing issues such as waste reduction, ethical sourcing, and corporate social responsibility. By doing so, it equips readers with the knowledge to contribute meaningfully to a more sustainable and equitable food system.

Another unique feature of this book is its global perspective. While it provides in-depth coverage of trends and practices relevant to specific regions, such as the United States, Europe, and emerging markets like India and China, it also offers a broader view of the interconnectedness of food systems worldwide. Case studies and examples from diverse cultural and economic contexts enrich the narrative, making the book a valuable resource for readers with varied interests and professional backgrounds.

The interdisciplinary approach of this book is one of its greatest strengths. By integrating insights from marketing, behavioural science, technology, and regulatory studies, it provides a comprehensive understanding of the food industry's multifaceted nature. Readers will find themselves equipped with not only theoretical knowledge but also practical tools and strategies to address real-world challenges. The inclusion of case studies, practical examples, and forward-looking analyses ensures that the book is both informative and actionable.

This book is more than a resource; it is an invitation to engage deeply with the complexities of the modern food industry. It challenges readers to think critically about the ethical and social dimensions of their decisions, whether as consumers, professionals, or policymakers. In a world where food choices have far-reaching implications for health, the environment, and society, such reflection is both timely and necessary.

I am honoured to introduce this work, which I believe will serve as a cornerstone for understanding and innovating within the fields of food marketing, labelling, and technology. The authors' dedication to capturing the nuances of these topics is evident on every page, and their commitment to bridging theory and practice makes this book an invaluable asset for all who seek to navigate and contribute to the evolving food industry.

The authors of this book bring a wealth of expertise and diverse perspectives to the subject. Dr. Debasish Biswas, an Assistant Professor at Vidyasagar University, specializes in business administration with a focus on consumer behaviour and food marketing. Prof. Debarun Chakraborty, an Associate Professor at the Indian Institute of Management Nagpur, contributes his deep understanding of branding and strategic management in the food sector. Finally, Atanu Manna, a Research Scholar in Environmental Studies, offers unique insights into the sustainability and ethical dimensions of the industry. Together, their combined knowledge and interdisciplinary approach enrich this work, making it a vital contribution to the literature on food marketing, labelling, and technology.

As you embark on this journey through the pages of food marketing, labelling, and technology, I encourage you to approach it with curiosity and an open mind. Whether you are a seasoned professional looking to refine your strategies, a student eager to learn about the industry's dynamics, or a policymaker striving to shape better regulations, this book has something to offer. It is my sincere hope that it will inspire new ideas, foster meaningful dialogue, and pave the way for innovations that benefit both businesses and consumers.

Prof. Stefano Bresciani
Full Professor of Innovation Management and Digital Transformation
Director of the PhD in Business & Management
Director of the Master in Business Administration
Co-Editor of the *British Food Journal*
Department of Management
University of Turin, Italy

CHAPTER 1

Food Marketing, Labelling, and Technology

An Overview

LEARNING OBJECTIVES

By the completion of this chapter, you should be able to:

- Comprehend the key concepts of food marketing, labelling, and technology and understand their significance in the modern food industry.
- Appreciate the nature, scope, and importance of studying food marketing, labelling, and technology.
- Identify the links between food marketing, labelling, technology, and other subjects.
- Understand the essential role of food labelling in providing transparent and accurate information to consumers, ensuring regulatory compliance, and fostering consumer trust.
- Grasp the key challenges faced by food marketers, and identify opportunities for innovative solutions.
- Comprehend the future trends and innovations in food marketing in shaping the industry's trajectory.

1.1 INTRODUCTION

In the modern era, the food industry has become a complex and evolving landscape that significantly serves as the foundation of human survival and well-being. However, food marketing, labelling, and technology play a crucial role in shaping and altering the food industry with the help of influencing customer choice, increasing sales, safeguarding product safety, and driving innovation. Food marketing is the priority for any business success, while labelling emphasizes the product content and its usage pattern, and technology helps to preserve, process, and distribute safe food. Food marketing is crucial for all stakeholders involved with the food system such as food producers, professionals, consumers, and regulatory bodies. The

DOI: 10.1201/9781003473305-1

chapter focuses on the scope, nature, and importance of studying food marketing, labelling, and technology. The chapter also focuses on the challenges, opportunities, and future trends of the food industry that will help to delve deeper into the modern food landscape.

1.2 FOOD INDUSTRY AND THE GLOBAL ECONOMY

The food industry is a crucial sector which sustains the world's population. It consists of a broad spectrum of interconnected activities, from food processing and agricultural production to distribution, marketing, and retailing. The food industry's significance in the present era goes far beyond accomplishing basic human needs since it plays a vital role in the global economy and impacts various social, environmental, and economic factors. Understanding the significance of the food industry and how it interacts with food marketing, labelling, and technology is crucial to understand the broader applications of these ideas in the modern world. The significant roles of the food industry in the global economy are as follows:

Providing Food for Everyone on Earth: The food industry plays a pivotal role in producing, processing, and distributing food to feed the growing global population. With a population of over 8 billion, it is essential to ensure food availability and security. The industry's ability to effectively meet this demand is vital for addressing global hunger, malnutrition, and poverty.

Economic Importance: The industry plays a significant role in national and regional terms as well as on a global scale. The food industry significantly contributes to a country's GDP, especially in lower- and lower-middle-income countries. For example, it has been calculated that in Myanmar the food and beverage industry alone accounts for 30% of the nation's GDP (Figure 1.1).

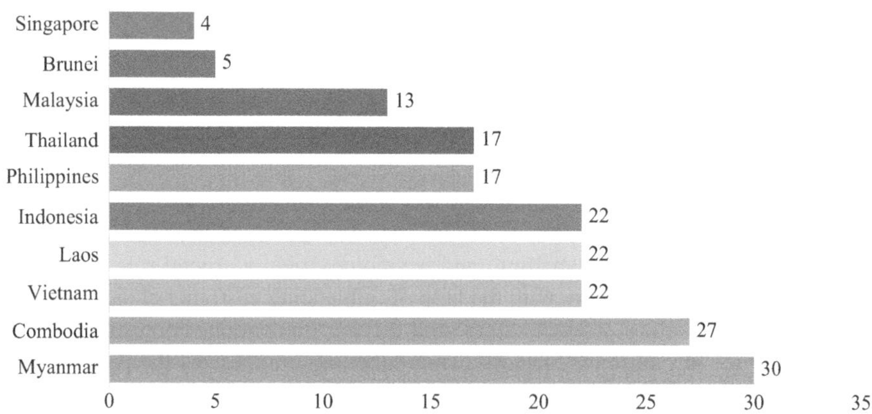

Figure 1.1 Food industry's contribution in the share of GDP among ASEAN countries.

Source: The Authors.

The industry also generates millions of employment opportunities, ultimately contributing to the national- and global-scale economy. For example, in the US the agriculture, food and allied industries provide 21.1 million jobs, nearly 10% of total employment (Figure 1.2).

Contribution to International Trade: The food industry plays a significant role in international trading as it helps to fulfil domestic demands and export surplus products. Exporting food products also helps to foster economic growth, employment generation, create new market opportunities, and improvise international bonding (Figure 1.3).

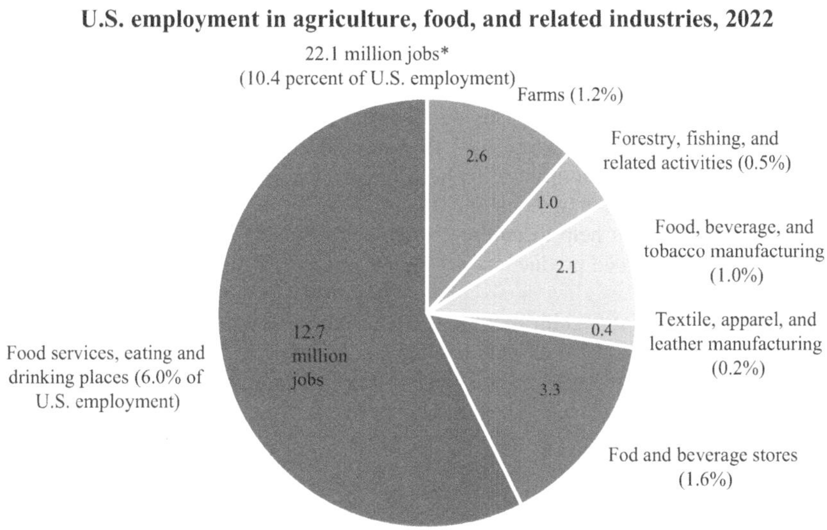

Figure 1.2 Employment in agriculture, food and related industries.

Source: The Authors.

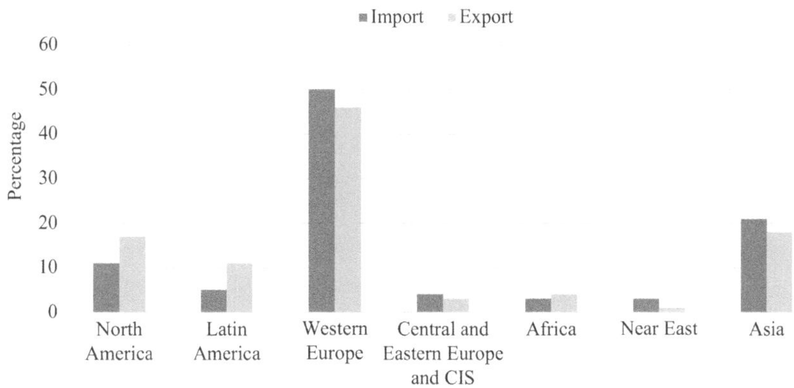

Figure 1.3 Import and export of food products in 1994.

Source: The Authors.

Impact on Rural and Agricultural Development: The food industry has a strong bond with agriculture and rural development. Farmers are the key players who supply raw foods or agricultural products for food processing and manufacturing ready-to-eat food. Food production helps their livelihood, and selling food products also becomes farmers' primary source of earning. The supply of food products significantly helps to improve the rural economy. Thus, empowering and supporting farmers can improve rural economic prosperity and agricultural sustainability.

Technological Advancements and Innovation: The food industry continuously adapts to technological advancements and innovations to improve production efficiency, food safety, and distribution. Technical integration in food marketing, labelling, and supply chain management drives innovation, enhances consumer experiences, and fosters sustainable practices.

Environmental Implications: The food industry plays a significant role in environmental sustainability (Figure 1.4). The stakeholders need to be concerned about the environmental footprint to reduce climate change and challenges relating to globalization. However, technological advancement, innovation, and sustainability awareness can help to minimize the environmental footprint by implementing sustainable food production, packaging, and distribution practices in the food industry.

Cultural and Social Significance: Food is, of course deeply intertwined with culture and society. The food industry influences dietary habits, culinary traditions, and cultural identities across the globe. Marketing professionals must understand

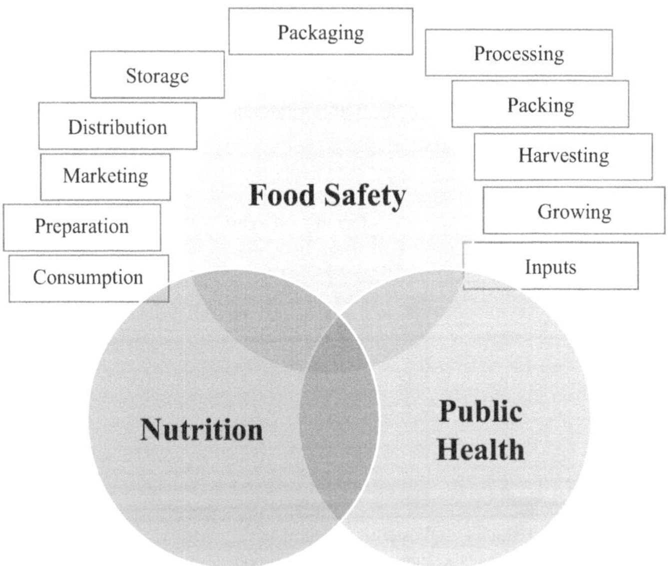

Figure 1.4 Food safety, nutrition, and public health relationship.

Source: The Authors.

socio-cultural aspects of different regions while designing effective food marketing campaigns and labelling strategies tailored to specific regions and communities to improve industry effectiveness.

Consumer Health and Nutrition: The food industry has a profound influence on consumer health and nutrition. Food marketing, labelling, and innovative technologies play a significant part in determining public health and government spending on the healthcare sector. The sector can improve public health and reduce healthcare costs by promoting healthy food choices through the provision of accurate nutrition labelling and responsible marketing.

It can be observed that the present-day food industry plays a significant role in the global economy, going well beyond meeting basic nutritional needs in the present era. At present, the food industry is an economic powerhouse, a key player in international trade, and a driving force behind innovation and technological advancements. However, the stakeholders need to understand the importance of the food industry and its interconnectedness with food marketing, labelling, and technology in order to achieve social, environmental, and economic development.

1.3 THE IMPORTANCE OF FOOD MARKETING, LABELLING, AND TECHNOLOGY

In the modern era, food marketing, labelling, and technology are essential components which have a significant effect on enterprises' sustainability and profitability. The interplay among marketing strategy, labelling techniques, and technological advancement plays a crucial role in shaping consumer preferences, building brand identities, ensuring food safety, fostering innovation, and meeting the demands of an evolving market landscape. It is therefore necessary to understand the importance of these concepts as they help to navigate the complexities of the global food supply chain.

Influencing Consumer Behaviour and Choices: Food marketing and labelling is a powerful tool, which has a paramount impact on consumer behaviour and purchase decisions. Effective marketing techniques can evoke feelings, build brand identification, and position food products in the market. The power of persuasion through advertising and promotion can increase sales and brand loyalty, ultimately impacting a company's growth. Thus, understanding the nuances of consumer psychology and preferences is crucial for professionals to adopt messages that resonate with their target audiences and foster enduring connections with consumers.

Empowering Consumers through Food Labelling: Food labelling plays a critical role in acknowledging and empowering consumers about product ingredients. However, transparent and accurate food labelling allows consumers to make decisions aligned with their health, dietary needs, and ethical values. When it comes to ingredients, nutritional value, allergens, and certifications, transparent and systematic labelling is essential in the present time when customers are looking for better and more sustainable food options. This transparency fosters trust and long-lasting relationships between consumers and brands, increasing demand for products that reflect consumer values.

Driving Innovation and Competitive Edge: In the modern era, technology plays a transformative role in the food industry, which drives continuous innovation and improved efficiencies. Integrating advanced technology in food marketing and supply chain management enables businesses to remain competitive, adjust to customer demand dynamics, and rationalize operations. Today, advanced techniques such as data analytics and AI-driven insights help to provide valuable consumer trends, which enables businesses to develop innovative products and niche marketing campaigns to satisfy the changing needs in the food market.

Ensuring Food Safety and Regulatory Compliance: In the present era, consumers are more concerned about food safety, while accurate food labelling and technology-driven traceability systems may helpful in maintaining compliance with stringent regulations. Proper labelling provides the consumers about the ingredients and other health-related information, while recent technology allows for immediate recalls and the efficient management of food safety incidents. Thus, proper food labelling is vital in the food industry as its over or under-statement may harm consumers' well-being and a brand's reputation.

Meeting the Demands of a Dynamic Market: The modern food industry is constantly changing due to the global market, shifting customer preferences, and the emergence of new trends. Food marketing, labelling, and technology allow businesses to operate with the agility to adapt and respond to these dynamic demands. Companies can customize their marketing strategy and product offerings to match the diverse needs of customers in various locations and demographics through efficient market research and data-driven insights.

From the above discussion, it can be noted that food marketing, labelling, and technology are paramount in navigating the multifaceted landscape of the modern food industry. They have a considerable influence on shaping consumer perceptions and purchasing decisions, as well as influencing business strategies, legal compliance, and overall rates of industry growth. However, understanding these concepts will help professionals to know their collective contribution to the vibrant food industry. Professionals can integrate these three concepts in order to develop sustainable practices, build consumer trust, and drive the food industry towards a more consumer-centric and forward-thinking future.

1.4 NATURE OF FOOD MARKETING, LABELLING, AND TECHNOLOGY

Food marketing, labelling, and technology is a multifaceted and dynamic subject which encompasses various interconnected aspects of the food industry. It involves strategically promoting and selling food products to consumers, regulating essential information on packaging of food products, and applying different scientific and engineering principles to improve food production, processing, and distribution. Several key attributes, detailed below, characterize the nature of this subject:

Interdisciplinary Approach: Food marketing, labelling, and technology include various disciplines such as marketing, regulatory compliance, consumer behaviour, nutrition, data analytics, and food science. The subject provides a comprehensive grasp of several disciplines to create effective strategies that appeal to consumers, adhere to regulations, and improve production techniques.

Consumer-Centric Focus: The core value of the subject is underlying consumer needs, preferences, and expectations. Effective food marketing and labelling strategies are centred on providing value to consumers, building trust, and engaging them both emotionally and rationally. Marketers and food producers need to consider the psychological influencers influencing consumer decision-making, including emotions, cultural, and social influences. However, technology-driven innovations in food processing and distribution also assist in shaping industry practices and supply chain management.

Evolving Trends and Innovations: The subject continuously evolves due to the dynamic nature of consumer behaviour, technological advancement, and emerging trends in the modern food industry. However, stakeholders must constantly be updated with emerging technologies to leverage new opportunities and address uncertainties.

Regulatory Landscape: The subject is connected to various national and international regulations and operates in a complex and dynamic regulatory landscape in order to ensuring consumer safety, transparency, and fair practices. Compliance with these regulations is essential for industry players to maintain credibility and trust among consumers. Governments and international agencies impose several rules to protect consumers from deceptive marketing, ensure precise and standardized labelling, and monitor the safety of technological advancements in the food industry. However, food producers, marketers, and technology developers must follow these regulations strictly to maintain market competitiveness in the modern and evolving food market.

Sustainability and Ethical Considerations: The subject faces growing concerns about sustainability, ethical considerations, and environmental impact in the modern food industry. Consumers are increasingly conscious of food products' ecological footprint, which drives businesses towards sustainable practices and transparent labelling for brand reputation. As food marketing and labelling have immense power in shaping consumer preferences, it is the responsibility of producers and business professionals to ensure sustainability, transparency, truthfulness, and social accountability.

Global Perspective: Food marketing, labelling, and technology in the modern food industry extend beyond national boundaries and adaptable to many geographic and socio-cultural settings. Food marketing, labelling, and technology deal with global supply chains, international marketing campaigns, and maintaining global labelling standards. However, brands must consider cultural differences, regional preferences, diverse regulatory frameworks, and local consumer behaviours when designing marketing strategies.

Technological Integration: Technology plays a crucial in the modern food industry. The subject shows the integration of technology in all aspects of the food industry, such as integrating blockchain techniques for supply chain management and integrating data analytics and artificial intelligence (AI) for effective personalized marketing and quick decision-making.

From the above discussion, it can be said that the nature of the subject of "Food Marketing, Labelling, and Technology" is dynamic. To understand the subject, scholars, food producers, and business professionals need a comprehensive understanding of marketing principles, consumer behaviour, technological innovations, and regulatory frameworks t. The subject is far-reaching, given that it significantly influences consumer preferences, industry practices, and the landscape of the global food market. However, business professionals can manage this complex scenario effectively and contribute to a more open, effective, and consumer-friendly food market through the adoption of adopting consumer-centric strategies, sustainability practices, and technology-driven innovations.

1.5 SCOPE OF FOOD MARKETING, LABELLING, AND TECHNOLOGY

The scope of this subject is broad and multifaceted from the perspective of the modern food industry. The subject holds significant importance in the technological advancement era as it discusses the convergence of marketing strategies, labelling practices, and the role of technology role in transforming the way of food production, distribution, and consumption. It aligns various aspects influencing the food industry's functioning and impact on consumers, businesses, and society. Thus, food marketing, labelling, and technology is the study of food marketing techniques and strategy, displaying product information and safe usage patterns, and applying technology in the selection, preservation, and processing of safe food. Food marketing applies marketing theories and principles to improve sales and mitigate consumer demand. Food labelling focuses on presenting information in a way that complies with regulatory frameworks while educating consumers on how to handle, prepare, and consume food safely. Food technology is primarily concerned with applying food science concepts and innovative technology for selecting, preserving, processing, packaging, distributing, and using safe food. The scope of this subject encompasses several key areas, including:

Food Product Development: Food marketing, labelling, and technology study provide a significant perception of the marketing strategy formulation, precise and law-oriented labelling technique, and innovative technology application for attracting consumers. Analyzing these elements helps food producers, labelling experts, and business professionals to build consumer-centric and appealing food product development based on identifying consumer needs and preferences.

Consumer Behaviour and Preferences: Studying food marketing, labelling, and technology helps professionals to understand how food marketing strategies and labelling practices influence consumer behaviour, perceptions, and purchasing decisions. The stakeholders need to understand the socio-cultural and geographical differences while formulating marketing strategies and labelling food products as the modern food industry deals with the global supply chain and dynamic regulatory framework.

Branding and Identity: Food marketing, labelling, and technology assist in building a brand by considering consumer preferences. Attractive packaging and branding assist in developing brand identity, attracting new consumers, and improving consumer loyalty. The stakeholders must provide attention to create a unique brand identity for their food products to establish consumer trust and loyalty.

Food Packaging and Labelling: Food packaging and labelling is a systematic procedure that helps in providing essential information to consumers, including nutrition facts, allergen declarations, and certification labels. Food labelling also offers several specific details, such as the ingredients present in the food, the percentage of those ingredients, the quantity (percentage or weight) of vitamins, nutrients, and minerals, any elements that can cause allergies, and environmental parameters under which some elements are produced such as organic or free range. This information helps consumers to determine whether or not they will buy the product.

Regulatory Compliance: The food industry professionals have a strong need to follow the standard regulations for producing, processing, and labelling food products. Most countries have their own food regulations and are trying to adjust their regulation within the broader global standards. For example, the European Union mandates 12 pieces of information that must be present on food labels: (i) product name, (ii) list of ingredients, (iii) any allergens present, (iv) quantitative ingredient declaration, (v) net quantity, (vi) durability dates, (vii) storage, (viii) instructions for use, (ix) business name and address, (x) country of origin, (xi) nutritional declaration, and (xii) alcoholic strength. However, global agencies such as the World Health Organization (WHO) and the Food and Agriculture Organization (FAO) have jointly worked with the Codex Alimentarius Commission to establish international standards for food labelling. Thus, business professionals need to understand industry standards and legal requirements before entering a target market.

Technological Advancements: In the modern era, the integration of technology in food marketing, labelling, and packaging significantly enhances the effectiveness of production. Today, several technologies, such as digital marketing, data analytics, machine learning and AI, e-commerce, and traceability systems are all paramount in production and decision-making. The technology also helps to achieve edible packaging with zero waste and sustainable packaging in order to reduce the environmental footprint.

Sustainable Practices: In the modern era, consumer awareness towards sustainability and technological advancement has increased their concerns about sustainability. The growing importance of sustainability and ethical considerations in food marketing and labelling drives professionals towards sustainable marketing, eco-friendly packaging, and socially responsible practices.

Online and Digital Marketing: In the digitalized era, consumers are now inclined towards purchasing food products online. The number of food delivery app users and internet users are both increasing rapidly, creating a favourable field for implementing digital marketing strategies in the modern food industry. For example, the number of food delivery app users in the U.S. was 36.4 million in 2019; by 2023, the figure had risen to 53.9 million (Statista, 2024). This trend allows professionals to explore the role of online platforms, social media, and influencer marketing in promoting food products and implementing them to engage consumers.

Global and Cultural Perspectives: The modern food industry operates globally and adapts to many geographic and socio-cultural settings. Incorporating global and cultural perspectives help marketers to build effective food marketing and labelling strategies for the successful penetration of diverse markets. The subject will enable us to recognize the cultural differences which allow brands to create relevant and appealing marketing campaigns for local people. Food producers, marketers, and tech developers need to take into account culturally sensitive approaches and international regulatory frameworks for establishing a robust global presence and fulfilling consumer demand.

Food Safety and Traceability: Food safety and traceability play crucial roles in food marketing, labelling, and technology study. Food safety refers to the procedures and safeguards put in place to stop contamination and food-borne diseases throughout the entire food production and distribution process. Hazard Analysis and Critical Control Points (HACCP), quality assurance, allergen management, testing and monitoring are among the crucial steps that help to ensure food safety. Technological innovations also play a pivotal role in addressing traceability concerns using bar codes, QR codes, and Radio Frequency identifiers (RFIDs). The management of food safety and tracking systems helps professionals to address challenges and uncover opportunities for market growth.

Health and Nutrition Trends: In the modern era, consumers are becoming increasingly health-conscious, and this development is creating a significant shift in consumer preferences towards the adoption of healthier and more nutritious food options. This aspect has had a considerable influence on the food marketing, labelling, and technology landscape. The stakeholders must adopt transparent labelling and effective marketing strategies to fulfil the demands of health-conscious consumers. However, this trend has shifted over time, and professionals need to be updated and responsive to changes in consumer preferences and regulatory requirements in order to improve customer loyalty and brand value.

Ethical and Social Responsibility: Ethical and social responsibility provides a strong foundation for building consumer trust, maintaining a positive brand image, and contributing to the creation of a sustainable and socially conscious food industry. Business professionals need to integrate ethical principles into marketing strategies and precise labelling information in the modern food industry to improve brand loyalty and consumers' well-being.

The subject will become increasingly important in the future, as the modern food industry is dynamic and constantly evolving. However, business professionals will be able to adapt to changing consumer preferences, technological advancements, and global trends for exploring the interconnectedness of food marketing, labelling, and technology. The subject will assist scholars, food producers, technology developers, and business professionals address the challenges and opportunities the modern food industry faces in accomplishing the demands of a diverse and conscious customer base.

1.6 LINKING WITH OTHER SUBJECTS

The subject of "Food Marketing, Labelling, and Technology" is an emerging subject within the modern food industry, and it is fundamentally linked with several other disciplines. Understanding these connections is vital for comprehending these elements' broader impact and implications in the food industry (Figure 1.5). Here are some of the key subjects that are linked to Food Marketing, Labelling, and Technology:

Figure 1.5 Subject links with food marketing, labelling, and technology.

Source: The Authors.

Nutrition and Health: The subject has a strong connection with nutrition and health. The marketing strategy, labelling, and presentation of food significantly impacts consumers' perceptions of product nutritional value and health benefits. Professionals must understand nutrition science in order to create informative and accurate food labels that help consumers to make healthier choices.

Consumer Psychology: Food marketing is heavily influenced by consumer psychology and behaviour. Food-related consumer behaviour, in turn, is affected by several factors extending beyond nutrition and sometimes, including social, cultural, and religious values. The study of consumer psychology and behaviour will help scholars and professionals to gain an understanding of the psychology behind different product labels and marketing strategies which attract consumers and improve sales and brand loyalty.

Food Safety and Quality Assurance: Food marketing, labelling, and technology are strongly connected to the topic of food safety and quality assurance, as food labelling and technology play vital roles in ensuring consumers have reassurances in this area. Food safety and quality assurance help to comprehend the behaviour of microorganisms in food, and this ultimately assists in efficient food quality control, food production, food safety, storage, and distribution. Business professionals need to understand the food safety protocols and quality control processes for maintaining consumer trust and loyalty in the food industry.

Environmental Sustainability: Food marketing, labelling, and technology are strongly related to environmental sustainability. This subject helps to comprehend sustainability perspectives in food production, processing, distribution, and quality control. Promoting sustainability with the help of eco-friendly food practices and packaging solutions and minimizing the industry's environmental effects can be accomplished by incorporating environmental-friendly methods into the modern food industry. This will help the stakeholders to promote sustainability and reduce environmental pollution.

Business and Marketing Management: Food marketing, labelling, and technology strongly connect with the subject of business and marketing management. Effective marketing strategy development is essential for food products to enhance sales. Thus, stakeholders need to understand the marketing principles necessary to create successful marketing strategies, brand recognition, and integrating technology to reach target audiences effectively. However, customer segmentation based on socio-cultural and religious dimensions is also crucial for targeting customers.

Data Analytics and AI: The integration of AI and data analytics in food marketing, labelling, and technology has brought about rapid changes in the modern food industry. Digitalization in marketing allows businesses to gather consumer insights, track consumer intention, and tailor marketing efforts accordingly in order to stay competitive in the digital era. Integrating AI and big data analytics help professionals investigate many customer reviews and feedback to improve food labelling and consumer attraction. The data-driven insights and automation technology help professionals to develop recipes, sensory analysis, personalized nutrition, and optimize flavours to enhance consumer happiness.

Food Laws and Standards: Food laws and standards play a significant role in marketing, labelling, and technology adoption. The food industry is subject to various national and international regulations regarding labelling, advertising, and marketing practices. The stakeholders must comply with these laws so as to avoid legal issues and maintain consumer trust.

Supply Chain Management: Supply chain management is closely linked with food marketing, labelling, and technology. An efficiently managed supply chain ensures product traceability, supports timely marketing efforts, promotes environmental sustainability, and enables the responsible utilization of technology. The integration of supply chain principles into food marketing, labelling, and technology is essential for branding, consumer loyalty, adherence to regulations, and achieving favourable outcomes for all parties involved in the food industry. Thus, business professionals need to understand supply chain logistics that help to safeguard food products' efficient movement from producers to consumers.

Economics: In the modern food industry, economic theories and principles provide a fundamental structure for understanding the interrelationship among food marketing, labelling, and technology. Economic principles help stakeholders to make effective decisions and shape the future of the food industry by looking at consumer behaviour, pricing dynamics, market structures, technological innovation, and sustainability. Business professionals need to understand the fundamental economic functions necessary for designing effective strategies and food labelling that address consumer demands, ensure ethical practices, and drive sustainable growth in the food market.

As can be seen from the above, "Food Marketing, Labelling, and Technology" is an evolving concept in the modern food industry due to evolving consumer preferences, technological advancement and innovations, and changes in food safety regulations. Different subjects contribute effectively and provide a holistic understanding of their impact on food marketing, labelling, and technology.

1.7 THE CONCEPT OF FOOD MARKETING, LABELLING, AND TECHNOLOGY

1.7.1 Concept of Food Marketing

According to Dr Philip Kotler (2009), "Marketing is a societal process by which individuals and groups obtain what they need and want through creating, offering, and freely exchanging products and services of value with others." However, food marketing mainly involves promoting, advertising, and selling food products and related services to consumers. It entails a variety of strategic initiatives aimed at raising awareness, stimulating demand, and influencing customer behaviour to boost sales and brand loyalty within the food sector. Food marketing also includes different advertising strategies for effective marketing, from traditional advertising methods to modern digital and social media campaigns. The principal focus of food marketing is to accomplish consumer demands, build brand recognition, and fulfil business goals while adhering to ethical and legal standards. Food marketing has several key aspects, which are detailed below:

- **Product Development**: Food marketing helps professionals to identify consumer needs and preferences to create consumer-centric food products that align with market demands and appeal to the target audience.
- **Pricing Strategy**: Food marketing assists in determining the optimal pricing strategy for food products by considering factors such as production costs, competitive pricing, and perceived value to attract customers while maintaining profitability.
- **Promotion and Advertising**: Promotion and advertising help to reach the audience effectively, utilizing diverse marketing channels.
- **Distribution and Retailing**: In modern food marketing, distribution channels and retailing play a crucial role in ensuring the timely delivery of food products to target markets and retail outlets that improves product availability to customers.
- **Branding and Packaging**: Branding and packaging is a crucial parameter in food marketing. It helps to create distinctive brand recognition and design attractive packaging to differentiate food products from competitors.
- **Consumer Research**: Consumer research helps professionals to gain insights into consumer behaviour, preferences, and emerging trends that ultimately assist in making effective decisions.
- **Ethical Considerations**: In the modern era, consumers are becoming aware of sustainability and ethical considerations. Thus, incorporating ethical and socially responsible practices in food marketing, such as sustainable sourcing, regulation-oriented labelling, and an effective marketing strategy, attract conscientious consumers and promote a positive brand image.

1.7.2 The Concept of Food Labelling

Food labelling primarily refers to providing transparent and regulation-guided essential information about a food product to make the consumer aware of the health benefit. Food labels include details about the product's nutritional content, ingredients, allergens, storage instructions, and other relevant information. The primary purpose of food labelling is to ensure transparency, safety, and accuracy in the representation of food products, empowering consumers to select products that align with their dietary needs, preferences, and values. The US Food & Drug Administration (FDA) provides particular instructions on conventional food that food labelling is mainly required for most prepared foods, such as drinks, desserts, canned and frozen foods, cereals, bread, and snacks, while nutrition labelling is very much essential for the raw product (such as fruits and vegetables). Labelling guidelines are not standardized across all countries; most countries have recently been amended to match international standards (Figure 1.6). In this regard, the WHO and the FAO jointly work with Codex Alimentarius Commission to establish global standards for food labelling (Figure 1.7). Food labelling has several common key elements, as follows:

- **Nutritional Information**: Food labels display the nutritional facts of the product, such as calories, macronutrients (such as carbohydrates, proteins, and fats), and micronutrients (such as vitamins and minerals).

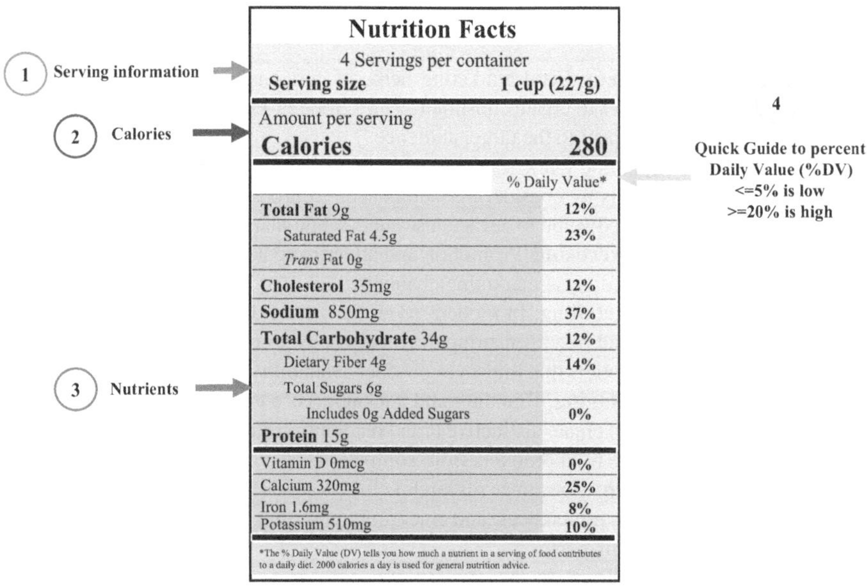

Figure 1.6 Sample label for frozen lasagna.

Source: The Authors.

Figure 1.7 Five types of information required on a food label.

Source: Callahan et al., 2020.

- **Ingredient List**: The ingredient list outlines all components of the food product in descending order of their predominance by weight. It helps consumers to identify specific ingredients and potential allergens present in the product.
- **Allergen Warnings**: Food labels highlight the presence of common allergens, such as peanuts, soy, wheat, dairy, and eggs, to protect consumers with allergies from unintentional consumption. For example, England enacted Natasha's Law in October 2021 to guide allergen labelling following the death of Natasha Ednan-Laperouse from allergens (sesame seeds) which had been present in a food item but not disclosed at the time of sale. There are 14 principal allergen contents that need to be highlighted if it is present in the food, and those are as follows: (i) Celery and celeriac; (ii) Cereals containing gluten; (iii) Crustaceans (lobster and prawns); (iv) Eggs; (v) Fish; (vi) Lupin; (vii) Molluscs (mussels and squid); (viii) Milk; (ix) Mustard; (x) Nuts (pecans, cashews, almonds, walnuts, and hazelnuts); (xi) Peanuts; (xii) Sesame seeds; (xiii) Soybeans; (xiv) Sulphur dioxide and sulphites at levels above 10 mg per kg or 10 ml per litre.

- **Serving Size**: It is very important to integrate the number of servings and size in the food labelling. The information about standard serving number and size (cups or pieces) helps consumers to understand the quantity of food for which the nutritional content is provided. It also enables them to better manage portion control and dietary intake.

- **Storage and Preparation Instructions**: Food labels may include guidelines on storage conditions (e.g., refrigeration), appropriate preparation methods, and cooking instructions for optimal product consumption (Figure 1.8).
- **Date Marking**: Date labelling is crucial in a food labelling process; it is an offence to sell the product after its "use by" dates. Date labels indicate the freshness and safety of the product, such as "best before" or "use by" dates, to guide consumers on product shelf life and quality (Figure 1.9).

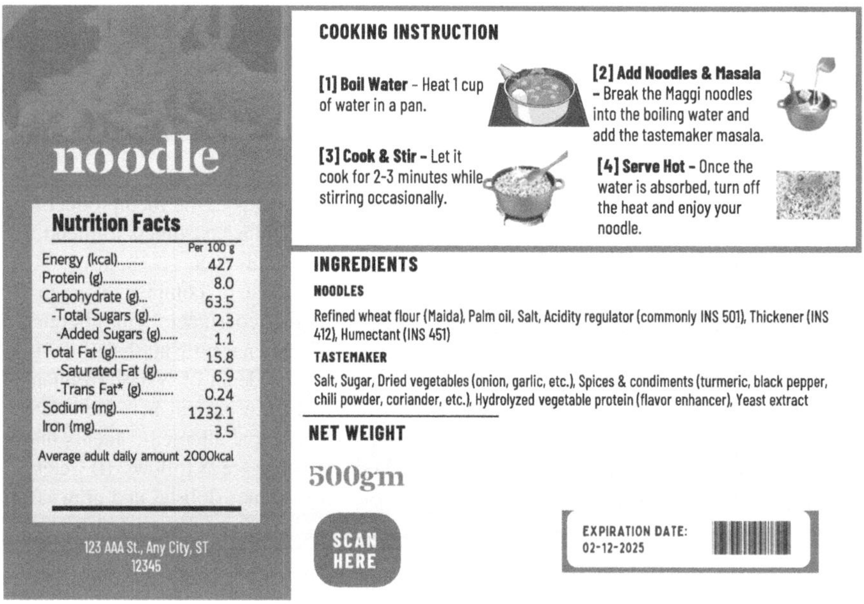

Figure 1.8 Sample cooking instructions in food labelling.

Source: Prusa & Gilbert, 2021 The Author.

Figure 1.9 Date labelling.

Source: The Authors.

- **Country of Origin**: The food product's origin country labelling is crucial for the manufacturer as it makes people aware of the source of the product and production location (Figure 1.10).
- **Certification and Claims**: Food labelling also includes the certification. Food labels may display certifications, such as ISO 9001 QMS and 22000 FSMS, to notify consumers about specific attributes or production practices. However, regulatory standards need to follow for claiming health and nutrient content in the food product.

1.7.3 Concept of Food Technology

The concept of food technology primarily encompasses the application of scientific and technological knowledge in food production systems, from raw material sourcing to food consumption. The Oxford Reference dictionary defines food technology as "the application of science and technology to the treatment, processing, preservation, and distribution of foods" (Oxford University Press, n.d.). It uses technology and advanced food processing, preservation, packaging, distribution, and marketing techniques to improve food safety, quality, efficiency, and sustainability.

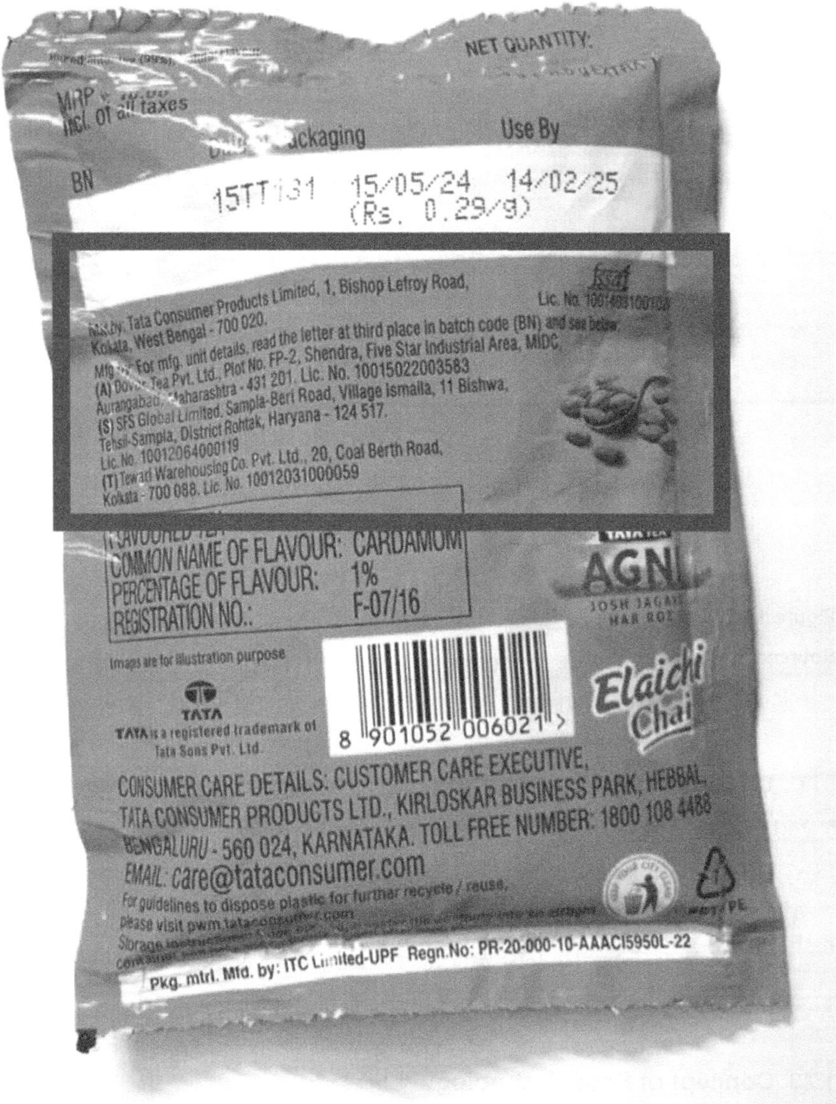

Figure 1.10 Sample of country of origin and certification.

Source: The Authors.

Food technology also significantly enhances food safety, extends shelf life, reduces food waste, and meets the demands of a rapidly changing market. Technology-driven insights through data analytics, AI, and digital marketing also allow businesses to understand consumer preferences. The critical elements of food technology applied in the modern food industry are listed below:

- **Food Processing**: Food technology significantly helps to transform raw food into processed food through various methods, such as heating, cooling, drying, fermenting, and preserving. These processes enhance the flavour, texture, and nutritional value of food products.
- **Food Preservation**: Food technology includes techniques to preserve food items to prevent spoilage and extend the product's life cycle. Several techniques can be applied to preserve food quality and safety, such as canning, freezing, pasteurization, and the use of a vacuum packing system.
- **Food Packaging**: Food technology is essential in designing and producing packaging materials that safeguard food products from taint, spoilage, and physical harm while giving consumers the information they need via labelling.
- **Food Safety and Quality Assurance**: Food technology conducts stringent testing, inspections, and surveillance at different stages of the production process to ensure that food products meet safety and quality standards.
- **Sensory Evaluation**: The use of sensory evaluation techniques (such as magnetic position sensor and humidity monitoring sensor) in the food industry allows for better flavour and a more satisfying overall shopping experience for consumers by evaluating the taste, texture, aroma, and visual appeal of food products (Figure 1.11).

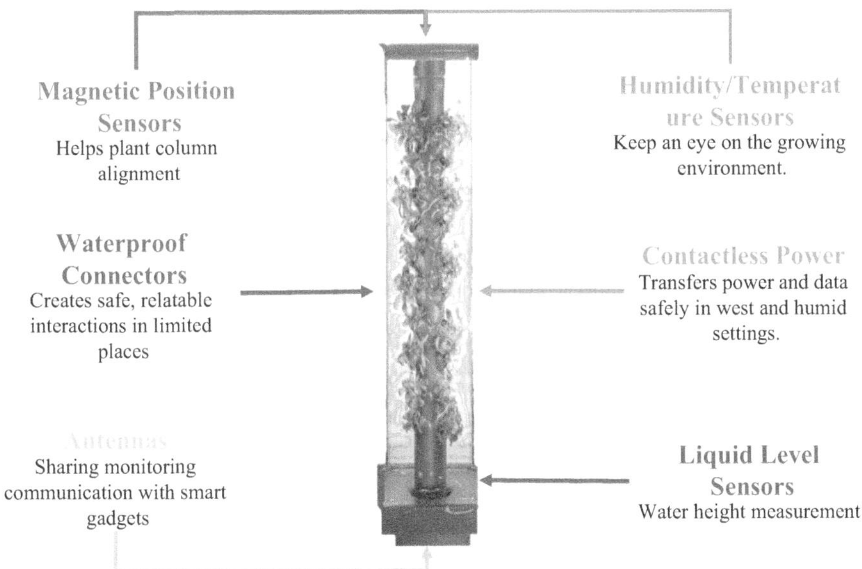

Figure 1.11 Different types of sensors used for monitoring ideal crop growth.

Source: The Authors.

- **Genetic Modification**: Food technology includes genetic engineering to create Genetically Modified Organisms (GMOs) that may have enhanced nutritional value, pest resistance, or other desirable traits for fulfilling consumer demand.

1.8 KEY CHALLENGES AND OPPORTUNITIES IN FOOD MARKETING, LABELLING, AND TECHNOLOGY

Food marketing, labelling, and technology offer business professionals many opportunities to meet consumer demand and drive growth. However, the modern food industry faces several challenges that need innovative solutions.

The key challenges are as follows:

- **Consumer Trust and Transparency**: In the digital era, branding and customer loyalty maintenance have become one of the most challenging tasks for the business professional due to intensive competition and increasing consumer awareness. However, false advertising, inaccurate labelling, and underhand sales technique can significantly harm a brand's reputation.
- **Health and Nutrition Concerns**: Food marketers must promote goods that consider consumers' changing dietary preferences while addressing worries about excess sugar, salt, unhealthy fats, and other nutrition-related issues as consumers become increasingly health-conscious.
- **Regulatory Compliance**: The regulations differ from country to country, creating a significant challenge for the stakeholders to manage the global food market. However, food producers, business professionals, and technology developers need always to consider the target market, focusing on international standard to meet major consumer demand.
- **Technological Integration**: In the modern era, the technological revolution creates several opportunities for the food market, helps stakeholders to enhance production and quality, and gains consumer insights. However, rapid technological change and integration into the organization is a primary challenge for food businesses, particularly for smaller or more traditional ones.
- **Environmental Sustainability**: Nowadays, consumers are aware of and inclined towards sustainable products. Consumers are looking towards socially responsible brands demonstrating sustainability efforts and reducing their ecological footprint.

However, effective food marketing, labelling, and technology provide several opportunities for businesses to enhance profitability:

- **Personalized Marketing**: The advancement in data analytics and AI integration helps professionals to gain insight into consumer preferences. It creates an opportunity for personalized marketing where brands can tailor content, offers, and product recommendations to individual consumer preferences.
- **Clean and Transparent Labelling**: Poor labelling may cause several issues, such as contamination and legal penalties. Emphasizing accurate labelling helps the organization to improve communication with consumers as well as improving customer loyalty and brand value. Today, AI helps label checks by considering standard labelling information and regulations.

- **Health and Wellness Messaging**: An effective food marketing strategy can take advantage of consumers' growing interest in health and wellness by highlighting the nutritional benefits of products, accommodating special dietary requirements, and disseminating informative health content.
- **E-Commerce and Direct-to-Consumer (DTC) Sales**: The development of e-commerce platforms and increasing Direct-To-Consumer (DTC) sales in the modern era presents opportunities for the food industry to reach a larger audience, cut costs associated with distribution, and strengthen direct relationships with customers.
- **Efficient Supply Chain Management**: Food storage and distribution is one of the most challenging tasks for businesses. The advancement in food technology provides the opportunity to manage the supply chain efficiently using blockchain technology. AI and other advanced technology help to manage stock, establish effective distribution channels, and track packages that ultimately enhance trust and safety.
- **Sustainability Initiatives**: Incorporating sustainability initiatives, such as eco-friendly packaging, ethical sourcing, and carbon footprint reduction, can resonate with environmentally conscious consumers and differentiate brands in the market.
- **Social Media Influencers**: Social media marketing plays a crucial role in amplifying sales of a product. Collaborating with social media influencers provides an excellent opportunity to reach a targeted audience quickly.

1.9 FUTURE TRENDS AND INNOVATIONS IN FOOD MARKETING, LABELLING, AND TECHNOLOGY

Future trends mainly suggest a development or change towards something new with progressing time. Directions and innovations in food marketing, labelling, and technology will likely shape the modern food industry in the coming years. The technological advancement and dynamics in consumer preferences lead the market towards an evolution in their nature. These changes will influence how food products are marketed, labelled, and delivered to consumers. Among the key future trends and innovations in this field are the following:

Personalization and Customization: At present, big data analytics, AI, and machine learning help to develop effective marketing strategies based on consumer feedback analysis. AI and data analytics developments will make it possible to personalize marketing strategies by adapting packaging, advertising, and food product recommendations to the dietary requirements and preferences of specific consumers. Among the other emerging technology trends are Augmented Reality (AR) and Virtual Reality (VR). These technology experiences will be integrated into food marketing campaigns, allowing consumers to virtually explore products, see nutritional information, and visualize how recipes are prepared.

Smart Packaging and Labelling: The packaging industry is currently undergoing an evolution, with competition in the market driving companies to seek competitive advantages. However, future packaging will become more interactive and informative through the incorporation of technologies like QR codes, NFC (Near Field Communication), and RFIDs to provide real-time product information, provenance details, and storage instructions. Businesses are now focusing on the consumer's sustainability awareness and trying to develop sustainable packaging techniques for food products. In future, sustainable packaging options, including biodegradable materials and edible coatings, will emerge as eco-friendly alternatives to traditional packaging in reducing global environmental concerns and attracting consumers.

Sustainability-Focused Marketing: Due to environmental awareness, brands will increasingly emphasize their commitment to sustainability through marketing campaigns highlighting eco-friendly practices, responsible sourcing, and a reduction in the carbon footprint. Food marketing will also focus on promoting plant-based and alternative protein products to cater to the rising demand for sustainable and ethical food choices.

Supply Chain Transparency: Supply chain management plays a crucial role in the modern food industry. There are several emerging technologies which are currently being used and whose employment is likely to be increased in future, such as blockchain technology, AI and the Internet of Things (IoT), 5G Networks, cloud-based products, robotics and automation to enhance supply chain transparency. Their combined use will allow consumers to trace the journey of food products from farm to fork, ensuring authenticity and ethical practices.

Health and Wellness Apps: Food marketing will increasingly rely on mobile-based apps that provide individualized health and nutrition advice, recipe recommendations, and tracking tools to support healthier lifestyles.

Influencer Marketing Evolution: Business professionals will deal with media influencers to reach target audiences as micro-influencers will gain in popularity in the future. Micro-influencers' popularity will help to develop influencer marketing as they provide more genuine and focused connections with interested audiences.

Voice Commerce: Voice commerce for food products will be made possible through the use of voice-activated devices such as smart speakers, ultimately enabling consumers to place hands-free grocery and recipe orders.

Nutrigenomics and Personalized Nutrition: Future food labelling techniques will incorporate more information about the nutritional impact of products based on individuals' genetic profiles and provide personalized nutrition insights.

Digital Payment Integration: Food retailers will incorporate digital payment options and loyalty programmes into their marketing strategies in response to the rise of cashless transactions.

Food Delivery and Drone Technology: Future food delivery services will use drone technology to make deliveries more quickly and effectively.

The future trends are dynamic for the modern food industry as consumer preferences and technological advancements evolve rapidly. Business professionals must prioritize the aforementioned trends and upcoming technologies to integrate them into their operations in order to gain a sustained competitive advantage in the food industry.

1.10 CHAPTER SUMMARY

The chapter mainly highlights the fundamental concepts of food marketing, labelling, and technology and their importance within the modern food industry.

The chapter begins by showing the food industry's contribution to the global economy. The relations highlight that the food industry plays a crucial role in the global economy, in areas such as international trading, contribution to GDP, feeding the world population, and public health. The chapter then discusses the importance of studying food marketing, labelling, and technology, as well as the nature and scope of the subject. Understanding these concepts is crucial not only for professionals in the food industry but also for consumers. Transparency in food labelling empowers consumers to make informed choices aligned with their health, dietary preferences, and ethical values. Knowledge of marketing strategies helps consumers to recognize and critically evaluate the influences shaping their purchasing decisions.

The chapter also delves deeper into the intricacies of food marketing, labelling, and technology, exploring the challenges businesses face and the opportunities for growth and sustainability. However, scholars are professionals can able to enhance their marketing practices by understanding these concepts navigate regulatory landscapes, and create meaningful connections with consumers. Comprehending these concepts will enable consumers to make conscious choices, supporting brands that align with their values and contributing to a more transparent and consumer-centric food industry. The chapter also discusses the new trends in the food industry, and it will assist marketers and scholars in shaping the future of food marketing and taking sustained competitive advantage.

CASE STUDY

Natasha's Law in England – Raising the Bar for Food Labelling Transparency

In October 2021, England enacted a pivotal food safety legislation known as Natasha's Law. This legislative change was a response to the tragic death of Natasha Ednan-Laperouse, a teenager who suffered an allergic reaction to undeclared sesame seeds in a pre-packaged sandwich. The law highlights the importance of transparent food labelling and sheds light on the role of technology and marketing in ensuring consumer safety.

Background

Natasha Ednan-Laperouse suffered from severe food allergies. In 2016, she purchased a sandwich from a Pret a Manger store at Heathrow Airport. Unbeknownst to her, the sandwich contained sesame, to which she was allergic. Her lack of awareness stemmed from the omission of allergen information from the package. Unfortunately, Natasha consumed the sandwich and subsequently suffered a fatal allergic reaction. Her death propelled a campaign led by her parents for stricter food labelling laws, culminating in the UK government's decision to revamp labelling regulations for prepackaged foods for direct sale.

Legislative Impact

Natasha's Law mandates businesses to provide a complete ingredients list with clear allergen labelling on all prepackaged food intended for direct sale. This includes various products, such as ready meals, ice cream, canned or packaged goods, restaurant and take-away meals, catering and event food, supermarket and grocery store offerings, as well as baked goods such as cakes, pastries, and bread. The goal of this law is to give consumers with allergies the information they need to make safe food choices.

Case Analysis

***Pre-Law Scenario*:** Before Natasha's Law, Non-Pre-Packed for Direct Sale (PPDS) food could be sold without a complete ingredients list, leading to potential risks for allergy sufferers.

***Technology's Role*:** Advances in labelling technology have made it easier for food businesses to include detailed allergen information on even the smallest packaging. Quick Response (QR) codes and digital labels allow consumers to access a wealth of information via smartphones.

***Marketing Considerations*:** With the enactment of Natasha's Law, marketing strategies had to adapt. Clear, transparent labelling became a selling point for health-conscious consumers and those with dietary restrictions.

***Benefits Realized*:** The implementation of the law led to increased consumer trust and safety. It also drove innovation in food labelling technologies and marketing approaches focused on transparency and consumer well-being.

Natasha's Law exemplifies the critical intersection of food marketing, labelling, and technology. It stands as evidence of how policy can influence and improve food safety standards, leading to greater consumer empowerment and industry responsibility. Through technological advancements and adaptive marketing, food producers and retailers in England can now offer safer products that cater to the needs of individuals with food allergies, thereby setting a new standard for food labelling worldwide.

Discussion Questions

1. How did Natasha's Law change the food labelling scenario in England and Wales, and what implications does this have for food marketing strategies?
2. What technological advancements have facilitated the implementation of Natasha's Law, and how might future technologies further impact food labelling?

3. Consider the impact of Natasha's Law on small food businesses versus large corporations. What are the challenges and benefits for each?
4. How can food businesses leverage the requirements of Natasha's Law in their marketing campaigns to differentiate themselves in the market?
5. What lessons can other countries learn from the implementation of Natasha's Law regarding food safety and labelling?
6. Discuss the potential role of smartphone apps and digital platforms in enhancing the effectiveness of Natasha's Law. How might they contribute to an increase in consumer awareness and safety?

BIBLIOGRAPHY

Belonax, J. J. (1997). *Food Marketing*. Simon & Schuster. https://www.google.co.in/books/edition/Food_Marketing/f0hEAAAAYAAJ?hl=en

Blanchfield, J. R. (2000). *Food labelling*. Woodhead Publishing.

Blume, A. (n.d.). *Food Date Labeling Act Would Tackle Confusion Over Safety and Quality Issues*. Food Safety Magazine. Retrieved August 12, 2023, from https://www.food-safety.com/articles/8579-food-date-labeling-act-would-tackle-confusion-over-safety-and-quality-issues

Callahan, A., Leonard, H., & Powell, T. (2020). *Nutrition: Science And Everyday Application (OER)*. Lane Community College. https://openwa.pressbooks.pub/nutritionscience2/chapter/1e-understanding-food-labels/

FAO. (1997). *The state of food and agriculture*. fao.org. Retrieved November 5, 2024, from https://www.fao.org/4/w5800e/w5800e.pdf

FDA. (2022). *How to Understand and Use the Nutrition Facts Label*. FDA. https://www.fda.gov/food/new-nutrition-facts-label/how-understand-and-use-nutrition-facts-label

Gunter, B. (2016). *Food Advertising*. Springer International Publishing. https://doi.org/10.1007/978-3-319-40706-7

Kokkoris, M. D., & Stavrova, O. (2021). Meaning of food and consumer eating behaviors. *Food Quality and Preference*, *94*, 104343. https://doi.org/10.1016/j.foodqual.2021.104343

Kotler, P. (2009). *Marketing Management*. India: Pearson Prentice Hall.

Mahmoud, B. (n.d.). *Food Safety Approaches to Achieve Better Public Health Outcomes in Developing Countries*. Food Safety Magazine. Retrieved August 12, 2023, from https://www.food-safety.com/articles/6357-food-safety-approaches-to-achieve-better-public-health-outcomes-in-developing-countries

Meixner, O., Riefler, P., & Schanes, K. (2021). Sustainable Consumer Behavior and Food Marketing. In *Sustainability* (Vol. 13, Issue 22). MDPI. https://doi.org/10.3390/books978-3-0365-2595-2

Oxford University Press. (n.d.). *Food technology*. Oxford Reference. https://doi.org/10.1093/oi/authority.20110803095827187

Petty, L. (2017). *Food Allergen Labels: Free Download & Advice*. High Speed Training. https://www.highspeedtraining.co.uk/hub/free-food-allergen-printable-labels/

Prusa, K., & Gilbert, K. (2021). *Food Product Development Lab Manual (Open Educa)*. Iowa State University Digital Press. https://iastate.pressbooks.pub/foodproductdevelopment/chapter/designing-a-package-label/

Rahman, M. S. (2007). *Handbook of food preservation*. CRC Press.

Rajendram, S. (2023). Translanguaging as an agentive pedagogy for multilingual learners: affordances and constraints. *International Journal of Multilingualism*, *20*(2), 595–622. https://doi.org/10.1080/14790718.2021.1898619

Statista. (2024, February 29). *Amount of smartphone food delivery app users in the U.S. 2019–2023*. https://www.statista.com/statistics/1189404/food-delivery-app-users-us/

von Kameke, L. (2023). *Share of GDP contributed by the food and beverage industry in ASEAN in 2018, by country*. Statista. https://www.statista.com/statistics/1124811/asean-gdp-contribution-food-and-beverage-industry-by-country/

CHAPTER 2

Evolution of Food Marketing

LEARNING OBJECTIVES

By the end of this chapter, you should be able to:

- Understand the historical phases of food marketing.
- Identify and describe the key characteristics of the different phases of food marketing (fragmentation, unification, segmentation, and digitalization and globalization phases).
- Explain how innovations have shaped food marketing strategies over time.
- Examine the role of branding and advertising in the evolution of food marketing.
- Understand the evolving pattern of food labelling and regulations and its effect on consumer trust, transparency, and marketing strategies.
- Understand how changes in consumer preferences have impacted marketing trends.
- Understand the role of globalization and digitalization in the modern era of food marketing.

2.1 INTRODUCTION

The food marketing industry has evolved significantly over the years, driven by technological advancements, shifting consumer preferences, and cultural changes, from a localized trade to a sophisticated global industry. Food marketing primarily includes the promotion, distribution, and sale of food products where modes and strategies change based on communication technology, new production, and consumption patterns. This chapter will attempt to explore the historical development of food marketing. The whole chapter is broken down into four main phases: (a) fragmentation phase – the markets were decentralized and regional; (b) unification phase – the time of mass production and a rise in national brands;

DOI: 10.1201/9781003473305-2

(c) segmentation phase –the marketing strategies directed towards specific consumer groups; and (d) globalization and digitalization phase – focused on healthy products and the rise of online platforms. The chapter will help to understand how the food industry evolved over time through these prominent phases. Along with highlighting the unique challenges and developments that each era brought forth, this chapter will also highlight the evolution of digital marketing strategies from the early days of fragmented local markets to the present. However, food marketing is a continuously evolving process, reflecting not only technology breakthroughs but also changing customer preferences and a growing focus on health and sustainability.

2.2 THE EVOLUTION OF FOOD MARKETING

The development of food marketing can be divided into four main categories: (a) the fragmentation phase; (b) the unification phase; (c) the segmentation phase; and (d) the digital and globalization phase. A brief detail of each phase is discussed below.

2.2.1 The Fragmentation Phase (Before 1870–1880)

The fragmentation phase is considered the initial stage of food marketing. The primary characteristics of this phase were decentralization – local markets were common, and branding and marketing initiatives were not very strong. Throughout this time, food production and marketing were primarily local and primitive in nature because they took place before the late 19th-century industrial revolutions. This phase can be seen as the initial stage in this process, laying the foundation for several procedures which led to the development of the modern food industry.

2.2.1.1 Local Markets and Direct Producer-Consumer Relationships

During this stage, food markets were highly localized, with the majority of food being acquired directly from small-scale farmers and producers. Most food was produced and consumed within a relatively small geographic area, frequently within the same town or region, due to the lack of convenient transportation options. Strong interpersonal bonds were upheld between producers and consumers, who regularly interacted in face-to-face encounters at neighbourhood markets. The absence of complex marketing techniques was attributable to direct contact, which cultivated a sense of loyalty and trust. Local markets were important hubs for the trade of food, especially in rural areas where most people lived. Farmers and traders brought fresh vegetables, dairy products, and meats for sale at these weekly markets. Due to the lack of technological advancements in food preservation, consumers purchase relatively little fresh food, making food marketing both urgent and crucial.

2.2.1.1.1 The Role of Trust in Food Marketing

The basis of food marketing was trust, since there was no branding, packaging, or mass media present at this time. Customers purchased from farmers who they either knew personally or who had a good reputation in the community, and informal connections rather than official laws frequently ensured the quality of the food. A farmer's business might succeed or fail based on their track record of growing and delivering fresh, high-quality food. Word-of-mouth referrals were crucial, and local trust networks were crucial in maintaining customer confidence during a period when food safety was a major worry.

2.2.1.1.2 Challenges of a Fragmented Market

Producers who wanted to grow business outside of their immediate geographic region faced difficulties due to the local character of marketplaces. Food, particularly perishables like dairy, meat, and fresh fruit, could not move very far without the modern-day transportation infrastructure. As a result, there was minimal regional variation in food availability, quality, or cost, and the food systems were extremely localized and fragmented. There were many different regional food systems that functioned independently of one another since it was practically hard to market food over long distances.

2.2.1.2 Early Technological Innovations in Food Preservation

Despite the limitations of this era, food marketing did begin to change gradually as a result of some noteworthy early advancements in food preservation. Food preservation was vital for life prior to the Industrial Revolution, especially in areas with severe winters or short growing seasons. To increase the shelf life of food, methods including pickling, smoking, drying, fermenting, and salting were used frequently. Even though these techniques were archaic in comparison to modern food technology, they were early attempts to preserve food for later use, enabling producers to promote their products beyond daily and instant sales.

2.2.1.2.1 The Role of Salting and Smoking

During this period, salting was a popular preservation technique. By removing moisture from food and stopping the formation of bacteria, salt not only improves flavour but also serves as a preservative. In order to increase the length of time that meat and fish could be sold in marketplaces further away from the point of production, these foodstuffs were frequently salted or smoked. In order to give meat and fish a unique flavour and further preserve it for transportation, smoking uses heat and smoke to dry them out (Figure 2.1). The reach of producers was widened by the frequent foot or horse-drawn cart transportation of salted or smoked meats to nearby cities and marketplaces.

Figure 2.1 Smoking used for food processing and preservation.

Source: The Authors.

2.2.1.2.2 The Advent of Canning and Its Marketing Implications

The invention of canning was the most important technical advancement made in the latter part of the fragmentation era (Figure 2.2). Food could now be preserved for an extended period of time in sealed containers thanks to the invention of canning in 1809 by the French chef Nicolas Appert. Canning was first developed to give soldiers access to preserved food during times of war, but it quickly spread to civilian markets. Despite its early lack of widespread adoption, canning was a breakthrough in food preservation that would eventually revolutionize the way in which food was marketed and sold. In this phase, canning was still in its infancy stage. However, it set the groundwork for later marketing campaigns that prioritized food safety, convenience, and shelf life. Canning not only made food items last longer, but also created the opportunity for suppliers to preserve and transport food for long distances, both of which led to a uniform market in the future. Furthermore, it also led to the popularization of packaging, which would become essential in allowing food marketers to distinguish their products from those of competitors.

2.2.1.3 The Lack of Formal Branding and Marketing

The concept of formal branding did not exist during this phase. The majority of food was sold without a label, with only a few food products having identifiable brands or logos. Food was mainly bought by consumers due to its visual appeal, freshness, and level of trust in the producer. Thus, trust in the producer played a significant role in consumer choices, and marketing efforts were predominantly transactional, with minimal emphasis on product differentiation among producers.

Figure 2.2 Food preservation through the canning process.

Source: The Authors.

2.2.1.3.1 Marketing by Reputation

In the early phase of marketing, branding was not as formalized as it is today. Instead, farmers and producers relied on their reputation within the community in order to attract customers. Maintaining a good reputation was crucial for long-term success, as news of low-quality goods or unfair practices spread rapidly in small and close-knit communities. In this sense, the producer's reputation served as an early form of branding, providing consumers with a basis to assess the quality and reliability of the product.

2.2.1.3.2 Visual and Sensory Appeal

During this period, another form of marketing was based on the visual and sensory attractiveness of the food itself. At local markets, merchants openly exhibited their food items, allowing consumers to visually assess, smell, and handle the products prior to making a purchase. This hands-on direct sensory experience replaced branding since consumers could immediately judge the freshness and quality of food. Thus, farmers show their foods in a good manner to attract consumers, so one will sometimes see attractive bright fruits and shiny vegetables shine above the rest.

2.2.1.3.3 Word-of-Mouth Marketing

In the past, word-of-mouth marketing held significant importance within the food industry, particularly in regions lacking widespread mass media or advertising. Producers heavily depended on satisfied customers to propagate information about their offerings. This form of organic marketing played a pivotal role in smaller communities where consumer loyalty was cultivated gradually through favourable interactions and robust interpersonal connections. The establishment of trust between consumers and producers fostered an informal web of endorsements, representing a fundamental marketing strategy.

2.2.1.4 The Emergence of Early Food Labelling Practices

The fragmentation period was characterized by the absence of formalized food labelling. Most of the products were sold in an unpackaged form, and the concept of labelling food by informing the people about ingredients, origin, or the quantity of nutritional value was still centuries away. However, the rudimentary labelling practice began to surface with the increasing demand to distinguish products, especially for large markets.

2.2.1.4.1 Handwritten Labels and Packaging

In urban markets, food was often sold in bulk or semi-packaged forms. Producers may use handwriting labels to indicate what was sold, or its origin (Figure 2.3). For example,

Figure 2.3 Sample handwritten packaging for honey.

Source: The Authors.

cheese or butter might be labeled with the farm or region it came from, giving consumers a basic sense of where their food was produced. Similarly, in preserved and cured goods, the label may indicate only that the product is "smoked" or "salted." The primary purpose of this early type of labelling was to aid consumers in making selections about what to buy by giving them basic information. But it also signalled the start of a trend that would eventually lead to comprehensive labelling regulations.

2.2.1.4.2 The Role of Packaging in Emerging Markets

As stated earlier, packaging was not common during the fragmentation phase. However, certain preserved foods, like canned goods, did need to be packaged. Canning was one of the first technologies to require packaging because it involved sealing food in glass jars or metal containers. Early canned goods often had simple labels showing what it contained along with (sometimes) the producer's name or location. This was the first time that packaging and labelling became important for marketing food products, laying the foundation for the future of branded, packaged goods.

2.2.1.5 The Transition to a Unified Market

The conclusion of the fragmentation phase, around the end of the late 19th century, marked the beginning of a more unified and centralized food market. Breakthroughs in transportation, including the expansion of railroads and steamships, enabled food products to be transported over longer distances, opening up the potential for national and international markets. In addition, the advancement of refrigeration technology assists the companies to extend the product's shelf life and also provides an opportunity to improve the food distribution channel over longer distances.

Modern food marketing was observed in the latter part of the 19th century. However, branding, advertisement, and labelling concepts were founded during this phase. As time went on, technological advancements and shifts in social mores caused the food system to become more complex and interdependent. This complicated and structured marketing system leads to the new phase of marketing, which is known as the unification stage.

KEY FEATURES OF THIS PHASE INCLUDE

- The development of local and regional markets
- Small-scale, fragmented markets focused on local and regional distribution
- Limited branding, with emphasis on personal relationships between producers and consumers
- Direct sales from farms to local markets or consumers
- Production and consumption
- Lack of mass production; consumers purchased goods produced close to their location
- Little or no advertising or promotion and packaging facilities.

2.2.2 The Unification Phase (1880–1950)

The unification phase became significant as it included the expansion of the transportation network, the advent of popular national brands, and rapid advancements in mass production. During this period, fragmented markets became national in extent as a result of industrialization and technological advancements. The advertising strategy adopted by the companies plays a crucial role in differentiating themselves from the competitors and leads to the standardization of branding, labelling, and marketing practices.

2.2.2.1 Mass Production and the Birth of National Markets

The Industrial Revolution played a significant role in advancing and integrating food markets. Technological advancement in the fields of food processing, machinery, and agriculture assists companies in manufacturing huge amounts of finished products in a short period. The huge production in a shorter period leads to the reduction of food prices in the food market.

2.2.2.1.1 Mechanization and the Rise of Processed Foods

The mechanization in food production includes lower labour intensity, factory-based food processing, higher productivity, and faster expansion in the market area. Several processed foods first introduced in this period, such as cereals, canned products, and snacks, have become mainstays in consumers' diets. The mass production of processed foods facilitated a more extensive geographic distribution of their sales. As a result of the transition in food production, food marketing underwent a significant transformation, enabling producers to provide standardized, homogeneous food to large urban populations.

2.2.2.1.2 Transportation and the Expansion of Distribution Networks

The advent of steamships and railroads revolutionized food delivery during this period. Food could now be moved over great distances without significantly spoilage, which made it easier to build both domestic and international food markets. The advent of refrigerated railcars and ships made it possible to transport perishable goods such as fruit, dairy, and meat, which increased the variety of food products that could be sold and commercialized over great distances. Food enterprises were able to penetrate new markets and extend their operations beyond local or regional limits for this enlarged distribution network.

2.2.2.2 Branding: Building Consumer Trust

Companies realized the value of branding as a means of fostering customer loyalty and trust as food production and delivery grew increasingly mechanized. As the gap between producers and consumers grew, the interpersonal connections that defined the fragmentation stage started to weaken. Companies responded by using branding as a means of setting themselves apart from the competition and trying to develop an image of dependability and consistency.

2.2.2.2.1 The Emergence of Iconic Food Brands

A number of recognizable food brands that are still well-known now first emerged during this unification period. Packaged goods such as morning cereals, canned soups, and sauces were first produced by companies like Kellogg's, Heinz, and Campbell's. To gain customers' confidence and to increase familiarity, they employed branding and advertising. These trademarks evolved into status symbols for excellence and dependability, giving customers the peace of mind that, even if they were no longer doing direct business with the manufacturer, they were nevertheless buying a reliable, safe product. For example, Kellogg's development of cornflakes in the late 19th century drastically transformed the morning food industry. The company addressed the growing consumer interest in health and nutrition with a marketing effort emphasizing the health benefits of a light, readily digestible breakfast. Kellogg's employed a highly effective branding approach to establish a strong connection between their product and the notion of healthy living.

2.2.2.2.2 Advertising and the Role of Mass Media

As national companies became more well-known, they had to use mass media ads more and more. Because of this, radio, magazines, and newspapers became the main locations for food-based advertisements. This allows companies to promote their goods all over the country to a large audience. Food ads from the beginning focused on the items' quality, ease of use, and safety in order to meet customers' needs for trustworthy food choices. In the late 1800s and early 1900s, advertising agencies grew in popularity. These agencies helped food companies to highlight their product's uniqueness through advertising. It was normal practice at the time to promote products using mascots, jingles, and themes. Such aspects allowed businesses to make ads that people would remember and that would make generate strong feelings. Some products also began to make use of famous people, with celebrities' names being used to sell food products.

2.2.2.3 The Evolution of Food Labelling

Industrialization changed the surrounding environment, leading to the quick growth and evolution of the idea of food labelling. However, there was also some concern raised about the possibility of food consumption and adulteration as the food purchases moved away from direct ties between consumers and producers. This concern led towards the establishment of food safety regulations and uniform rules for labelling.

2.2.2.3.1 *Early Labelling Laws*

In the late 1800s and early 1900s, national governments introduced different rules and regulations concerning food labelling to uniform the labelling practice and provide essential information about the product to the customer. In 1906, for example, the USA enacted the Pure Food and Drug Act, which marked a big step forward in terms of controlling food labels. The rule was enacted to safeguard the consumer from contaminated and dangerous foods. The act also ensures that the provided information on the product label should be accurate; otherwise, penalties would be imposed. The law set the groundwork for more advancements in the decades that followed. This regulation eventually gave rise to the creation of the Food and Drug Administration (FDA), which has the responsibility to ensure food safety and proper label application.

2.2.2.3.2 *Nutritional Information and Health Claims*

Food companies started to provide nutritional facilities in the food products in order to attract health and nutrition-conscious customers. Similarly, several breakfast foods were advertised as containing extra vitamins and minerals. However, food companies would sometimes claim and advertise their products as nutritious without any scientific proof. This concern became a major challenge for the government and consumers to trust the company blindly.

2.2.2.4 *Technological Advancements in Food Production and Marketing*

The unification phase was known for its significant advancement in technology concerning food production, preservation, and distribution. These advancements also revolutionized food marketing and consumption behaviour.

2.2.2.4.1 *The Invention of Refrigeration*

The advent and widespread use of refrigeration in food preservation is a significant turning point for the food industry. Before the widespread use of refrigeration, perishable products such as meat, dairy, and fruit could only be transported over small distances. However, long-distance perishable commodity transportation underwent a significant change with the widespread use of railcars and refrigerated ships in the late 19th century. This invention greatly expanded the global accessibility of food items, hence increasing consumer access worldwide. Moreover, refrigeration technology facilitated the development of entirely new food categories, particularly frozen foods. Companies such as Birds Eye played a pioneering role in the development of the industry by introducing frozen vegetables and meats as convenient, time-saving alternatives to fresh produce. The widespread adoption of frozen foods in modern kitchens was driven by marketing efforts which emphasized their convenience, quality, and ability to retain essential nutrients.

2.2.2.4.2 The Canning Industry and Processed Foods

During the fragmentation phase, the canning industry emerged and subsequently underwent rapid expansion in the unification phase. Technological advancements in canning facilitated the production of a diverse range of canned products, including fruits, vegetables, soups, and meats. Canned foods were marketed as convenient and long-lasting. This is also a practical solution for busy homes, attracting customers who wanted to keep their pantries well-stocked. Campaigns for canned products frequently highlighted their safety and convenience. For example, companies like Campbell's utilized marketing strategies to promote their canned soups as quick, easy-to-prepare meals that preserved the flavours and nutrients typically associated with home-cooked food. The canning processed foods allows marketers to extend their geographical areas for selling the food products as it extends the food's shelf life.

2.2.2.4.3 Mechanization and the Standardization of Food Products

The application of mechanization in food manufacturing leading food standardization and uniformity which enhanced the effectiveness of marketing strategies. Customers can expect consistent quality in every package from the same brand due to the standardized food processing system. This standardization of product quality assists the brands to maintain brand loyalty. Furthermore, the homogeneity of food products allowed businesses to conduct nationwide marketing campaigns, guaranteeing a constant taste regardless of the product's location.

2.2.2.5 Advertising and Consumer Culture

Advertising became an important strategy for food companies throughout this phase as the companies attempted to stand out in a competitive market. Early in the 20th century, after World War I, consumer culture really took off, giving food marketers a fantastic chance to meet customer needs for novelty, prestige, and convenience.

2.2.2.5.1 The Role of Radio in Food Marketing

The introduction of radio in the 1920s revolutionized the advertising industry and allowed companies to reach a wider audience through it. Radio commercials featured famous jingles, catchy slogans, and the endorsements of radio celebrities. Thus, as a result of their intensive radio advertising campaigns, brands like Wheaties and Cheerios became well known at the local and national levels.

2.2.2.5.2 The Expansion of Supermarkets and Consumer Choice

The expansion of supermarkets during the 1930s and 1940s significantly influenced the consolidation of the food market. By offering a diverse array of food products in a single location, supermarkets revolutionized consumer choices. This diversification prompted food companies to innovate their marketing strategies to secure shelf space and capture consumer attention. Consequently, packaging design became a pivotal element of food marketing, as companies utilized vibrant colours, bold typography, and compelling imagery to differentiate their products within the competitive landscape of supermarket shelves. For instance, in 1934, the renowned baseball player Lou Gehrig became the first athlete to be featured on the back of the Wheaties box.

KEY FEATURES OF THIS PHASE INCLUDE

- Mass production and distribution
- Introduction of mass production techniques due to the Industrial Revolution
- National markets began to form as transportation improved (railroads, automobiles)
- Brands started to emerge (e.g., packaged goods, standardized products)
- The rise of department stores and supermarkets as centralized points for food distribution
- Early food advertising
- Emergence of food advertisements in newspapers and magazines
- Growth of branded goods and consumer trust in large, consistent brands
- Development of food labelling regulations

2.2.3 The Segmentation Phase (1950–2000)

The segmentation phase is the most significant and innovative period in the history of food marketing, which started after World War II and continued until 2000. During this phase, companies were more concentrated on segment-based food products to target specific demography to meet customer demand. Food companies now have unprecedented opportunities to engage with broad audiences because of the growth of mass media, particularly television, and the revolutionary changes in technology that have transformed food production and delivery techniques. The increasing health-conscious customer provides another opportunity for companies to lead to the creation of creative marketing campaigns that highlight aspects such as convenience, sustainability, and nutritional value.

2.2.3.1 Post-War Consumerism and the Growth of Processed Foods

From the aftermath of World War II, the global food consumption pattern and economic growth undertook many significant changes, particularly in the western regions. Along with that, the growth and transformation of the food industry also changed with the pace of socio-economic status. At this stage, processed foods have become a huge part of contemporary diets because they are convenient and affordable.

2.2.3.1.1 The Role of Television in Food Advertising

The advent of television in the 1950s marks another revolutionary step in the history of food marketing. Companies soon began to realize how vital television was for advertisement purposes to convey the product message to a wider audience through visual media. TV ads, therefore, became a constituent part of brands' marketing strategies. Ads were planned taking care that they highlighted the simplicity of usage, family values, and fast living lifestyles so that they found resonance with the steadily increasing numbers of middle-class consumers. The biggest companies, such as Coca-Cola, Kraft, and General Mills, made heavy use of television advertising to anchor their product into becoming household brands. To make the value of their product memorable, they frequently used catch characters, slogans, and jingles in their TV commercials. Some of the best-known commercial figures include Tony the Tiger, Ronald McDonald, and the Pillsbury Doughboy, all of which became much more than simply become popular commercials. They soon became household names for the companies that represented them and allowed them to get in touch with their audiences on a more emotional level. These catchphrases and mascots then created a sense of comfort and trust with the products they represented while also assisting in brand awareness.

2.2.3.1.2 The Rise of Convenience Foods

People are now literally introduced to a pre-war and inter-war period when frozen dinners, quick soups, and ready-to-eat snacks become prominent in the post-war period. Companies like Birds Eye and Swanson became prominent during this time, popularizing the frozen and canned food industries by offering convenient meal options for busy families. The group of customers was able to popularize these products to the housewives, who were the primary decision-makers with regard to family food purchases. The products were favoured by advertisements that have thus portrayed them as a time-saving convenience for women in terms of their varied responsibilities and modern lifestyles.

2.2.3.2 The Emergence of Market Segmentation

In this phase, food firms realized that they needed to target certain consumer segments with tailored marketing strategies in order to deal with diversified and complex food markets. This was the beginning of the market segmentation strategy. In this strategy, markets were split into numerous divisions based on psychographics, geography, demography, and behaviour and separated the market into multiple submarkets. Companies started to build tailored strategies for the specific target group by taking into account the unique demands and preferences of these groups.

2.2.3.2.1 Demographic Segmentation

The demographic segment primarily includes different demographic characteristics such as age, gender, income, and family size which were recognized as essential factors in the development of food marketing strategies. The baby boomer cohort, born between 1946 and 1964, emerged as a significant target market for food manufacturers upon reaching maturity and establishing their own families. Through television and printed media, the companies focused on these groups in order to promote instant-ready foods such as breakfast cereals, soft drinks, and snack items. However, the companies also focused on the younger age customer segment and developed several items with particular focus on these segments. Popular companies such as Cap'n Crunch, Froot Loops, and Kool-Aid used vivid packaging and captivating characters to attract youngsters, while concurrently highlighting convenience and nutritional benefits to appeal to parents.

2.2.3.2.2 Psychographic Segmentation

Psychographic segmentation divides markets according to customers' lifestyles, attitudes, and values. This segmentation-based company's strategy emphasizes the emotional and psychological needs of the customer instead of selling goods based on price and convenience. For example, the companies sell healthy and organic foods to customers who are health-conscious. Some brands, Stonyfield and Whole Foods, for example, had discovered that consumers were willing to pay more for products that align with their values. Luxury foods like gourmet candies, fine wines, and handmade cheeses were sold to wealthy people who wanted to eat foods that were of high quality and less available. These items are marketed in magazines and TV programmes. Thus, the establishment of a marketing strategy based on consumer preference becomes essential to gain brand loyalty from consumers.

2.2.3.3 Health and Nutrition Trends

A growing consciousness regarding health and nutritional foods was noted among customers during the last stage of the segmentation phase. Food firms started to modify their marketing techniques accordingly to deal with the growing consciousness.

2.2.3.3.1 The Rise of Health-Conscious Consumers

The market observed an upsurge in healthy food products during the 1960s and 1970s as people became more aware of obesity, heart disease, and high cholesterol levels. Companies started to develop innovative products that are healthy to fulfil these rising demands. Some popular brands such as Weight Watchers and Special K took advantage by making low-calorie meal options available and selling them in specific ways to those who are more interested in losing weight. For example, brands such as Diet Coke and Splenda focused mostly on diet food and sugar-free products, which became highly popular during subsequent decades such as the 1980s and the 1990s. The products were especially created for the consumers who wanted to cut down on sugar but refused to give up the taste of sweets. Food labels, for example, today use words such as "low-fat," "sugar-free," and "all-natural" that offer further proof that the public are making increasing demands for healthy diet foods.

2.2.3.3.2 *Nutritional Labelling and Regulation*

Food labelling regulations have indeed changed much today since most people are becoming interested in health and nutrition. In 1990, the US government implemented the Nutrition Labeling and Education Act (NLEA), requiring food manufacturers to put a lot of nutritional information about their products on their packages based on standard guidelines. These details included the amount of portions of their foods, calorie counts, fat types, and vitamin levels. This law was intended to enlighten people on the facts needed to make wise decisions about what they eat. The act was introduced in order to lead transparency in marketing. Instead of this, companies used this as a chance to disclose their level of nutrition and made the act an instrument for marketing. They put information on labels that are easy to see about foods that are low in calories, high in fibre, or high in nutrients. Several companies redesigned their product to meet the guidelines and attract the targeted segment. They also added low-fat, low-sodium, and organic versions of their current lines of products to attract the health-conscious customer segment.

2.2.3.4 *Technological Innovations in Food Production and Marketing*

In this stage, the technological advancement made tremendous progress in food industry concerning its marketing and manufacturing procedure. Advancements in the processing, delivery, and food safety help businesses create new products and improve the whole experience.

2.2.3.4.1 *The Growth of Convenience Foods and Packaging Innovation*

The 1960s and 1970s showed a significant rise in the accessibility of convenience foods, including microwavable meals, prepackaged snacks, and quick drinks. This development is mostly due to increased female labour-force involvement in the industries and it resulted in a higher demand for fast lunch choices. For this reason, companies started to develop new products that were faster and easier to prepare. In the 1970s, the invention of microwave ovens changed the way people cooked at home. As a result, food companies started to make versions of popular meals that could be cooked using this new technology. The sales of convenient foods were expanded as the packaging technology became advanced. Businesses developed pop-top cans, single-serving portions, and resealable packaging, all of which were intended to make it simpler for customers to store and use food goods. Technological developments in packaging have significantly assisted in increasing the use of frozen foods. Several companies, such as Stouffer's and Lean Cuisine, became familiar for providing frozen meals that are quick to cook and consume.

2.2.3.4.2 *The Impact of Digital Marketing*

With the widespread adoption and rapid growth of the internet in the 1990s and 2000s, digital marketing came to the fore, disrupting more traditional food marketing strategies. At this stage, the e-commerce sites assisted consumers with buying foodstuffs without moving out of the comfort zones and helped food manufacturers reach more people through virtual channels. Today, food companies increasingly recognize that websites, social media, and email marketing are all key weapons in actively engaging consumers, expanding offers of promotion, and building brand preference. Social media influencers and bloggers, in particular with regard to niche industries like organic, vegan, and gluten-free food products, have shaken up the food marketing industry. Companies started to engage with influencers to reach the specific target groups of people they wanted to focus on and to build a real connection with their followers.

KEY FEATURES

- Post-war consumerism and marketing
- The growth of consumer culture in the post-World War II era
- The advent of television and mass media advertising (jingles, brand mascots, celebrity endorsements)
- The expansion of supermarkets, offering more choices and a wider variety of products
- Rise of segmentation and targeted marketing
- Companies began to segment markets based on demographics, lifestyles, and preferences
- Introduction of product differentiation, niche marketing, and premium branding
- Health-conscious trends emerged (low-fat, low-sugar, organic)

2.2.4 The Digital and Globalization Phase (2000–Present)

The phase of food marketing during the digital and globalization phase represents rapid technical advancement and the growing interconnectedness of global marketplaces. This phase is characterized by the omnipresence of digital platforms, the expansion of e-commerce and also the increasing role of data-based approaches in marketing. Global companies started providing both global as well as local food preferences to attract customers and fulfill their needs. Customer preferences are also time-poor and, at this stage, customers are focusing much on convenience, sustainability health, and transparent products. The meaning of food marketing has increased with time to target the needs of a tech-savvy as well as socially conscious consumer market.

2.2.4.1 The Rise of Digital Marketing and E-Commerce

Digital marketing and e-commerce platforms gain utmost prominence in this phase of digitization. There are various robust communication channels that exist to connect with consumers, such as online advertising, social media campaigns, influencer marketing, email marketing, and other digital channels.

2.2.4.1.1 Online Shopping and E-Commerce Platforms

The most significant penetration of e-commerce platforms gradually changed the way people could buy foods. Major online channels such as Amazon, Instacart, and Walmart have allowed customers to purchase groceries and more through direct orders from their homes. Food manufacturers have come to realize this and focused on establishing a solid e-presence while simplifying their product portfolios to make them suitable for internet sales. E-commerce assists consumers in delivering a much more tailored purchase experience with data analytics and enhances the overall buying experience. Companies can formulate customized plans according to the purchase intentions and proper recommendation of the products, offers, and discounts.

2.2.4.1.2 Social Media and Influencer Marketing

The technological advancement of social media made it feasible for food manufacturers to have real-time communication with their customers. Moreover, other famous social media sites such as Instagram, Facebook, YouTube, and TikTok are more relevant for food manufacturers to highlight their products. All this is achieved through collaborations with influential people, attractive materials, and viral marketing. Influencer marketing practices become a vital tool of marketing. People consider the recommendations given by the influencers who are particularly in the field of food and health industries. Food companies attempted to connect with the influencers to reach out to their targeted customers to build trust. Some of the most popular brands hire social media influencers to market their product to health-conscious consumers. The low-calorie ice cream brand Halo Top is one such brand. Another example is Blue Apron, a company that offers meal kits. Influencers put more vigour and sales behind these products by posting pictures, reviews, and recipes made from these products in their respective channels.

2.2.4.2 Data-Driven Marketing and Personalization

The emergence of a data-driven strategy has substantially changed the way marketing was practiced in those decades. Companies can now collect and analyze data about their customers using data-driven modelling to develop very specific marketing efforts based on demographics, lives, and habits of groups of people.

2.2.4.2.1 Big Data and Consumer Insights

Data analytics is a mathematical technique that is used by organizations in order to carefully study the preferences of customers, analyze trends in the market, and evaluate the efficiency of marketing efforts. Companies can precisely segment their targeted customers and prepare products that meet the changing tastes of their customers with the help of data analytics. Based on the data-driven modelling, the companies can evaluate the pattern of consumer preferences and the changes in preferences over time. For example, Nestle and PepsiCo use consumer data to track consumer behaviour. Thus, companies can take advantage of the market by looking at what their customers want and coming up with new lines of products or making current ones better. In this way, they can meet their buyers' expectations.

2.2.4.2.2 Personalization and Customization

The consumer's preferences changed over time and presently they are looking for tailored experiences from the companies. In response, organizations in the food sector have begun using data to provide personalized meal plans, product recommendations, and promotional offers. There are several companies, such as HelloFresh and Blue Apron (mentioned above), which have started providing meal kits based on the data analytics outcomes to create customized meal plans that cater to the nutritional choices (vegan, keto, or gluten-free diets) of their customers. This customization technique helps the companies to meet the customer needs to improve overall buying experiences.

2.2.4.3 Globalization and the Rise of International Food Markets

In the globalization phase, companies can easily start expanding their businesses in cross-border countries. Food companies started to operate internationally and reach customers in a range of cultural and socioeconomic contexts. As a result, this tendency has led to the launch of novel tastes, components, and goods into international marketplaces.

2.2.4.3.1 The Globalization of Food Brands

Some companies, such as McDonald's, Starbucks, and Coca-Cola, become global brands as their presence is observed in various countries. These brands have skillfully customized their marketing strategies to align with local preferences while maintaining a unified global brand identity. For instance, McDonald's have changed their menus based on local preferences. For example, they provide vegetarian selections, such as the McAloo Tikki Burger in India, while in Japan, the company features offerings such as Teriyaki Burgers. The success of the companies in this phase relied heavily on the ability of the company to maintain a strong global presence while adapting to local markets.

2.2.4.3.2 The Influence of Global Food Trends

The global food trend accelerated and modified after globalization. The worldwide customers are showing interest towards superfoods, plant-based diets, and sustainable eating. This extensive acceptance is the result of social media, food blogs, and food industry leaders. The food companies also adopted these new trends and started to offer healthy products. Adopting this trend, some companies, such as Beyond Meat and Impossible Foods, become global leaders in providing plant-based items. By aggressively marketing their plant-based meat alternatives on social media, these firms have spearheaded the sustainable food movement.

2.2.4.4 Health, Sustainability, and Transparency in Food Marketing

The growing health-conscious customer forced companies to modify their marketing methods to prioritize sustainability, health, and transparency. However, the customers also focus on the firms that are attempting to be transparent about their sourcing methods, ingredients, and environmental impact.

2.2.4.4.1 The Rise of Health and Wellness Products

The health and wellness market is booming as customers are prone to invest in products that are healthy for the body. For this reason, the market observed a growing trend of health-advantageous foods such as non-GMO, organic, gluten-free, and keto-friendly alternatives. Prominent companies like RXBAR, which is well-known for its protein bars, and Chobani, which is renowned for its Greek yogurt, have benefited from this health-conscious trend by promoting their goods as natural, pure, and devoid of artificial ingredients. Their marketing approaches emphasize the use of straightforward, instantly identifiable components to appeal to customers who value transparency and nutritional value in the meals they choose to eat.

2.2.4.4.2 Sustainability and Ethical Marketing

Consumers start to care about how their food choices affect the environment. Manufacturers begin highlighting eco-friendly packaging, responsible labour practices, and sustainable sourcing. As such, in this phase, sustainability and ethical marketing also became important because of their lesser impacts on the environment and society. Good examples of that trend come from companies such as Ben & Jerry's and Patagonia Provisions, who consciously brand their companies around sustainability and social responsibility. For example, Ben & Jerry's has fair trade ingredients and environmentally friendly packaging, while Patagonia Provisions centres on ethical sourcing and regenerative agriculture. These businesses can convey their sustainability pledge through digital channels such as the social media in which like-minded customers seek out branding stories.

2.2.4.4.3 The Importance of Transparency

The opposite would state that transparency in food marketing becomes significant because it can help businesses gain trust, empower and educate consumers, and advocate for sustainable practices. Customers are now expecting the firms to speak to what they source, how they produce it, and to be transparent about how and where they source their ingredients. This shift encourages the companies to describe the product's ingredients, chemicals, and other nutritional content within its labelling. As an example, Panera Bread started a programme named the "No No List," listing all the fake chemicals and pesticides extracted from food items. The entire purpose of the campaign was to build trust with the consumers, to give the business an image of transparency and to promote healthy eating.

2.2.4.5 *The Future of Food Marketing: AI, Automation, and New Technologies*

Food businesses have been presented with a whole array of possibilities and problems, as a result of the changes in the food marketing industry resulting from digitization and globalization. The changing environment compels companies to continue negotiating complexities in customer expectations, regulatory frameworks, and the improvements in technology to remain in the competition. Companies always need to focus on the concerns faced by the customers in improving buying experiences. The digital world has brought many problems, but it has also led to the development of new ways in which to communicate with customers. Food companies are now focusing on giving personalized experiences to their customers through the use of direct communications channels, making customers more loyal to the brand. For instance, companies began to utilize social media platforms for direct communication with customers and started responding to feedback, solving problems, and establishing closer ties with clients. Future innovations in automation, AI, and other emerging technologies will go a long way in shaping the course of food marketing into the future. This is expected to personalize clients further, increase the efficiency of food production and delivery, and open new channels.

2.2.4.5.1 AI and Personalization

Companies began to apply AI to food marketing to enrich personalization. Based on data about consumers, food manufacturers began to use AI algorithms to predict preference, provide suitable recommendations, and increase the effectiveness of marketing techniques. Of interest is that businesses such as Amazon and Nestlé have started to use AI to tailor product recommendations according to past purchase history and browsing behaviour. AI also allows companies to analyze current trends and come up with potential future development for innovative products. While gauging consumer data, AI may even identify the market gaps and come up with unique product ideas targeting the needs of the individual customer.

2.2.4.5.2 Automation in Food Production and Delivery

Automation has a significant implication on the transformation of food marketing in consumption, production, storage, and delivery. It also allows companies to maintain quality and safety. This case will assist companies to enhance efficiency since their efforts reduce, the cost of labour reduced, and human errors reduced. Application of a smart delivery system on food marketing automation reduces the cost and time of delivery. Automated checkout processes and drone deliveries are becoming increasingly common in the retail sector, providing consumers with fast and simple methods to purchase food. As part of its attempts to improve customer satisfaction and expedite delivery, Domino's Pizza, for instance, has started to test autonomous delivery cars and AI-powered order systems. Automation also helps companies to optimize supply logistics using machine learning and AI technologies. Using mobile apps, tablets, or smart goggles employees can monitor and adjust the production equipment to improve productivity and quality.

KEY FEATURES OF THIS PHASE

- Internet and e-commerce.
- Shift to digital marketing, e-commerce platforms, and online grocery shopping.
- Data-driven marketing strategies.
- Social media and influencer marketing.
- Growth of influencer-led food marketing.
- Sustainability and health trends.
- Emphasis on organic, sustainable, and ethical food production.

2.3 CHAPTER SUMMARY

The evolution of food marketing is a result of changes in societal, technological, and consumer behaviour. The chapter divided the whole evolution into four phases to highlight the major changes observed in each phase. The evolution started from the fragmentation phase, characterized by local markets and direct ties between consumers and producers, to the unification phase, characterized by mass manufacturing and the emergence of national brands. To meet the needs of a more complex and expanding market, food marketing has seen ongoing development.

The market segmentation started in the segmentation phase, where companies started to customize their ads and goods for certain psychographic and demographic groups. Wellness and health-related trends had a greater impact on consumer choices during this time, leading to the rise of processed foods and convenience foods.

In the era of globalization and digitalization, the emergence of new technologies such as social media, digital marketing, and e-commerce has completely changed the way in which food commodities are marketed and sold. Thanks to this change, food producers now have more opportunities to engage with a wider range of customers in the global market. Simultaneously, consumers of today want products to meet increased levels of sustainability, personalization, and transparency.

In summary, it can be concluded that food marketing development is a continuous and dynamic process. Food firms will need to constantly innovate and modify their marketing tactics in response to new trends, technology, and customer expectations to be relevant in a constantly changing environment. Nonetheless, the fundamental tenets of comprehending client needs, fostering trust, and offering value—which have traditionally directed the enterprise—will continue to be crucial.

CASE STUDY

Coca-Cola's Marketing Evolution: From Local Tonic to Global Icon

For over a century, Coca-Cola, which is now a globally recognized brand, has adeptly navigated and evolved through several stages of food marketing. From its humble beginnings in 1886 as a regional medicinal subject, the brand has developed into a formidable competitor in the beverage business all over the world. Coca-Cola's success may be attributed, in large part, to the company's capacity to experiment with new marketing strategies and make necessary adjustments. The purpose of this case study is to investigate the development that Coca-Cola has made in the food marketing industry through the stages of fragmentation, unification, segmentation, and digitization. It sheds light on the ways in which the corporation maintained its dominance in the very competitive worldwide market by using branding, advertising, and technology breakthroughs.

Fragmentation Phase (Early Beginnings): The first Coca-Cola factory opened in Atlanta, Georgia, in 1886. Initially, the beverage was sold as a regional medicine, and sales relied heavily on trust and personal relationships between customers and pharmacies in the area. When Coca-Cola's unique writing mark was formed at the end of the 1800s, it was a big part of giving the brand an early visual identity.

Unification Phase (National Expansion): In the early 1900s, Coca-Cola grew well beyond its local market. It did this by using mass production and delivery networks to make its products available all over the United States. During this time, Coca-Cola spent a lot of money on ads in newspapers and, later, on the radio. Coca-Cola's early motto, "Delicious and Refreshing," made it seem like an easy-to-find drink that could be drunk on a daily basis. Coca-Cola employed renowned personalities to endorse its goods, as well as notable visual branding, such as the adoption of the red-and-white image and the introduction of the contour bottle in 1915, to differentiate itself from competitors. During the unification phase, these tactics helped Coca-Cola become a more accepted and well-known brand.

Segmentation Phase (Market Segmentation): Coca-Cola turned its attention to segmented marketing after World War II, primarily focusing on different customer groups. The business invented television advertising in the 1950s, producing catchy jingles and commercials with themes of joy, refreshment, and youth. Coca-Cola persisted in broadening its range of products to cater to various customer tastes. In 1982, for example, it introduced Diet Coke to appeal to customers who were more health-conscious. Coca-Cola strengthened its worldwide footprint during this time by tying its brand to significant international events like the Olympics and FIFA World Cup, capitalizing on the expanding trend of lifestyle marketing.

Digital and Globalization Phase (Adapting to the Digital Era): In the 2000s, Coca-Cola entered the digital age by embracing social media marketing and e-commerce to reach a global audience. The business communicated with customers directly by using social media sites like Facebook, Instagram, and YouTube. Digital platforms were employed by campaigns like "Share a Coke," which allowed consumers to customize Coca-Cola bottles with their names in order to promote customer involvement and produce viral social media content. Coca-Cola has also modified its marketing to appeal to consumers' demands for transparency and sustainability by endorsing eco-friendly behaviours and using recycled packaging. Additionally, the company focuses on localized marketing, tailoring its campaigns to reflect regional tastes and values.

Coca-Cola has achieved long-standing success by adapting its marketing strategies to different phases of food marketing. By understanding consumer trends, leveraging emerging technologies, and consistently innovating its marketing approach,

Coca-Cola has maintained its status as one of the most recognizable and beloved brands in the world.

Discussion Questions

1. How did Coca-Cola's marketing strategies during the unification phase (1880–1950) help build consumer trust and loyalty?
2. In what ways did Coca-Cola adapt its marketing during the segmentation phase to appeal to different consumer groups?
3. What role did technological advancements, such as the rise of television and social media, play in Coca-Cola's success during the segmentation and digital phases?
4. How did Coca-Cola's "Share a Coke" campaign reflect the principles of modern food marketing in the digital era?
5. In the context of Coca-Cola's global marketing, how does the company balance its global brand identity with localized marketing efforts?

BIBLIOGRAPHY

O'Hagan, L. A., & Eriksson, G. (Eds.). (2024). *Food Marketing and Selling Healthy Lifestyles with Science: Transhistorical Perspectives*. Taylor & Francis.

CHAPTER 3

Consumer Behaviour and Decision-Making in Food Choices

LEARNING OBJECTIVES

By the end of this chapter, you should be able to:

- Understand the significance of consumer behaviour in the food industry.
- Analyze the influence of consumer decision-making on food product development.
- Identify and evaluate the various factors that influence food choices.
- Apply the seven-step decision-making process to understand how consumers make food-related decisions.
- Examine modern trends in consumer food choices, such as health-conscious eating, ethical and sustainable consumption, and the demand for convenience foods.
- Explore the role of marketing in shaping consumer preferences and decision-making processes.
- Assess the impact of technology on consumer behaviour.
- Recognize the challenges associated with understanding consumer behaviour in the food industry.

3.1 INTRODUCTION

The academic discipline of Consumer behaviour examines the decision-making processes and actions undertaken by individuals, groups, and organizations in the acquisition, utilization, and disposal of goods, services, ideas, or experiences with the aim of fulfilling their needs and desires. Within the realm of food choices, consumer behaviour comprises a diverse array of factors that exert influence over individuals' decision-making processes regarding their dietary selections, meal timings, and portion sizes. There are various factors that exert influence over individuals' decisions regarding food

DOI: 10.1201/9781003473305-3

consumption. These factors encompass a wide range of elements, including personal preferences, cultural background, socioeconomic status, and health considerations.

A comprehensive comprehension of customer behaviour is crucial for organizations operating within the food industry. By comprehending the various aspects that impact individuals' food choices, enterprises can devise marketing tactics that have a higher probability of achieving success. As an illustration, a company enterprise engaged in the promotion of a novel health-conscious food item may prioritize highlighting the various health advantages associated with the product. This strategic approach aims to attract consumers who are primarily driven by health-related considerations. The phenomena of consumer behaviour is intricate and subject to constant evolution. By comprehending the various aspects that exert an influence on individuals' food choices, enterprises can design marketing tactics that possess a higher probability of achieving success.

3.2 IMPORTANCE OF UNDERSTANDING CONSUMER BEHAVIOUR IN THE FOOD INDUSTRY

Comprehending consumer behaviour holds considerable significance within the food industry due to various factors:

- By comprehending consumer preferences, enterprises can create items that have a higher probability of achieving success. For instance, if a company possesses knowledge regarding the growing consumer inclination towards healthier food options, it can strategically direct its efforts towards the creation and enhancement of products that cater to this demand.
- Through comprehensive comprehension of consumer demographics, psychographics, and media consumption patterns, firms can optimize the precision of their marketing strategies. For instance, if a company possesses knowledge on their target market comprising young adults, they might concentrate their advertising efforts on media platforms that enjoy significant popularity within this specific demographic.
- The establishment of competitive pricing for items can be achieved by a comprehensive grasp of consumer perceptions of value by firms. For instance, if a corporation possesses knowledge regarding consumers' inclination to pay a higher price for organic food items, they can strategically determine the pricing of their organic products to align with this consumer behaviour.
- The development of effective promotional campaigns can be facilitated by a comprehensive understanding of consumer motives and attitudes. This understanding enables firms to create promotional efforts that possess a higher likelihood of resonating with their intended target audience. For instance, if a company possesses knowledge regarding consumer motivation driven by health-related concerns, it might strategically devise promotional initiatives that centre on the advantageous health attributes of its products.
- Through a superior comprehension of consumer behaviour relative to their rivals, enterprises can attain a competitive advantage. For instance, when a corporation

possesses knowledge regarding a novel trend in customer behaviour, it might gain a competitive advantage by being the pioneer in introducing a product that aligns with this trend.

3.3 THE IMPACT OF CONSUMER DECISION-MAKING ON FOOD PRODUCT DEVELOPMENT, MARKETING, AND SALES

3.3.1 The Influence of Consumer Decision-Making on the Development of Food Products

- The process of consumer decision-making is of paramount importance in the realm of food product creation. By comprehending the various aspects that exert an influence on individuals' food choices, enterprises can formulate and produce items that possess a higher probability of satisfying the preferences and desires of consumers.
- As an illustration, when a corporation possesses knowledge regarding the growing consumer inclination for healthier food options, it can strategically direct its efforts towards the creation and refinement of products that exhibit reduced levels of fat, sugar, and sodium.
- Furthermore, enterprises have the ability to utilize consumer input as a means to enhance their current product offerings. As an illustration, in the event that a business is provided with input indicating that consumers perceive a specific product to possess excessive saltiness, the business has the option to undertake a reformulation process aimed at lowering the sodium content of said product.

3.3.2 The Influence of Consumer Decision-Making on Food Marketing

Consumer decision-making has a substantial impact on food marketing. By gaining insights into the decision-making processes of consumers regarding food, firms can devise marketing efforts that have a higher probability of achieving their desired outcomes. For instance, if a company possesses knowledge regarding the impact of health concerns on consumer behaviour, they might strategically direct their marketing communications towards highlighting the health advantages associated with their products. Furthermore, enterprises have the ability to utilize customer data in order to formulate and implement focused marketing initiatives. For instance, if a corporation possesses knowledge regarding its target market consisting of young adults, it can strategically allocate its marketing endeavours towards social media sites that enjoy significant popularity within this specific demographic.

3.3.3 The Influence of Consumer Decision-Making on Food Sales

Consumer decision-making has a crucial role in influencing food sales. By comprehending the various aspects that exert an influence on individuals' food selections, enterprises can enhance their prospects of generating sales. For instance, if a corporation possesses knowledge regarding consumers' inclination to pay a higher price for organic foods, they can strategically adjust the pricing of their organic products to align with this consumer behaviour. Furthermore, enterprises have the ability to utilize consumer feedback as a means to enhance their customer service. For instance, in the event that a firm is provided with feedback indicating that consumers are encountering challenges in locating their products within retail establishments, they can collaborate with retailers to enhance the positioning of their products.

In summary, the process of consumer decision-making exerts a substantial influence on various facets of the food sector. By comprehending the cognitive processes involved in consumers' decision-making regarding food, businesses can enhance their decision-making processes pertaining to product creation, marketing strategies, and sales techniques.

3.4 FACTORS INFLUENCING FOOD CHOICES

Food decisions are influenced by a multitude of circumstances, encompassing a broad spectrum of variables.

3.4.1 Individual Factors

- **Age**: The dietary preferences of individuals may undergo alterations as they advance in age, owing to modifications in metabolic processes, taste inclinations, and health-related considerations.
- **Gender**: The dietary choices of individuals may vary based on gender, influenced by both biological and cultural variables.
- **Income**: Individuals with higher salaries are more likely to possess the means to acquire a diverse range of dietary options, including organic or speciality foods.
- **Education**: Education plays a significant role in influencing individuals' awareness of the health benefits associated with various food options, potentially leading to a greater propensity to choose healthier dietary selections.
- **Lifestyle**: Individuals who engage in regular physical activity may exhibit distinct dietary requirements compared to those who lead a predominantly inactive lifestyle.
- **Health status**: Individuals with specific medical issues may be required to adhere to specialized dietary regimens.
- **Individual preferences**: It is widely acknowledged that individuals possess distinct tastes and may exhibit a preference for specific types of cuisine over others.

3.4.2 Social Factors

Social variables refer to the various elements and influences inside a society that shape individuals' behaviours, attitudes, and interactions. These aspects encompass:

- **Family unit**: The family unit has a significant role in shaping individuals' dietary preferences through the transmission of knowledge about food and the demonstration of either health-promoting or detrimental eating behaviours.
- **Friends**: The influence of friends on food choices might manifest through their ability to either promote or discourage specific eating behaviours.

- **Culture**: The influence of culture on dietary choices can be observed via the lens of customs, traditions, and religious beliefs.
- **Social norms**: Social norms exert a significant influence on individuals' food choices by establishing societal standards that determine the acceptability or unacceptability of various eating behaviours.

3.4.3 Psychological Factors

Psychological factors refer to the internal influences that impact an individual's thoughts, emotions, and behaviours. These factors encompass several aspects:

- **Motivations**: Individuals may exhibit several motivations for engaging in eating behaviours, including physiological hunger, psychological boredom, or emotional tension.
- **Attitudes**: The attitudes individuals hold towards food have the potential to impact the decisions they make regarding their dietary selections. An individual who holds the belief that maintaining a healthy diet is of utmost significance is inclined to have a higher likelihood of selecting nutritious food options.
- **Beliefs**: The beliefs held by individuals regarding food have the potential to exert an influence on their decision-making processes pertaining to food selection. For instance, an individual who holds the belief that all processed meals are detrimental to health is less inclined to consume such goods.
- **Emotion**: The influence of emotions on individuals' eating choices has been widely observed. For instance, individuals experiencing stress may exhibit an increased propensity to consume comfort foods.

3.4.4 Situational Factors

Situational factors refer to the external circumstances or conditions that influence a certain situation or event. These factors can include several elements, such as:

- Time limitations can influence individuals to opt for convenient food options, such as fast food or frozen meals.
- The lack of availability of nutritious food options may contribute to an increased likelihood of individuals opting for bad eating choices.
- The influence of individuals' moods on their dietary choices is a significant factor to consider. For instance, individuals experiencing a state of happiness may exhibit an increased propensity to opt for lavish eating choices.

3.4.5 Economic Factors

Economic issues exert a substantial influence on individuals' decision-making processes about food selection. Several significant economic factors have been identified as influential in shaping individuals' eating choices.

- **Income**: Individuals with lower incomes may have greater challenges in accessing nutritious dietary options. This phenomenon can be attributed to the fact that the cost of nutritious food options tends to exceed that of less healthy alternatives. As an illustration, it is commonly observed that unprocessed fruits and vegetables tend to exhibit higher price points compared to their processed counterparts.
- **Food prices**: The cost of specific food items can also impact individuals' decisions regarding their food consumption. In the event that there is a rise in the cost of meat, individuals may exhibit a greater inclination towards selecting more economical substitutes, such as beans or lentils.
- **Food subsidies**: Food subsidies have the potential to enhance the affordability of nutritious food options. An illustration of this can be seen in the Supplemental Nutrition Assistance Program (SNAP), which offers benefits that can be utilized for the acquisition of nutritious food items such as fruits, vegetables, and other healthful provisions.
- **Food deserts**: Food deserts is the term given to geographical regions characterized by restricted availability and accessibility of nutritious food options. This phenomenon may be attributed to an insufficient presence of grocery stores or farmers' markets within the vicinity. Individuals residing in areas classified as food deserts are potentially inclined to opt for unhealthy food options due to their increased accessibility.
- **Food marketing**: Food businesses frequently employ persuasive strategies to promote the consumption of nutritionally deficient products, often resorting to tactics such as targeting youngsters or leveraging the influence of celebrities. This phenomenon has the potential to enhance the attractiveness of unhealthy food products to consumers.

In conjunction with the aforementioned economic variables, additional determinants that may exert an influence on individuals' dietary decisions encompass personal inclinations, cultural conventions, and health-related considerations. Nevertheless, economic variables can exert a significant influence on the dietary decisions of individuals with lower socioeconomic status. An investigation conducted by the United States Department of Agriculture (USDA) revealed that households with lower incomes exhibit a higher propensity to acquire food items that are deemed unhealthy, such as beverages with high sugar content, snacks with excessive salt levels, and processed meat products. This phenomenon can be attributed to the relatively lower cost of certain food items compared to their healthier counterparts. The economic determinants that exert influence on individuals' food preferences might yield substantial consequences for the overall well-being of the population. For instance, empirical research has demonstrated a positive correlation between residing in food deserts and a higher propensity for obesity among individuals. This phenomenon can be attributed to their restricted availability of nutritious food options. To facilitate the promotion of a nutritious dietary pattern, it is imperative to acknowledge and examine the economic determinants that exert an influence on individuals' selection of food items. Potential strategies to address the issue at hand encompass implementing food subsidies, enhancing the availability of nutritious food options in marginalized regions, and imposing regulations on food marketing practices.

3.5 SEVEN-STEP DECISION-MAKING PROCESS ON FOOD CHOICES

a. The first step in the process is to identify the need or problem that requires attention.
   - Do you have feelings of hunger, thirst, or a general need for something?
   - What are the anticipated benefits of consuming food or beverages?
   - Are there any particular nutrients or constituents that you are seeking?

b. Collect relevant data and information
   - One should engage in the practice of examining food labels and menus in order to get knowledge regarding the nutritional composition of various choices.
   - Seek suggestions from others within your social network, including friends, family members, or healthcare professionals.
   - Conduct an investigation of the potential health advantages or hazards associated with various dietary items and beverages.

c. The options should be identified.
   - What are the complete range of food and beverage choices at your disposal?
   - Are there any innovative or nutritious variations that might be applied to well-known culinary preparations?
   - Is it possible to prepare a meal or beverage at home rather than purchasing it?

d. Carefully consider the available evidence.
   - It is advisable to evaluate the advantages and disadvantages associated with each available alternative.
   - What are the potential health advantages and dangers associated with a particular phenomenon or intervention?
   - How does the inclusion of each option align with your overarching dietary objectives?
   - What are the potential financial and environmental consequences associated with each alternative?

e. Engage in the process of decision-making.
   - Please select the meal or beverage item that you deem to be the most optimal choice for your own preferences and dietary needs.
   - Do not hesitate to modify your selection if necessary.

f. Initiate appropriate measures and interventions.
   - Acquire and ingest the selected meal or beverage.
   - Appreciate the experience and derive pleasure from consuming your meal or beverage.

g. Assess the outcome of your decision-making process.
   - Provide a reflection on your level of pleasure with the food or beverage you have selected.
   - Would you go for the same decision if given the opportunity to do so?
   - What alternative actions could be implemented in future instances?

By adhering to the prescribed seven-step process, individuals can enhance their decision-making abilities and cultivate a heightened sense of awareness when selecting and consuming food and beverages (Figure 3.1).

Figure 3.1 Seven steps involved in decision-making process on food choices.
Source: The Authors.

3.6 TRENDS IN MODERN CONSUMER FOOD CHOICES

Contemporary consumer food preferences are progressively influenced by many prevailing patterns, encompassing:

- **Nutritious food**: The significance of consuming nutritious foods for the purpose of sustaining overall well-being is increasingly recognised by consumers. The current trend is generating an increased need for items that possess reduced levels of sugar, fat, and salt, while exhibiting elevated quantities of protein, fibre, and other vital nutrients.
- **Sustainability**: There is a growing consumer awareness of the ecological consequences associated with their dietary preferences. The aforementioned trend is resulting in an increasing need for sustainable food items that are cultivated within the local region, obtained through ethical means, and subjected to minimal processing.
- **Convenience**: In contemporary society, individuals with demanding schedules and a mobile lifestyle seek food and beverage alternatives that offer convenience and need minimal preparation. The aforementioned factors are contributing to an increase in the market demand for snacks, meal kits, and other conveniently prepared meals.
- **Personalized option**: The need for personalized food and beverage options that cater to individual dietary demands and preferences is on the rise among consumers. The current trend is resulting in an increasing need for products that are free from

certain components, such as gluten, lactose, and animal-derived ingredients, thereby catering to the preferences and dietary restrictions of consumers.

Several distinct patterns are currently influencing the decision-making process of contemporary consumers when it comes to food selection.

- The growing popularity of plant-based diets is evident in the increasing demand for meat alternatives, dairy alternatives, and plant-based egg replacements. This phenomenon can be attributed to various elements, encompassing apprehensions pertaining to human well-being, ecological viability, and the ethical treatment of animals.
- The escalating consumer preference for clean labelling is evident as individuals are progressively seeking food products that are characterized by the use of uncomplicated, identifiable, and organic constituents. The increasing scepticism towards processed foods and the aspiration to adopt a better diet are among the contributing factors to this phenomenon.
- The rising popularity of ethnic cuisines can be attributed to the growing consumer need for novel and stimulating culinary encounters. The increasing desire for genuine ethnic components and culinary offerings is fueling the need.
- The expansion of the functional food industry: Functional foods refer to food products that have undergone scientific validation to demonstrate their ability to provide targeted health advantages. The industry is experiencing growth because to customer demand for dietary improvements that enhance health and well-being.

Several trends are now influencing contemporary consumer food preferences. With the increasing awareness among customers regarding the health, environmental, and social implications associated with their food preferences, it is anticipated that the food and beverage industry will witness a surge in innovative practices in the foreseeable future.

3.6.1 The Rise of Health-Conscious Eating

The global phenomenon of health-conscious eating has gained significant traction for a multitude of influential causes.

- The level of awareness about the association between diet and health has witnessed a notable rise. Presently, consumers possess a heightened understanding of the significance of consuming nutritious meals to uphold general wellness and mitigate the risk of enduring chronic ailments such as heart disease, stroke, cancer, and diabetes.
- The increasing accessibility of nutritious food choices: The proliferation of supermarkets, health food stores, and internet sellers has contributed to the expanded availability of healthy food options in recent years.
- The impact of social media and celebrities: The utilization of social media platforms and the endorsement of celebrities have contributed to the promotion of healthy eating practices through the dissemination of their personal dietary habits and recipes.

Several significant phenomena are contributing to the increasing prevalence of health-conscious eating:

- The trend towards the consumption of whole, unprocessed foods: There is a growing inclination among consumers to choose whole, unprocessed foods rather than processed food products. The rationale for this assertion is that complete meals provide higher nutritional value and are less prone to incorporating detrimental components, such as excessive sugar, salt, and harmful fats.
- The increasing prevalence of plant-based diets: The adoption of plant-based diets is experiencing a notable surge in popularity, driven by consumer preferences for healthier and environmentally sustainable dietary choices. Research has demonstrated that the adoption of plant-based diets has several health advantages, such as mitigating the likelihood of developing cardiovascular disease, stroke, neoplastic conditions, and type 2 diabetes.
- There is a noticeable increase in consumer demand for organic and locally sourced meals. The rationale behind the preference for organic foods is in their cultivation methods which exclude the application of synthetic pesticides and herbicides. Additionally, locally sourced foods are frequently favoured due to their enhanced freshness and sustainability.
- The prominence of gut health: There is an increasing recognition of the significance of gut health in relation to general health and welfare. There is a growing trend among consumers to seek out food products that promote the health of the gastrointestinal tract, including fermented foods, foods rich in probiotics, and foods rich in prebiotics.
- The increasing prevalence of health-conscious dietary practices is exerting a substantial influence on the food and beverage sector. Food firms are actively addressing the prevailing trend by engaging in the development of novel goods that exhibit enhanced health attributes, increased sustainability, and heightened convenience. One illustration of this phenomenon may be observed in the current market, where numerous food corporations have begun to introduce a wide array of plant-based alternatives, including meat substitutes, dairy replacements, and various other plant-based items. In addition, they are providing an expanded selection of organic and regionally sourced items.

The emergence of health-conscious dietary practices is a favourable progression, as it engenders a shift towards more nutritious and environmentally sustainable eating habits among individuals. It is probable that this will yield a favourable outcome for public health in the long run.

3.6.1.1 Organic Food

Organic foods refer to food products that are cultivated or produced by agricultural practices that aim to minimize the utilization of synthetic fertilizers, pesticides, and other artificial substances. The primary objective of organic farming practices is to safeguard the environment and foster biodiversity. The perception commonly held is that organic foods possess superior health benefits and exhibit more sustainability compared to conventional meals. The following are among the advantages associated with the use of organic foods:

- Organic fruit exhibits a potential increase in specific nutrients, including antioxidants and other flavonoid compounds.
- Organic produce has reduced levels of pesticide residues in comparison to conventional produce, resulting in diminished exposure to pesticides.

- Organic farming practices contribute to environmental preservation through pollution reduction and the promotion of biodiversity.
- Organic farming practices are commonly seen as possessing greater ethical value compared to conventional farming practices due to their emphasis on safeguarding animal welfare and environmental preservation.

Nevertheless, it is crucial to acknowledge that organic foods do not inherently provide superior health benefits or exhibit more sustainability compared to conventional meals. The cost of organic foods tends to be higher compared to conventional foods, and their availability may be limited in certain regions. Moreover, many research has indicated that there exists minimal disparity in the nutritional composition of organic and conventional food products. Ultimately, the determination of whether or not to consume organic foods is a subjective matter. The use of organic foods entails both advantages and disadvantages, necessitating consumers to thoroughly evaluate these aspects before to reaching a conclusion.

3.6.1.2 Veganism

Veganism is a lifestyle that aims to minimize, to the greatest extent feasible and practical, the utilization of animals for food, clothing, or any other purpose, with the intention of eliminating exploitation and cruelty towards them. Vegans adhere to a dietary and lifestyle choice that excludes the consumption of meat, fish, poultry, eggs, and dairy products. Additionally, they abstain from utilizing animal-derived materials, such as leather, fur, wool, and silk. Veganism is driven by a diverse range of ethical, environmental, and health considerations:

- Ethical considerations arise from the perspective of vegans who maintain that the act of exploiting and terminating animals for the purposes of sustenance, attire, or any other utilization is morally unjustifiable. The authors contend that animals possess sentience and hence should be afforded the entitlement to life without experiencing injury or exploitation.
- Environmental concerns: Animal husbandry has been identified as a significant contribution to climate change, deforestation, and water pollution. Vegan proponents assert that the mitigation or eradication of animal agriculture is vital in safeguarding the environment.
- Health considerations: The adoption of a vegan diet has the potential to promote optimal health and provide a rich source of essential nutrients. Individuals who adhere to a vegan diet exhibit a propensity for reduced incidence of heart disease, stroke, cancer, and type 2 diabetes in comparison to those who consume animal-derived products.

The following are several advantages associated with adhering to a vegan lifestyle:

- The adoption of a vegan diet has been associated with a decreased susceptibility to chronic ailments, including but not limited to cardiovascular diseases, cerebrovascular incidents, malignancies, and type 2 diabetes. The reason for this is that vegan diets generally consist of a high intake of fibre, fruits, vegetables, and whole grains, while being low in saturated fat and cholesterol.

- Enhanced gastrointestinal well-being: Vegan diets frequently exhibit a notable abundance of prebiotics and probiotics, both of which provide advantageous effects on the gut microbiota. The presence of a well-functioning gut microbiome has been associated with enhanced digestive processes, strengthened immune responses, and improved emotional well-being.
- Weight loss and management: Vegan diets have been found to be beneficial for weight loss and management. The reason for this is that vegan foods are generally characterized by their low caloric content and high fibre content.
- The environmental impact of animal husbandry is significantly diminished: it serves as a prominent catalyst for climate change, deforestation, and water pollution. Vegan diets exhibit a reduced environmental footprint in comparison to diets that incorporate animal-derived items.

For individuals who express an interest in adopting a vegan diet, a plethora of tools exist to facilitate the initiation of this dietary choice. In addition, there exists a plethora of delectable vegan dishes that are readily accessible through online platforms and cookbooks, boasting simplicity in their preparation.

3.6.1.3 Dietary Supplements

Dietary supplements refer to goods designed to augment the nutritional intake by providing additional nutrients that may be deficient or insufficiently present in one's diet. Dietary supplements are available in a diverse range of formats, encompassing capsules, pills, powders, liquids, and gummies. Commonly consumed dietary supplements include:

- Vitamins are indispensable nutrients that are crucial for the optimal functioning of the human body. Several commonly consumed vitamin supplements include vitamin A, vitamin C, vitamin D, vitamin E, and the B vitamins.
- Minerals are vital nutrients that are necessary for the correct functioning of the human body. Calcium, iron, magnesium, and potassium are among the most often-consumed minerals in the form of dietary supplements by individuals.
- Herbs and botanicals refer to plant species or extracts that are utilized for their therapeutic attributes. Several often consumed herbs and botanicals that individuals commonly utilize as dietary supplements encompass ginseng, turmeric, ginger, and Echinacea.
- Additional supplements: Probiotics, omega-3 fatty acids, and protein shakes are among the commonly consumed nutritional supplements.

Individuals consume dietary supplements for a multitude of purposes. There is a subset of individuals who opt to consume dietary supplements with the intention of enhancing their general state of health and their overall sense of well-being. Some individuals opt to consume dietary supplements in order to target particular health issues, such as weariness, anxiety, or joint pain. Additionally, a number of individuals opt to consume dietary supplements with the intention of enhancing their athletic capabilities or facilitating weight loss.

3.6.2 Ethical and Sustainable Choices

Ethical and sustainable choices encompass decision-making processes that take into account the potential consequences on the environment, society, and the overall welfare of the planet. These options can be exercised across various domains of human existence, encompassing dietary preferences, sartorial selections, and consumer product acquisitions. There exist numerous justifications for the significance of adopting ethical and sustainable decision-making practices. Firstly, these decisions have the potential to contribute to environmental conservation. Numerous commodities and amenities that are utilized by individuals are dependent on the utilization of natural resources, encompassing vital elements such as water, energy, and land. Through the adoption of ethical and sustainable practices, individuals can actively contribute to mitigating their environmental footprint and safeguarding invaluable resources.

Making ethical and sustainable decisions can contribute to the enhancement of individuals' well-being. A significant number of individuals globally are employed in various industries responsible for the production of goods and services utilized by society. By engaging in ethical decision-making, individuals can contribute to the promotion of fair treatment and the provision of a livable wage for these people. Additionally, we have the capacity to provide assistance in promoting sustainable enterprises that demonstrate a steadfast dedication to both social and environmental accountabilities.

Ultimately, opting for ethical and sustainable choices aligns with moral principles and is considered morally justifiable. It is incumbent upon each individual to assume the responsibility of environmental stewardship and to make conscientious decisions that will contribute to the advancement of a more sustainable and prosperous future for humanity as a whole.

Below are a few recommendations for making ethical and sustainable decisions:

- It is advisable to purchase locally sourced and seasonally available vegetables. This practice aids in mitigating the environmental consequences associated with transit and storage.
- It is advisable to opt for organic products wherever they are available. The implementation of organic agricultural techniques yields benefits in terms of both environmental sustainability and human health.
- It is advisable to decrease one's intake of meat and dairy items. The production of meat and dairy products constitutes a significant factor in the exacerbation of climate change and other environmental challenges.
- It is advisable to reduce one's use of clothing and opt for garments that are manufactured using sustainable materials. The fashion sector plays a significant role in environmental degradation and the exacerbation of climate change.
- It is advisable to engage in the practice of repairing and reusing objects wherever feasible. This practice aids in the mitigation of waste generation and the preservation of valuable resources.

- Advocate for enterprises that demonstrate a strong dedication to social and environmental responsibility. Numerous enterprises are actively engaged in endeavours aimed at improving global conditions. By providing support to these enterprises, we may contribute in effecting a positive change.

Navigating the complexities of ethical and sustainable decision-making might present difficulties, although it is crucial to bear in mind that even incremental efforts contribute significantly towards positive change. Through implementing minor modifications in our everyday routines, each individual has the capacity to contribute positively to the environment and subsequently benefit future generations.

3.6.2.1 Fair Trade

Fair trade food goods refer to food items that have obtained certification indicating their compliance with fair trade norms. The primary objective of fair trade standards is to provide equitable remuneration for farmers and workers in developing nations, as well as to ensure their labour is conducted under safe and healthy conditions. Fair trade food goods encompass a diverse range of commodities, including, but not limited to, coffee, chocolate, cocoa, tea, sugar, bananas, rice, honey, nuts, and many more.

The act of purchasing fair trade food goods entails providing support to farmers and workers in developing nations, thereby facilitating the enhancement of their livelihoods. Additionally, your efforts contribute to the advancement of sustainable agriculture and the promotion of ethical business practices. The acquisition of fair trade food goods offers several advantages, which are outlined below:

- The promotion of fair trade standards serves to provide support to farmers and workers in developing nations by guaranteeing equitable compensation for their products and ensuring the provision of safe and healthy working environments.
- Advocate for sustainable agriculture: Fair trade norms incentivize farmers to adopt environmentally friendly farming practices aimed at safeguarding the ecosystem.
- Advocate for ethical business practices: The implementation of fair trade norms serves to foster ethical business practices, encompassing principles of transparency and accountability.
- Enhanced product quality: Fair trade farmers frequently exhibit heightened motivation to cultivate superior products due to their awareness of receiving equitable compensation.
- Supermarkets and health food stores frequently provide fair trade food products. Fair trade food goods can also be sourced through online platforms.

Below are a few recommendations for selecting and purchasing fair trade food items:

- It is advisable to seek out the Fairtrade International certification label when examining food goods. The presence of this label signifies that the product has obtained certification, indicating that it has been manufactured in adherence to fair trade principles and regulations.
- It is advisable to procure fair trade products in large quantities wherever feasible. This practice has the potential to yield cost savings and mitigate the issue of excessive packaging waste.

- One can demonstrate support for fair trade firms by directly purchasing their items or by making purchases through fair trade shops.

Through the deliberate adoption of ethical and sustainable practices, individuals possess the capacity to effectuate positive change for both the environment and forthcoming cohorts.

3.6.2.2 Cruelty-Free

Cruelty-free food goods refer to food items that have been manufactured or cultivated without causing any harm or suffering to animals. This implies that the animals utilized in the production of the food were not subjected to any superfluous discomfort or distress. The accessibility of cruelty-free food products is gradually improving, although they may still provide a greater challenge to locate than is the case for normal food goods. Among the food products that are often recognized as cruelty-free are the following:

- Plant-based milks, including almond milk, oat milk, and soy milk, present a viable substitute for conventional cow's milk. Plant-based milks are manufactured in a manner that ensures no harm is inflicted upon animals, rendering them ethically superior. Furthermore, these milks exhibit positive environmental attributes.
- Plant-based meats, including tofu, tempeh, and seitan, present a viable substitute for animal-derived meat products. Plant-based meats are manufactured in a manner that ensures the absence of suffering inflicted upon animals. Furthermore, they frequently exhibit reduced levels of saturated fat and cholesterol when compared to their animal-derived counterparts.
- Vegan eggs serve as a viable substitute for traditional chicken eggs. Vegan eggs are produced using plant-derived components, such as chickpea flour or flaxseed meal.
- Cage-free eggs refer to eggs sourced from chickens that have been reared in spacious enclosures, allowing them the freedom to wander about unrestrictedly. Raising hens in this manner is considered to be a more compassionate approach compared to the conventional practice of confining them in battery cages.
- Free-range eggs refer to eggs derived from chickens that have been granted unrestricted access to outdoor areas, enabling them to wander freely. This approach represents the most ethical and compassionate method for raising hens.
- Grass-fed beef refers to beef derived from cattle that have been nourished primarily on a diet consisting of grass, as opposed to grain. The practice of raising cattle on grass-based diets is widely regarded as more ethically sound and environmentally beneficial compared to the conventional method of feeding them grains.
- Pasture-raised pork refers to pork derived from pigs that have been reared in open pastures, as opposed to being confined in restricted spaces. Pasture-raised pork is widely acknowledged as a more ethically sound and environmentally friendly alternative to pork produced under confinement methods.

When making selections of food goods that are cruelty-free, it is imperative to exercise caution and make a thorough examination of the labelling. Certain items may be designated as "vegetarian" or "vegan"; nonetheless, it is still possible for these products to contain components derived from animals. Furthermore, it is imperative to provide backing to enterprises that demonstrate a steadfast dedication to the well-being and ethical treatment of animals. Below are few recommendations for selecting and purchasing food products that adhere to cruelty-free principles:

- It is advisable to seek out items that have obtained certification from cruelty-free organizations, such as PETA or Leaping Bunny.
- It is advisable to make purchases from companies that possess a well-defined animal welfare strategy.
- It is advisable to thoroughly examine product labels in order to identify and abstain from purchasing items that incorporate components produced from animals.
- Advocate for enterprises that demonstrate a steadfast dedication to the well-being of animals.

By opting for food products that are cruelty-free, individuals can contribute to the mitigation of animal suffering. Individuals have the capacity to contribute towards the advancement of more compassionate and environmentally responsible methods of food production.

3.6.2.3 Environment-Friendly Products

Environment-friendly food items are food products that are manufactured and consumed in a manner that minimizes their adverse effects on the environment. Achieving this objective can be accomplished by the implementation of sustainable farming techniques, the mitigation of food waste, and the preference for locally sourced and seasonally available food items. The following are illustrative instances of food products that are environmentally sustainable.

- Organic food refers to agricultural produce that is cultivated without the application of synthetic pesticides, fertilizers, and herbicides. This practice is more environmentally friendly and has potential benefits for personal well-being.
- Plant-based foods, encompassing a variety of nutrient-rich sources such as fruits, vegetables, whole grains, legumes, and nuts, exhibit a comparatively reduced ecological footprint when contrasted with animal-derived food products. This phenomenon can be attributed to the fact that plant-based foods necessitate a comparatively smaller amount of land, water, and energy for their production.
- The environmental impact of locally produced food is comparatively smaller when contrasted with food that undergoes long-distance transportation. The reason for this is that food produced within the immediate vicinity necessitates a reduced amount of fuel and packaging for transportation.
- Seasonal food refers to the type of food that is cultivated and gathered during its respective season. Seasonal cuisine exhibits a tendency to possess superior freshness and nutritional value in comparison to food items that are not in season. Additionally, it possesses a diminished environmental footprint due to its reduced energy requirements for transportation and storage.
- Fair trade food refers to food products that have been cultivated, processed, and traded in compliance with established fair trade criteria. Fair trade norms are implemented to guarantee that farmers and workers are remunerated fairly for their products and are provided with safe and healthy working conditions.

When making decisions regarding environmentally sustainable food products, it is crucial to take into account the subsequent factors:

- **Production methods**: What were the techniques employed in the production of the food? The environmental benefits of organic agricultural practices surpass those of conventional farming practices.
- **Ingredients**: What are the components of the food product? The environmental impact of plant-based foods and locally sourced meals is comparatively lower when contrasted with animal-based foods and those that undergo long-distance transportation.
- **Packaging**: What is the method employed for food packaging? The adoption of minimal packaging is more environmentally beneficial.
- **Transportation**: What was the method employed for the transportation of food? Food that is produced within a local region has a reduced environmental footprint in comparison to food that undergoes long-distance transportation.
- **Social impact**: What were the methods employed in the production of the food? Fair trade norms are implemented to guarantee that farmers and workers are remunerated fairly for their products and are provided with safe and healthy working conditions.

By opting for food products that are environmentally friendly, individuals may contribute to mitigating their ecological footprint and fostering a more sustainable future for the global community.

3.6.3 Convenience and Ready-to-Eat Foods

Convenience and ready-to-eat foods refer to food items that have been processed, packed, and promoted with the intention of facilitating effortless and prompt consumption. The increased consumption of these food items can be attributed to shifting societal patterns, the prevalence of hectic routines, and a growing inclination towards convenience. Meal delivery services have numerous benefits, including time efficiency, decreased meal preparation efforts, and enhanced accessibility. However, it is important to acknowledge that these services also entail significant disadvantages pertaining to health and nutritional aspects. The following are many salient factors to contemplate regarding the convenience and accessibility of pre-prepared food items.

There are several advantages to consider.

- **Time Efficiency**: Convenience foods offer a rapid and effortless preparation or consumption process, rendering them very suitable for persons who own demanding schedules.
- **Accessibility**: Legumes are readily accessible in a range of formats, such as frozen, canned, packed, and takeaway, thereby ensuring their widespread availability in numerous locations.
- **Food waste reduction**: The reduction of food waste can be achieved by the implementation of portion control and the extension of shelf life, as these measures effectively avoid excessive purchasing and rotting.
- **Portion Control**: Numerous convenience foods are available in pre-portioned servings, facilitating the management of calorie intake and portion sizes.
- **Diversity**: A broad array of choices is accessible, accommodating diverse palates, dietary inclinations, and cultural heritages.

There are several drawbacks associated with this phenomenon:

- **Cost**: Certain convenience foods exhibit a higher per-serving expense in comparison to the act of preparing meals from raw ingredients.
- **Health Concerns**: The regular intake of convenience foods has been linked to the development of obesity, cardiovascular disease, and various other health complications.
- **Restricted Customization**: Pre-prepared meals may not consistently cater to individual dietary requirements or limitations due to restricted options for customization.

Illustrations of convenience and readily consumable food items encompass frozen dinners, microwaveable meals, quick food, canned soups, instant noodles, pre-cut veggies, and pre-packaged salads. It is imperative to strike a balance between the convenience offered by these food products and their nutritional value, while integrating them into a comprehensive dietary regimen that encompasses fresh, unprocessed foods. The use of measures such as the careful examination of nutrition labels, the selection of healthier alternatives, and the adoption of moderation techniques are essential for integrating convenience foods into a well-rounded and balanced dietary regimen.

3.6.3.1 The Impact of Urbanization and Fast-Paced Lifestyles

The phenomenon of urbanization and the adoption of fast-paced lives exert a notable influence on individuals' dietary preferences. Typically, individuals residing in metropolitan regions exhibit a higher propensity for consuming processed foods, fast food, and beverages with high sugar content when compared with their rural counterparts. This phenomenon can be attributed to various variables, including the following:

- Individuals residing in urban regions frequently experience limited time availability for preparing meals at home as a result of demanding job schedules and long commutes. This phenomenon may result in an increased dependence on convenience foods and fast food options.
- Urban areas have a more pronounced assortment and accessibility of processed goods in comparison to their rural counterparts. This phenomenon can be attributed to the existence of supermarkets, convenience stores, and fast food establishments.
- Urban populations typically have higher income levels than their rural counterparts. This grants individuals the capacity to get higher-priced and industrially manufactured food products.
- The prevalence of marketing and advertising for processed foods and sugary drinks is higher among urban dwellers compared to their rural counterparts. This factor has the potential to impact individuals' dietary preferences.
- The adoption of processed foods and fast food has been associated with several adverse health outcomes. Processed foods frequently contain elevated levels of harmful fats, sugar, and sodium. Additionally, these foods exhibit a deficiency in crucial nutrients, such as dietary fibre and vital vitamins. Fast food is commonly characterized by its elevated levels of harmful fats, sugar, and salt. Moreover, it frequently exhibits a deficiency in nutrient density.

- There exists a significant association between the use of processed foods and fast food and the incidence of several chronic diseases, such as obesity, cardiovascular disease, stroke, type 2 diabetes, and some forms of cancer.

Nevertheless, it is important to acknowledge that urbanization and fast-paced lives can also have good implications for food choices. For instance, urban regions frequently exhibit a wider array of fresh fruit and healthier eating choices in comparison to rural locations. The existence of farmers' markets, health food stores, and ethnic grocery stores accounts for this phenomenon. Moreover, it is frequently seen that individuals residing in urban areas exhibit a higher level of consciousness regarding the need of maintaining a nutritious diet compared to their counterparts in rural areas. This phenomenon can be attributed to the increased accessibility of nutrition education and resources in urban regions. In general, the influence of urbanization and fast-paced lifestyles on dietary preferences yields a varied outcome. The impacts of a given phenomenon might be both advantageous and disadvantageous, with the equilibrium being contingent upon the particular circumstances. The following recommendations provide guidance for individuals seeking to make nutritious dietary selections within urban settings:

- It is advisable to engage in proactive meal planning. This strategy can assist individuals in making healthier decisions when faced with time constraints.
- It is advisable to prepare meals in one's own residence whenever possible. This affords the user autonomy in determining the selection of ingredients and the application of cooking techniques.
- It is advisable to prioritize the consumption of fresh produce above processed foods. Fresh produce exhibits a higher concentration of essential nutrients and a reduced presence of detrimental fats, sugar, and salt.
- It is advisable to restrict consumption of fast food and beverages high in added sugars. Fast food and sugary beverages are commonly characterized by their elevated levels of harmful fats, sugar, and sodium.
- It is important to exercise caution regarding the marketing and advertising strategies employed for processed foods and beverages high in sugar content. One should use caution when confronted with assertions of "healthy" or "natural."
- Utilize the diverse range of nutritious dietary choices accessible in metropolitan environments. One can explore farmers' markets, health food stores, and ethnic grocery stores as potential venues for sourcing fresh vegetables and other nutritious food items.

3.7 THE ROLE OF MARKETING IN SHAPING CONSUMER BEHAVIOUR

The field of marketing exerts a substantial influence on consumer behaviour pertaining to food goods. Food marketers employ a diverse range of tactics to shape consumers' perceptions, attitudes, and behaviours in relation to food goods. Several strategies are encompassed by this list:

- The process of branding involves the establishment of a distinct and individualized identity for a food product or service, so enhancing its capacity to be retained in the memory of consumers and increasing its attractiveness. Branding is employed by food marketers as a strategic tool to establish robust emotional bonds with consumers and cultivate brand allegiance.

- Food advertising serves as a means of conveying information and promoting food goods to consumers. Advertising is employed by food marketers as a means to generate consumer awareness, establish brand associations, and ultimately influence purchasing behaviour. Food advertising frequently employs persuasive tactics and employs a diverse range of psychological strategies to sway consumers.
- Food promotions encompass several strategies employed by food marketers to generate interest and boost consumer demand for food items. Promotional activities may encompass a variety of incentives, such as coupons, discounts, contests, and sweepstakes. Food promotions have demonstrated significant efficacy in exerting influence on customer behaviour, particularly when they are temporally constrained.
- Public relations (PR) refers to the strategic management of a food company or brand's public image and reputation. This practice involves carefully shaping and influencing the way the general public perceives and interacts with the said entity in the context of food-related matters. Food marketers employ public relations strategies to cultivate favourable media coverage and establish mutually beneficial connections with influential individuals, including journalists, bloggers, and influencers. Food public relations serves as a strategic tool for the promotion of novel food products, accentuating the advantageous attributes of such products, and managing the adverse publicity around them.
- Direct marketing refers to a marketing strategy that involves direct communication with consumers, specifically in the context of food products. Direct marketing is a strategy employed by food marketers to effectively communicate with consumers by delivering personalized messages and offers. Food direct marketing serves as a strategic approach to promote novel food products, extend discounts, and get valuable customer feedback.

Food marketers employ a diverse range of psychological principles in order to manipulate consumer behaviour. Food marketers often employ the idea of scarcity as a strategic tool to cultivate a perception of limited availability, thus instilling a sense of urgency within consumers and motivating them to make a purchase of a particular food item before its potential depletion. Alternatively, marketers can employ the concept of social proof to influence consumers by presenting a food product as widely accepted and highly sought after. It is imperative to acknowledge that food marketers do not consistently engage in efforts to deceive customers. In numerous instances, companies endeavour to provide consumers with comprehensive information regarding their food products, thereby facilitating well-informed decision-making. Nevertheless, it is crucial to acknowledge the mechanisms via which food marketing can exert an impact on individual behaviour, particularly in the context of consumer purchasing decisions related to food items. The subsequent instances delineate how food marketing might influence customer behaviour in relation to food products:

- Advertising can be employed by a food company to establish a favourable correlation between its brand and a specific lifestyle or value. As an illustration, a cereal firm could employ advertising strategies to establish a connection between its brand and a lifestyle that promotes well-being and physical activity.

- A food corporation may employ a promotional strategy to incentivize people to sample a novel food item. As an illustration, a snack food corporation might provide complementary samples of a novel product to customers in a convenience store.
- A food company may employ public relations strategies to cultivate favourable press for its food items. As an illustration, a beverage corporation might engage in sponsorship of a health-related occasion or contribute to a charitable organization that advocates for the promotion of nutritious dietary practices.
- Direct marketing is a strategy employed by food companies to disseminate personalized messages and exclusive offers to individual consumers. As an illustration, an e-commerce entity might dispatch an electronic mail to a consumer, offering a price reduction on a comestible item that has been previously procured by said consumer.

Food marketers have the ability to employ diverse strategies in order to exert impact on consumers' purchasing choices throughout the point of sale. These tactics encompass:

- The practice of positioning food products at eye level.
- Utilizing visually appealing packaging.
- Providing discounted prices or promotional offers.
- Employing compelling sales strategies.

Consumer awareness of the impact of food marketing on their behaviour is of paramount importance, as it enables them to exercise informed decision-making when purchasing food goods.

3.8 THE IMPACT OF TECHNOLOGY ON CONSUMER BEHAVIOUR

The influence of technology on consumer behaviour regarding food choices is substantial. Indeed, the advent of technology has brought about a significant transformation in the manner in which individuals discover, purchase, and engage with food. The following are certain manners in which technology is influencing consumer behaviour regarding food preferences:

- The proliferation of technology has facilitated enhanced accessibility to information for consumers regarding food goods, encompassing pertinent details such as nutritional composition, constituent components, and provenance. This information has the potential to enhance customers' ability to make well-informed decisions regarding the food products they purchase and eat.
- The advent of technology has significantly enhanced the convenience factor for consumers in terms of food purchasing and consumption. As an illustration, individuals now have the ability to place meal orders via internet platforms or mobile applications, subsequently receiving doorstep delivery of their chosen items. This alternative can prove to be a convenient choice for individuals with busy schedules who are unable to allocate time for cooking or visiting a grocery shop.
- The utilization of technology is facilitating a heightened level of personalization in the realm of food selection. For instance, individuals now have the ability to

utilize food monitoring applications in order to monitor their caloric and macronutrient consumption. This information has the potential to assist consumers in making more educated decisions regarding their dietary choices and in attaining their fitness objectives.

- The utilization of technology is additionally enhancing transparency inside the food supply chain. Consumers are now able to utilize blockchain technology as a means to trace the provenance of their food items, thereby ensuring their ethical sourcing and production.

3.8.1 Online Shopping and Food Delivery Platforms

In the contemporary digital era, the advent of online shopping and meal delivery platforms has revolutionized the consumer experience. These platforms offer exceptional convenience and accessibility, enabling consumers to effortlessly explore a wide range of items and dining choices from the comfort of their residences. The advent of online shopping platforms such as Amazon and e-commerce websites has significantly facilitated the process of purchasing a wide range of products, spanning from electronics to clothing. Concurrently, meal delivery applications like Uber Eats, Zomato, and Swiggy have emerged, providing consumers with a vast array of culinary options that can be conveniently delivered to their homes. These services have shown to be quite beneficial during periods of social separation and lockdowns, since they provide a secure and effective means of meeting vital requirements and fulfilling gastronomic desires. The increasing prevalence of these platforms not only confers advantages to customers but also enables businesses to expand their customer reach and adapt to the dynamic digital economy. With the continuous advancement of technology, it is anticipated that online shopping and meal delivery platforms would assume a more important position in our daily routines.

The advent of online shopping and meal delivery platforms has brought about a significant transformation in the manner in which individuals engage in shopping and dining experiences. E-commerce platforms provide a convenient and user-friendly avenue for purchasing goods and services, thus facilitating the entry of emerging enterprises into the market and enabling them to access a broader clientele. The following are few notable advantages associated with online shopping and meal delivery platforms:

- Convenience is a prominent feature of online shopping and meal delivery platforms, as they allow individuals the opportunity to engage in the purchase of goods and services from the confines of their residences, unrestricted by temporal constraints. This has the potential to significantly reduce time and exertion, particularly in situations where individuals are occupied or possess restricted physical mobility.
- The advent of online shopping and meal delivery platforms has significantly expanded the assortment of items and services available to consumers, surpassing the offerings typically seen in brick-and-mortar stores. The reason for this phenomenon is attributed to the ability of online retailers to maintain a significantly greater inventory, while meal delivery platforms have the capacity to consolidate menus from a diverse range of eateries.

- The utilization of online shopping and food delivery platforms presents an opportunity for individuals to economies by facilitating the comparison of pricing across various stores, hence enabling the identification of optimal offers. Additionally, coupons and discounts can be utilized to achieve enhanced cost reductions.
- Online shopping and meal delivery platforms offer a plethora of information pertaining to the products and services under our consideration. Prior to making a purchase, individuals have the ability to engage in various activities such as reading reviews, comparing features, and examining product photos and videos.

Although online shopping and meal delivery platforms provide numerous advantages, it is essential to acknowledge the potential downsides associated with these services. As an illustration,

- Impulse purchases can be readily made when engaging in online shopping or placing orders for food delivery. This phenomenon can be attributed to the absence of actual presence within the business or restaurant, leading to a potential decrease in our conscientiousness regarding expenditure.
- The expenditure associated with shipping and delivery can accumulate significantly, particularly when procuring a substantial quantity of things or engaging with numerous vendors.
- Return and exchange policies: Certain online shops and food delivery platforms implement stringent return and exchange policies. This can provide challenges when attempting to initiate the process of returning or exchanging things that fail to meet our satisfaction.
- Security and privacy concerns are of paramount importance, necessitating the selection of trustworthy online shopping and food delivery platforms, as well as the implementation of measures to safeguard our security and privacy. This encompasses the utilization of robust passwords and exercising caution over the dissemination of personal information on the internet.

3.8.2 Augmented Reality (AR) and Virtual Reality (VR) in Food Choices

As discussed in previous chapters, augmented reality (AR) and virtual reality (VR) are nascent technological advancements that possess the capacity to fundamentally transform the processes of food selection and consumption. AR involves the superimposition of digital information onto the physical world, enhancing the user's perception and interaction with their surroundings. On the other hand, VR entails the creation of a fully simulated environment that totally immerses users in a virtual realm, detached from the physical world. Both of these technologies possess the capacity to offer consumers a more informative and engaging experience in the process of selecting food choices. There exist several distinct manners in which AR and VR can be effectively employed in the context of food selection.

- AR has the potential to be utilized as a tool for delivering nutritional information to consumers regarding food goods. As an illustration, consumers have the ability to utilize their smartphones to direct their attention towards a food product, thereby enabling them to access pertinent details such as the calorie count, fat content, and further nutritional information, which will be visually presented on the device's screen.

- AR has the potential to be utilized for the purpose of delivering consumers with comprehensive information pertaining to the provenance of food goods. As an illustration, consumers possess the capability to utilize barcode scanning technology to ascertain the origin or production location of a given food product.
- VR has the potential to be utilized in providing consumers with a simulated tour of a restaurant or food production facility. This could enable people to make more educated decisions regarding the food they consume.
- VR has the potential to be utilized in the creation of immersive culinary experiences. As an illustration, individuals have the potential to employ such technology in order to virtually visit a farm or vineyard and acquire knowledge pertaining to the intricacies of food production.

3.8.3 Artificial Intelligence (AI) and Personalized Recommendations

The utilization of artificial intelligence (AI) is anticipated to have a substantial impact on the customization of food recommendations. AI-driven systems have the capability to analyze diverse data points such as an individual's dietary limitations, food preferences, lifestyle, and health objectives in order to produce customized food suggestions. There exist several distinct methodologies via which AI can be effectively employed to tailor personalized meal recommendations.

- AI has the capability to detect and classify dietary allergies and intolerances. Through the examination of a user's medical history and genetic information, AI has the capacity to detect potential instances of food allergies and intolerances. The aforementioned data can thereafter be utilized to develop individualized dietary suggestions that circumvent the presence of these allergenic substances and intolerances.
- AI has the capability to provide recommendations for food choices that align with certain dietary objectives. AI has the capability to provide tailored nutritional recommendations to users based on their individual objectives, such as weight loss, muscle gain, or chronic disease management.
- AI has the capability to provide recommendations for food choices that align with cultural norms and practices. It has the capability to consider a user's cultural background and personal tastes in order to generate culinary recommendations. This can be particularly advantageous for individuals who are unfamiliar with a specific culinary tradition or who possess dietary limitations that are culturally specific.
- AI has the capability to provide recommendations for sustainable dietary options. It has the capability to consider the environmental consequences associated with the production and delivery of food when making suggestions for dietary choices. This feature has the potential to assist users in making more environmentally-conscious food selections.

Several companies have already implemented AI-powered food recommendation systems:

- Noom is a mobile application designed for weight management, utilizing AI algorithms to produce individualized dietary suggestions.

- Spoonacular is a recipe discovery tool that employs AI to provide users with personalized recipe recommendations, taking into account their specific dietary restrictions and food preferences.
- Yummly is a culinary exploration platform that employs AI to provide personalized recommendations for dining establishments and recipes, taking into account users' dietary limitations and gastronomic inclinations.

3.9 CHALLENGES IN UNDERSTANDING CONSUMER BEHAVIOUR

Comprehending consumer behaviour about food choices presents a multifaceted undertaking, owing to various influential elements.

- The diversity of individual choices and needs. Individuals possess varying eating preferences, dietary limitations, and health objectives. This phenomenon can provide challenges in the development of universally applicable observations regarding consumer behaviour in relation to food preferences.
- The impact of social and environmental factors. Food decisions are subject to a multitude of social and environmental influences, encompassing cultural norms, familial and peer influences, media exposure, and the accessibility of food options. The presence of multiple factors can pose challenges in isolating the precise influence of individual elements on dietary choices.
- The intricate intricacy of dietary decision-making. The decision-making process regarding food choices is often characterized by a lack of rationality and simplicity. Individuals may select their food options based on emotional factors, established patterns, or convenience, rather than prioritizing nutritional content or health-related factors.

In addition to the aforementioned overarching difficulties, there exist a multitude of distinct obstacles pertaining to comprehending consumer behaviour in relation to food decisions, including:

- The challenge associated with quantifying food consumption. Accurately quantifying individuals' dietary intake, particularly over extended periods, is a formidable challenge. This can provide challenges in examining the correlation between dietary preferences and several variables, including physical health and overall welfare.
- The impact of unconscious prejudices. Individuals may possess little knowledge on the various aspects that exert an influence on their decision-making processes when it comes to selecting food items. This phenomenon poses challenges in obtaining precise data regarding individuals' dietary preferences and devising efficacious strategies for promoting nutritious eating habits.
- The dynamic nature of the food environment. The food environment undergoes continuous transformation, characterized by the introduction of novel food products and marketing methods on a regular basis. This circumstance can pose challenges in staying abreast of contemporary trends and devising efficacious strategies to foster the adoption of healthy dietary practices within a dynamic milieu.

Notwithstanding these obstacles, there exists an expanding corpus of scholarly literature pertaining to consumer behaviour in relation to food selection. The present study contributes to the advancement of knowledge regarding the determinants that impact individuals' dietary preferences, ultimately aiding in the formulation of efficacious strategies aimed at fostering the adoption of healthful eating behaviours. Outlined below are several solutions aimed at addressing the obstacles associated with comprehending consumer behaviour in relation to food choices:

- Employ a mixed-methodologies approach in the research methodology. Mixed-methods research is an approach that integrates both quantitative and qualitative methodologies in order to gather data pertaining to consumer behaviour. This approach can contribute to a more thorough comprehension of the various aspects that exert effect on individuals' eating choices.
- Utilize longitudinal studies. Longitudinal studies are designed to track and gather data on participants' eating choices and other relevant characteristics over an extended period of time. This methodology facilitates the identification of the enduring impacts of many factors on individuals' dietary preferences.
- Utilize technological advancements. The utilization of technology enables the collection of data pertaining to dietary choices in a manner that is characterized by enhanced efficiency and accuracy. One illustrative instance involves the utilization of smartphone applications to monitor and record dietary consumption as well as levels of physical activity.
- Engage in collaborative partnerships with a diverse range of stakeholders. The comprehension of consumer behaviour about food choices necessitates a collaborative effort among researchers, policymakers, industry stakeholders, and community organizations. Through collaborative efforts, it is possible to cultivate a more holistic comprehension of the various determinants that impact individuals' dietary preferences and subsequently devise efficacious strategies to encourage the adoption of nutritious eating habits.

3.10 CHAPTER SUMMARY

The phenomenon of consumer behaviour and decision-making in food choices is intricate and multifaceted. Food choices are shaped by various elements, encompassing human tastes, social and environmental influences, as well as the inherent characteristics of the food options available. Gaining insights into consumer behaviour about food choices is crucial in order to devise efficacious interventions aimed at promoting healthy eating habits. Nevertheless, undertaking this endeavour is a formidable challenge owing to the diverse range of personal inclinations and requirements, the impact of societal and environmental elements, and the intricate intricacies inherent in the decision-making process around food selection.

Notwithstanding these obstacles, there exists an expanding corpus of scholarly literature concerning consumer behaviour in relation to food selection. The present study contributes to the advancement of knowledge regarding the determinants of dietary preferences and facilitates the formulation of efficacious strategies aimed at fostering the adoption of nutritious eating habits. The following are notable insights derived from the discourse on consumer behaviour and decision-making pertaining to food selection:

- Consumer food choices are subject to several influences, encompassing individual preferences, societal and environmental issues, as well as the inherent characteristics of food choices.
- Gaining insight into consumer behaviour about food choices is crucial in order to devise efficacious interventions aimed at promoting healthy eating habits.
- The field of consumer behaviour regarding food choices has witnessed a notable expansion in research endeavours. This surge in scholarly investigations has contributed to a deeper comprehension of the various determinants that shape individuals' food preferences. Moreover, these research efforts have played a pivotal role in the development of efficacious treatments aimed at fostering healthier dietary practices.

By acknowledging the complexities associated with comprehending consumer behaviour in relation to food choices and implementing efficacious interventions, it is plausible to foster the adoption of healthier and more sustainable dietary preferences among individuals.

CASE STUDIES

Case Study 1

An Examination of Consumer Behaviour and Decision-Making Processes in the Context of Food Choices

This case study aims to investigate the intricate relationship between consumer behaviour and decision-making processes in the context of food choices. By analysing various factors that influence consumers' preferences and behaviours, this study seeks to shed light on the underlying mechanisms that drive individuals to make specific food choices. The food sector is subject to continuous evolution, characterized by the frequent introduction of novel goods and marketing methods. This phenomenon might pose challenges for customers in making well-informed decisions on their food selection.

One viable approach for comprehending consumer behaviour and decision-making in the context of food choices involves the implementation of case studies. Case studies provide researchers with the opportunity to meticulously analyse the behaviour of a limited sample size or an individual. This analysis can yield valuable information regarding the determinants of individuals' dietary preferences, as well as the obstacles and limitations they encounter in the process.

Example

Researchers at the University of California, Berkeley conducted a case study examining consumer behaviour and decision-making in the context of food choices.

The researchers conducted interviews with a sample of 20 mothers from low-income backgrounds to gather information regarding their decision-making processes related to food selection.

The study revealed that the mothers encountered many obstacles when making decisions on food selection. The challenges encompassed the following:

- Limited financial resources posed a challenge for the women, compelling them to make arduous decisions regarding the allocation of their constrained financial means. This frequently entailed opting for food options that were less nutritious but more cost-effective.
- The mothers resided in regions characterized by the limited availability of nutritious food options, with a scarcity of grocery shops and a prevalence of fast food establishments. This was a challenge in terms of accessing nutritious dietary options.
- Time constraints: The mothers were occupied with professional responsibilities and the care of their children, resulting in limited availability for the preparation of nutritious meals.

Notwithstanding these obstacles, the mothers expressed a desire to prioritize nutritious dietary selections for their households. The study revealed that the mothers exhibited motivation stemming from many causes, which encompassed:

- The mothers expressed a desire for their children to possess good health and experience happiness. The parents were aware of the significance of adopting a nutritious diet in promoting the physical and mental well-being of their offspring.
- Maternal well-being: The mothers expressed a desire to prioritize their own health and decrease the risk of developing chronic illnesses.
- The mothers aimed to impart knowledge about nutritious food choices and demonstrate positive health-related actions to their children.

The implications of the subject matter under discussion are of significant importance and warrant careful consideration. This case study offers valuable insights into the various aspects that exert an effect on consumer behaviour and decision-making processes when it comes to food selection. The research additionally underscores the difficulties encountered by low-income households in their pursuit of nutritious dietary options.

The consequences of the study's conclusions extend to politicians, industry stakeholders, and community organizations. Policymakers has the capacity to formulate policies that facilitate the provision of accessible and affordable healthy food options. The industry has the potential to enhance the nutritional quality of food products and refine marketing methods to promote healthier choices. Community organizations have the potential to offer educational resources and assistance aimed at facilitating informed decision-making around healthy food choices for families.

The phenomenon of consumer behaviour and decision-making in food choices is multifaceted and intricate. Nevertheless, through comprehending the various elements that impact individuals' decisions regarding food selection and acknowledging the obstacles encountered by consumers, it becomes possible to devise efficacious interventions aimed at fostering a culture of nutritious consumption among the general population.

Questions

1. How do financial constraints influence the food choices of mothers from low-income backgrounds, as revealed in the case study conducted by researchers at the University of California, Berkeley?
2. What role do community organizations play in addressing the challenges of limited availability and time constraints faced by mothers when making nutritious food choices, according to the findings of the study?

Case Study 2

An Examination of Consumer Behaviour and Decision-Making Processes in the Context of Organic Food Selection

This case study aims to investigate the intricacies of consumer behaviour and decision-making processes in relation to the selection of organic food products. By analysing the factors that influence consumer choices, this study seeks to shed light on the motivations and preferences that drive individuals to opt for organic food items. Through a comprehensive examination of existing literature and empirical evidence, this research aims to contribute to the understanding of consumer behaviour in the context of organic food consumption. Organic food refers to food that is cultivated without the application of artificial pesticides, herbicides, and fertilizers. Furthermore, it is manufactured without the utilization of genetically modified organisms (GMOs). The popularity of organic food has experienced a notable surge in recent years, driven by heightened consumer consciousness regarding the potential advantages for both human health and the environment associated with organic farming practices. Nevertheless, it is frequently observed that organic food tends to have a higher price point compared to conventional food. This phenomenon can provide challenges for certain consumers in terms of their ability to afford organic food.

The present study examines the consumer behaviour and decision-making processes in relation to food choices, with a specific focus on organic food. The present study conducted by scholars affiliated with the University of California, Davis aimed to investigate the various determinants that impact consumer behaviour and decision-making processes in relation to food preferences, with a specific focus on the selection of organic food products. The study conducted by the researchers revealed that customers exhibit a multitude of motivations when it comes to purchasing organic food. These motivations encompass a wide range of factors and considerations:

- Health concerns: It is often believed by consumers that organic food possesses superior health benefits compared to conventional food. The reason for this is that organic food is cultivated without the application of artificial pesticides, herbicides, and fertilizers, which have been associated with a range of health issues.
- Environmental concerns: Consumers exhibit apprehension regarding the ecological ramifications associated with conventional agricultural practices. The sustainability of organic agriculture surpasses that of conventional agriculture due to its avoidance of synthetic pesticides, herbicides, and fertilizers.

- Ethical considerations: These arise as consumers' express apprehension regarding the well-being of animals reared for the purpose of food production. Organic agriculture imposes more stringent criteria on animal care compared to conventional agriculture.

The researchers additionally discovered that customers encounter many obstacles when confronted with the task of making selections regarding organic food. These issues encompass:

- Cost: Organic food frequently carries a higher price tag than that of conventional food. The affordability of organic food can pose challenges for certain consumers.
- Limited availability: The accessibility of organic food may be restricted in certain grocery stores. This phenomenon poses challenges for customers in their quest to locate organic food.
- Insufficient knowledge: A proportion of customers exhibit uncertainty on the definition and distinctions between organic and conventional food. This phenomenon can provide challenges for customers in making well-informed judgements regarding organic food.

The implications of the subject matter are significant and warrant further investigation and analysis. The consequences of the study's conclusions extend to politicians, industry stakeholders, and community organizations. Policymakers have the capacity to formulate policies that facilitate the advancement of organic agriculture, thereby enhancing the affordability and accessibility of organic food. The industry has the potential to enhance the accessibility of organic food items by implementing strategies that emphasize their advantages and effectively target consumers. Community organizations play a crucial role in offering educational resources and assistance to consumers, enabling them to acquire comprehensive knowledge regarding organic food and facilitating informed decision-making in relation to their dietary preferences. The complexity of consumer behaviour and decision-making in the context of food choices is evident. Nevertheless, via comprehending the various elements that impact individuals' decisions regarding food consumption and the obstacles they encounter, it becomes possible to devise efficacious measures aimed at encouraging the adoption of nutritious and environmentally-friendly dietary preferences among the general population.

Questions

1. How do health concerns influence consumer decisions when choosing between organic and conventional food products, according to the findings of the study conducted by scholars at the University of California, Davis?
2. What strategies can industry stakeholders implement to address the challenges of cost and limited availability associated with organic food, as highlighted in the case study?

CHAPTER 4

Food Product Development and Innovation

LEARNING OBJECTIVES

By the end of this chapter, you should be able to:

- Identify and describe the major steps in the food product development process, from idea generation to commercialization.
- Understand how technological advancements contribute to creating new and innovative food products.
- Evaluate consumer preferences that influence food product innovation.
- Understand the key factors for success in product development.
- Understand the challenges associated with the product development.

4.1 INTRODUCTION

Food product development is a very dynamic and critical process for the food industry as it keeps producing new and improved products that meet changing consumer demands and preferences. The food innovation process thus forms essential support for companies aiming to remain competitive in the market, which is becoming increasingly diverse and demanding. It ties in food science, marketing insight, and technological advancement for possibly attractive, but scalable and market-ready products.

Continuous innovation is required in the dynamic world where consumer preferences change with time. Thus, the company needs to innovate new products that are trending. For example, companies are focusing on sustainable and nutritional foods to accomplish the upsurge in demand for consumer health consciousness and sustainability products. Companies are also continuously innovating products to suit the modern consumer through plant-based alternatives, clean-label ingredients, or personalized nutrition solutions. However, the new product must additionally meet the requirements of regulatory bodies, achieve consumer acceptance, and remain profitable. This chapter contains an overview of the several phases of food product development, innovation motivations, and obstacles businesses encounter while launching new goods on the market.

DOI: 10.1201/9781003473305-4

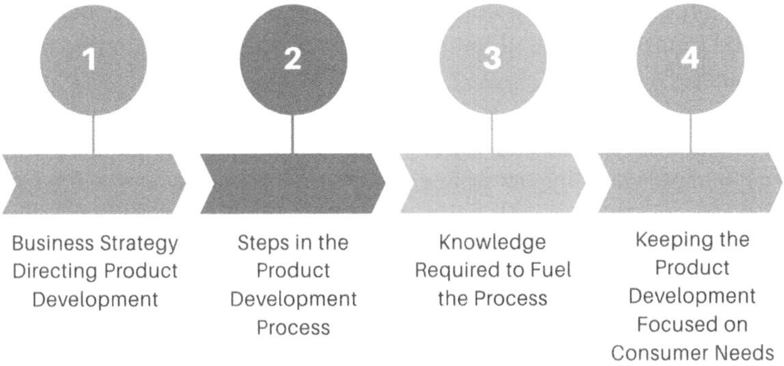

Figure 4.1 Elements of food product development.

Source: The Authors.

4.2 THE CORE ELEMENTS OF FOOD PRODUCT DEVELOPMENT

The manufacture of a new food product is an ever-changing and complicated process. It involves many strategic, practical, and consumer-focused factors. Thus, a company must focus on the factors that drive successful food product innovation before introducing new products in the market. This chapter will give an insight into four critical elements that form the backbone of food product development: the alignment of the process with business strategy, structured steps for development, the knowledge requirements to fuel the process, and, finally, putting consumer needs and aspirations at the heart of the process. These are the building blocks of a sound food product development strategy that ensures market-ready products but guarantees long-term success (Figure 4.1).

4.2.1 Business Strategy Directing Product Development

Every new product development attempts to thoroughly align with the organization's overall business strategy. Effective business strategies serve as a roadmap for innovation; they ensure that new products directly support broader, long-term goals, such as revenue growth, market share, sustainability, or brand positioning. In such scenarios, companies face a high likelihood of failing to advance their business agendas or losing the competitive capability of responding dynamically to changing markets while developing new products.

4.2.1.1 Aligning Product Development with Business Goals

Aligning business strategy in product development to the company's broader objectives is essentially the management of a firm by making sure that every new product fits into the company's overall objectives. From this alignment, businesses can use product development to attain long term goals. For instance, a business targeting a health-conscious customer will produce nutritious low-calorie food items to ensure alignment with their market positioning as a health-centric brand.

Another point of having a business strategy is that it dictates the use of resources for developing products. A company should combine budgetary, human, and technological resources according to strategic priorities. For instance, a food company whose strategy underscores international expansion will use its resources to develop products that meet different tastes in global markets and comply with the global requirements of regulation. Innovation efforts also have to be aligned with strategies for sustainability, as consumers are gradually becoming more environmentally sensitive. Hence, companies such as Impossible Foods have been able to incorporate sustainability elements into product development. Producing plant-based alternatives while keeping in view the mission of reducing the environmental impact created by regular food products, Impossible Foods regards itself as sustainable and caters to consumer demand for innovative products.

4.2.1.2 Strategic Market Positioning and Competitive Advantage

Strategic market positioning assists the marketer in positioning a product in the marketplace as well as differentiating the product from the competitors. For example, Chobani, a Greek yogurt brand, brought about an incredible revolution in the yogurt market with its high protein and low-fat Greek yogurt at a time when consumers were progressing towards the consumption of healthier options. It has contributed to a major market share for the company when its strategic focus was on health and wellness. In the food industry, a highly competitive market, companies that use product development to innovate and seize unmet consumer needs are the ones that benefit. A clear business strategy helps companies constantly introduce products that matter in consumer lives and thus differentiate themselves from others in a competitive marketplace.

4.2.2 Steps in the Product Development Process

Food product development includes several steps to ready the product for the market. In this multi-step approach, each stage requires careful planning, collaboration, and innovation towards the final product to meet consumer expectations and be feasible from manufacturing and commercial perspectives (Figure 4.2).

4.2.2.1 Idea Generation and Concept Development

New food product development is started from idea generation, whereby companies generate concepts based on consumer needs, market trends, and possible technologies. It is a foundational stage in the whole process of product development. The objective is to identify ideas that meet, or correspond to, the company's strategic purpose while filling gaps found within the marketplace.

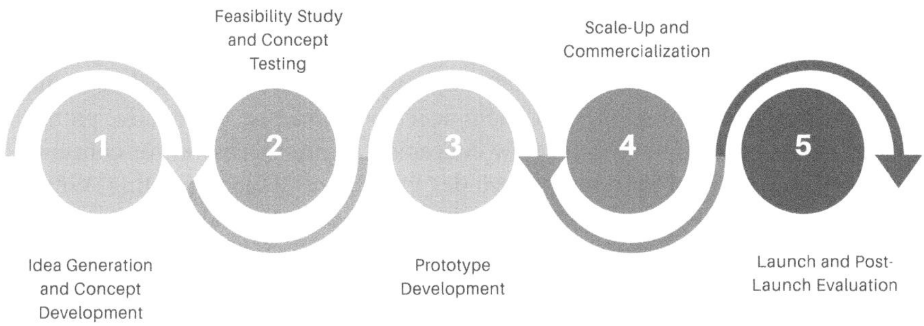

Figure 4.2 Steps involved in product development process.

Source: The Authors.

4.2.2.2 Identifying Consumer Needs

Consumer needs and preferences are the driving forces behind successful food innovation. For example, there has been an ever-steeper demand for plant-based products over the past few years based on consumer health, animal welfare, and environmental concerns. As a result, companies such as Impossible Foods and Beyond Meat have been able to innovate plant-based meat alternatives that deliver taste and texture mirroring actual meat. A company can determine its consumers' needs through various methods, for example, surveys, focus groups, and social media analysis. The outcomes of such steps allow companies to discover ways in which they could capitalize on untapped needs by innovating new products. For example, the growing need for gluten-free diets and low-carb food has resulted in the production of innovative, high-quality products: snacks, baked goods, and replacement meals.

4.2.2.3 Market Research and Trend Analysis

Companies in the idea generation phase also conduct market research and industry trend analysis in order to understand competitive forces. Research has discovered areas of growth, new trends, and shifts in consumer behaviour. For instance, in response to customer expectations, companies have been compelled to reformulate current goods and launch new ones due to the increased demand for clean-label products, or products created with straightforward, identifiable components. Through data analyses in reports such as Mintel, Nielsen, and Euromonitor, companies gain an understanding of which products are performing well and where there are opportunities for innovation. Market insights will allow food developers to spot niche markets or openings for new products to thrive, particularly in fast-growing sectors such as functional foods, organic products, and sustainability-focused products.

4.2.2.4 The Role of Innovation in Concept Development

Innovation is a good source for a company's successful concept development. Companies rely on food science, technological advancements, and creative problem-solving skills to create innovative new concepts that are successful for consumers. One of the most recent innovations includes the area of 3D food printing. With 3D food printing technology, it is now possible to make custom foods in unique shapes with specific textures and flavours. This could pave the way for personalized nutrition for people with specific dietary requirements or preferences. Along with new product conceptualization, the development process should also consider other aspects of development such as sensory appeal (taste, texture, aroma), nutrient content, and convenience. Consumers increasingly value healthy beverages that please not only their taste buds but also promote well-being. Nowadays, there is a need to tap into this market by developing functional beverages containing vitamins, probiotics, or adaptogens.

4.2.2.5 Feasibility Study and Concept Testing

From the concept, we then establish if it is viable technically and commercially. The feasibility study covers items such as production cost, possible sourcing of the ingredients, and compliance with regulations. Through this, we draw whether the concept is valid and can be scaled up to the mass production level.

4.2.2.6 Technical Feasibility

Technical feasibility deals with the possibility of manufacturing a new product. This case deals with whether or not the ingredients for the product are available, whether it is possible for it to be manufactured using existing equipment, and whether the entire process is efficient and cost-effective. For example, a company that specializes in the development of plant-based meat alternatives must ensure that it can rely on the availability of pea protein or soy protein and produce a product with a texture and flavour equivalent to meat. In some instances, the quality characteristics required in the product may require the development of new technologies or processes. A good example of this is the production of lab-grown meat, also commonly known as alternative meat. This involves the growing of animal cells in a laboratory setting. The technology is still comparatively novel, and companies such as Memphis Meats and Mosa Meat are striving to establish scalable processes that would be capable of allowing their lab-grown meats to be produced at a price that is competitive with conventional meats.

4.2.2.7 Regulatory and Compliance Considerations

The feasibility aspect also ensures that the product complies with all applicable food safety and labelling standards. Ingredients, additives, nutrition labelling, and health claims differ from country to country. For example, in the European Union, products manufactured through genetic modification (GMOs) must be appropriately labelled. In the United States, by contrast, the labelling of GMO products is still developing. Companies should ensure that any new product complies with all the regulations applicable to a business company before finally proceeding with product development. This process may start with consulting the law experts regarding the products proposed and the ingredients for the same; afterwards, it is submitted for approval to such regulatory bodies as the U.S. Food and Drug Administration or the European Food Safety Authority.

4.2.2.8 Consumer Testing and Concept Validation

Once the technical feasibility of the product is confirmed, the company will test concepts with consumers to either reject or accept the product idea. Pilot research tends to entail showing a focus group the concept or asking if the group members are interested in buying the product. Consumer testing enables the fine-tuning of the product concept, such that it reaches the target level. For example, when developing a new snack, the company shows consumers flavour profiles and seeks their input on preferences; therefore, it can refine the product before proceeding to other stages of development. In addition, there are frequently conducted sensory tests that test taste, texture, and appearance to ensure that the products meet consumer expectations.

4.2.2.9 Prototype Development

A prototype would be developed after the validation of a product concept. The prototype is the preliminary version of the product that works as an outline for the final version. The step is where food scientists and product developers, fine-tune the formulation, recipe, and processing method used in developing the product to obtain desired characteristics.

4.2.2.10 Ingredient Selection and Recipe Formulation

Formulation and ingredients are really important in prototype development, specifically on which ingredient mix and recipe will provide enough flavour, nutritional quality, and cost. The functionality of the final product quality depends on the quality of the ingredients; for example, it would be critical to find effective alternatives in gluten-free baked goods to reproduce the same sensory profile as wheat-based products. Ingredients such as almond flour, coconut flour, and tapioca starch often find their places into gluten-free formulations, but have unique impacts on the final product's texture and taste. During this stage, the food developer considers the product's nutritional profile. If a product is promoted as healthy, the developers may reduce the sugar, fat, or salt present without compromising on flavour and overall satisfaction. Snack foods are one area wherein consumers view healthier options for indulgence without having an adverse impact on health.

4.2.2.11 Process Optimization

The production process would be optimized to develop consistent and effective products. It means determining the best processing method, including baking, frying, freezing, or drying, along with the equipment and technologies necessary. For example, in frozen pizzas, the freezing process should be optimized in such a way that the product quality of the dough, sauce, and toppings is retained and has a prolonged shelf life. In optimizing the process, the companies also care that the product reaches the consumer within the stipulated time in such a manner that quality is not compromised. Therefore, the companies may opt for packing materials that guard the product against spoilage moisture, or contamination. For example, most snack foods, including potato chips and crackers, make use of Modified Atmosphere Packaging (MAP). MAP substitutes some or all of the oxygen in the headspace of their packaging with nitrogen in order to prolong shelf life.

4.2.2.12 Scale-Up and Commercialization

Once the prototype has been successfully developed, the next stage is to scale up production for commercialization. This involves transitioning from small-scale production to large-scale manufacturing while ensuring that the product retains its quality, flavour, and texture. Scaling up can present technical challenges, particularly when trying to replicate the characteristics of the prototype on a larger scale.

4.2.2.13 Manufacturing and Production Challenges

Scaling up introduces more complexity for food companies to use mass production techniques in terms of the capacity of the equipment, the sourcing of ingredients, and the uniformity of production. A recipe that works well in a test kitchen may not adapt as well to factory production. Ingredient ratios, cooking times, or processing could be varied while maintaining the characteristics of the desired product. Automation and standardization are also critical elements of mass production. They bring about efficiency and low costs in conditions of large-scale production. Firms will embrace technologies such as automated mixing systems, robotic packaging, and real-time quality monitoring to reduce the possibility of human error and to streamline the production process. For example, automated quality control systems are used by companies to detect defects, such as broken cookies or misshapen candies, in order to assure the quality of the product.

4.2.2.14 Packaging and Shelf-Life Considerations

During the scaling-up phase, firms must finalize the product packaging. Packaging serves critical functions, including safeguarding against contamination, maintaining freshness, and communicating essential information to customers. The growth in ready-to-eat meals requires packaging that can endure microwave warming while preserving freshness and delicacy. Another important consideration is the shelf life for commercialization. Food items should be safe and palatable for some months, or even years. To enhance shelf life without sacrificing quality, organizations use different preservation techniques, such as pasteurization, dehydration, or freeze-drying. For example, freeze-dried instant coffee is also conserved for flavour and aroma for long periods without refrigeration.

4.2.2.15 Launch and Post-Launch Evaluation

After the product has been successfully scaled up and manufactured, it is ready for a commercial launch. The launch phase involves bringing the product to market through various distribution channels and marketing strategies. A successful product launch requires careful planning and coordination across multiple departments, including marketing, sales, and supply chain management.

4.2.2.16 Launch Strategies and Marketing

The marketing strategy for a new food product plays a crucial role in its success. Rather than going it alone, businesses use a variety of strategies, such as influencer alliances, in-store sampling, social media promotions, large advertising campaigns, and more, to generate buzz for newly released products. For example, when Lay's launches a new flavour of potato chips, it can sponsor a social media campaign encouraging consumers to sample the new flavour and to share their views on the web. It helps to create buzz and word-of-mouth promotion. Distribution is another important element at the launch stage. The companies then have to decide which channel is necessary to market the product in a grocery store, speciality retailer, internet, or food service. For example, all Chobani yogurt ranges are found within a broad distribution channel from supermarkets to door-to-door online grocery delivery.

4.2.2.17 Post-Launch Evaluation and Continuous Improvement

After the product launch, a company follows the sales data and consumer feedback to determine how that product has performed. For example, if the consumers claim the snack item tastes too salty or, alternatively, that it has no real flavour, the company will adjust the particular recipe of the product to suit the consumers' needs. Another strategic aspect of post-launch evaluation entails the assessment of marketing strategy effectiveness. Companies would review statistics available on market penetration, brand awareness, and customer retention to know whether the product hit its sales mark. In most instances, companies may readjust their marketing strategies or alternative distribution channels to boost sales.

4.2.3 Knowledge Required to Fuel the Process

Successful food product development requires a diversity of knowledge, encompassing scientific, technical, and consumer insights. Integrating various fields of knowledge ensures that products are not only innovative, but also meet safety standards, nutritional requirements, and consumer preferences.

4.2.3.1 Food Science and Technology

The advancement of food products is contingent on the evolution of food science and technology. An understanding of the chemical, physical, and microbiological properties informs the development of novel products. Food scientists work together to optimize formulations, ensuring the final product achieves the intended flavour, texture, and shelf life. For example, the success of gluten-free products required significant investigation into substitute flours and how they behave in baking operations. Another area where technology is fundamental is in the process of making. Through automation and robotics, companies can multiply production without losing product quality. Indeed, AI and machine learning maximize this process by predicting how different ingredients will interact, optimizing recipes for taste, flavour, or other criteria, and even trying to anticipate what consumers would require.

4.2.3.2 Market Research and Consumer Insights

An understanding of consumer behaviour is critical to successful product development. Companies need to stay one step ahead of the markets and how consumer preferences would change products over time. Through surveys, focus groups, and social media data analysis, market research works out what consumers look for in new food products. Analytical tools for huge amounts of consumer data enable examining patterns and predicting future trends. Examples include renewed demand for healthier products, which has been driving innovation in low-sugar snacks, plant-based alternatives, and functional foods enriched with vitamins and probiotics. These trends make Chobani benefit from collecting consumer data and continuously innovating to build and expand its assortment to keep up with such trends.

4.2.3.3 Regulatory Knowledge

In the regulated food industry, a food company should keep up to date with numerous food safety standards and requirements. To ensure that products are safe and not unlawful, the company must comply with all local and international food laws. The knowledge of regulations regarding ingredients, labelling, and advertising claims helps to avoid legal pitfalls and ensures consumer trust in products.

4.2.4 Keeping the Product Development Focused on Consumer Needs

At the heart of food product development lies the consumer. Understanding and addressing consumer needs and aspirations are vital for the creation of successful food products. Companies must adopt a consumer-centric approach throughout the product development process to ensure that their products not only meet but also exceed consumer expectations.

4.2.4.1 Understanding Consumer Preferences

Today, consumers are more health-conscious and also knowledgeable than any previous generation. Their preferences are inclined towards transparency, sustainability, and personalized products. It could be the growing demand for plant-based alternatives or desires related to sustainable packaging; consumer preferences will take the lead for future developments of food products. Consumer research is what will keep the company ahead of this curve. Rising health needs are prompting increased demand for products tailored to an individual's requirements. As a result, firms are increasingly turning to data analytics and AI to offer customized food products aligned with individual segments or specific data related to one's health. The popularity of such customized products is also expected to soar as consumers seek products that conform to their specific dietary preferences and health goals.

4.2.4.2 Continuous Feedback and Adaptation

Successful companies also realize that product development involves continuous feedback and adjustment. Consumer preferences change day in and day out; what could be a hot sell today may be of little use tomorrow. Companies should be agile in the products they provide, continuously changing them based on consumer opinion, market trends, and technology changes. For instance, when Chobani introduced Greek yogurt, it targeted health-conscious consumers. However, with this change in the market, Chobani keeps innovating with new flavours, diverse product lines, and even non-dairy alternatives to keep changing according to the needs of its customers. This agility has thereby kept the company competitive in the crowded yogurt market.

4.2.4.3 Consumer-Centric Innovation

Innovative companies always keep the consumer at the forefront of every decision. For example, products such as Impossible Foods' plant-based burgers emerged primarily due to consumer pressure for sustainable alternatives to meat consumption. It is focused on delivering products that reflect the values of its consumers, which has helped Impossible Foods exist very successfully in the alternative protein market space. Consumer-centric innovation is also found in packaging and branding. Sustainable packaging is no more a trend, but a requirement. It lines with an increasing consumer demand for responsibility towards nature. Companies that invest in green packaging meet consumer expectations and contribute to global efforts toward waste reduction.

4.3 KEY FACTORS IN SUCCESSFUL PRODUCT DEVELOPMENT

The success of a new food product depends on a combination of factors, ranging from understanding consumer preferences to utilizing the latest technological advancements. Companies that excel in food product development can adapt to changing market conditions, innovate with new ingredients and processes, and ensure their products align with current trends like health, sustainability, and convenience. In this section, we'll explore the major factors that contribute to the success of food product development.

4.3.1 Consumer Trends and Preferences

The driving force behind most successful food products is a deep understanding of consumer trends and preferences. Companies need to develop products based on consumer trends to accomplish their preferences. Over the past few decades, the food industry has seen significant shifts in consumer behaviour, from the desire for convenience and indulgence to the growing interest in health, wellness, and sustainability.

4.3.1.1 Health and Wellness

The trend towards health-conscious eating had a profound influence on food product development. Consumers began to seek foods that taste good but also provide health benefits. Consumed functional food products are fortified with probiotics for gut health, antioxidants for immune support, or protein for muscle recovery. For instance, there is a development in plant-based proteins, which include those used in products from Beyond Meat and Impossible Foods, which cashed out from the rising perception of the bad impact of dietary products on the environment and health. The products are succeeding, not only because they cater to the higher-end vegetarian and vegan consumers but also because they address the flexitarian – consumers who are decreasing the proportion of meat in their diets but not necessarily eliminating it from their plates. Another big wave in this regard is clear labelling, which pushes for the ease of natural ingredients without any artificial additions, preservatives, or colors. Consumers are increasingly paying attention to ingredient labels. Companies that respond to this trend by manufacturing clean-label products can gain a competitive advantage. For instance, RXBAR has built its brand off the back of transparency, listing its core ingredients: egg whites, dates, and nuts – on the front of the packaging.

4.3.1.2 Convenience and Time-Saving Solutions

Modern consumers also look for convenience, especially in fast-moving modern life. Time-saving foods in the kitchen, ready-to-eat meals, pre-packaged snacks, and meal kits have become ever more popular. As discussed earlier, companies like HelloFresh and Blue Apron pioneered the meal-kit delivery service. The pre-portioned ingredients and simple recipes are now delivered to a consumer's doorstep, promising an easier and more convenient way of cooking at home. In addition to meal kits, there has been demand for "on-the-go" snacks as busy consumers look for convenient, quick, and healthy portions that suit their lifestyles. Kind Bars has capitalized on this, offering a whole ingredient-only healthy snack product using nuts, fruits, and seeds. These bars are convenient, and they market them as being superior to traditional snack foods.

4.3.1.3 Ethical and Sustainable Eating

Present-day consumers have become more sensitive to food purchasing decisions because they are concerned that their consumption may affect the environment and, therefore, request more information on producing these items to be ethically sourced. As a result of this awakening, more growth is observed for the labels or certifications of Fair Trade, Non-GMO Project Verified, and Organic. For instance, the ice cream manufacturer Ben & Jerry's has been focusing on sustainable sourcing and fair-trade ingredients as being part of a company that fits the values of an ethical consumer, thus placing it in the right position within ethical consumerism. Similarly, Patagonia Provisions, a food extension of the outdoor apparel company, sells its food products to consumers as sustainably sourced and environmentally friendly in their pursuit of food products that resonate with their ecological values. The success of food products depends upon how fast the firm keeps up with these ever-evolving trends. Thus, companies can develop products that will meet the demand now but for the future based on observing consumer behaviour and preferences.

4.3.2 Technology and Innovation

Technological advancements play a crucial role in food product development, offering new possibilities for creating, producing, and marketing food products. Companies that embrace innovation in food science, production processes, and digital tools are better positioned to develop successful products.

4.3.2.1 Food Science and Ingredient Innovation

One of the major areas where research and innovation takes place concerning the development of new food products is food science, enabling companies to produce new ingredients and formulations that offer better taste, texture, nutritional quality, and longer shelf life. For example, advancements in emulsification and encapsulation technologies allowed companies to manufacture alternative versions of traditional products, such as low-fat salad dressings or fortified beverages. Another area of innovation is the development of alternative ingredients that can replace conventional ones. Plant-based proteins such as pea protein and soy protein, and mycoprotein, have gained popularity, offering alternatives to animal-based proteins. Apart from addressing consumer health concerns, such ingredients also point towards environmental concerns over the production of meats.

Yet another emerging technology is 3D food printing, which companies can uses to create complex customized food products. It will revolutionize food production, especially in restaurants, hospitals, or space missions, where nutrition can sometimes depend on being tailored to an individual's needs. Still in its infancy, 3D food printing may one day produce foods whose textures, nutritional profiles, and appeal cannot be achieved in any other way.

4.3.2.2 Automation and Production Efficiency

Advances in technology concerning automation have also changed food production into a better and more efficient way of producing it. Companies can now produce large batches of food products mainly due to automated mixing systems, robotic assembly lines, and AI-driven quality control systems. Using automation, products are packaged readily, with few errors and a labour-intensive cost.

Mass production goes particularly well with automation, ensuring product quality consistency from small prototype batches to large production runs. Automated systems also prevent food waste by optimizing the use of ingredients and keeping products packaged and stored according to optimal shelf-life storage guidelines. In addition to production efficiency, companies continue to look for ways and means to utilize technology to better their supply chain management. A powerful example of blockchain technology is its ability to trace, from farm to table, exactly where the ingredients came from. Companies can ascertain transparency and traceability. Hence, firms must prove the ethical sourcing of ingredients in the supply chain or authenticate claims like "organic" or "fair trade."

4.3.2.3 Digital Tools and Data-Driven Innovation

The development of big data and AI has significantly impacted food product development. Companies can currently gain access to large volumes of consumer data, which can be used to recognize market trends and consumer preferences and identify gaps in the market. Through the analysis of this data, companies will position themselves better to develop new products with more probable appeals to consumers. For example, AI is implemented for consumer preference prediction and creates novelty in product formulation. Large companies such as IBM Watson can analyze flavour combinations and give innovative flavour pairing ideas for new product launches. McCormick & Company, on the other hand, employs AI to create novel seasoning blends based on data-driven insights into flavor trends.

Digital tools also facilitate better market research and consumer testing for companies. Online survey analysis, social media analytics, and data analytics sites can also give immediate feedback on consumer opinions. Thus, companies will be guided by factual data at any product development phase.

4.3.3 Sustainability and Ethical Considerations

Sustainability and ethics have become central to food product development as consumers increasingly prioritize products that align with their environmental and social values. Incorporating sustainable practices and ethical considerations into product development can lead to greater consumer loyalty, brand differentiation, and long-term success.

4.3.3.1 Sustainable Sourcing and Ingredient Selection

Sustainable sourcing implies that the raw materials used for the product should be eco-friendly or less harmful to the environment. Companies are increasingly looking for ingredients produced in ways that consume fewer natural resources, such as organic farming practices, water conservation, and lower carbon footprint transportation. Rainforest Alliance Certified and Fair Trade Certified products are examples of the ways in which companies can guarantee their ingredients meet certain environmental and social standards. One of the major emerging areas in sustainable food product development is alternative proteins, whether these be plant-based or lab-grown. These alternatives promise to be far more sustainable than animal-based proteins, with significantly fewer natural resources and lower greenhouse gas emission levels. This growth of plant-based offerings from companies such as Beyond Meat and Impossible Foods is legitimate, not just on consumer preference for healthier fare but also on the necessity for something more sustainable to be used as a food source.

4.3.3.2 Ethical Sourcing and Fair Trade

Consumers purchasing products also demand that these products not only be good for the environment but also have an ethical manufacturing process. Fair Trade certification ensures proper remunerations to the farmers and producers operating in developing countries are paid adequate wages, work under safe conditions, and have access to resources as needed. Coffee, chocolate, and bananas carry the label "Fair Trade Certified," which indicates that they uphold such standards. Ethical considerations in food products can help set a brand apart and eventually build consumer trust relating to their social responsibility. For example, Divine Chocolate Company bases its declaration on the point that it is co-owned by the cocoa producers who produce the chocolate; thus, all dividends are accorded to the farmers, including those involved in the decision-making process. This appeal to ethical sourcing resonates well with consumers who would like their purchases to have a positive impact on society.

4.3.4 The Importance of Cross-Functional Collaboration

Successful food product development relies on collaboration between multiple departments within a company, as well as external partnerships. From research and development (R&D) to marketing and supply chain management, a cross-functional approach ensures that all aspects of the product are considered and optimized.

4.3.4.1 Internal Collaboration

New products are developed in a company with a cross-functional team, including R&D, marketing, finance, operations, and quality assurance. Here, various departments offer something different within the process. For example, the R&D team focuses on ingredient selection and formulation, while the marketing team ensures that the product aligns with consumer preferences and branding strategies. Cross-functional teams also ensure that products are feasible from production and cost perspectives. The finance team is in close contact with R&D to ensure the production at cost, and the operations ensure that the product can be done on a large scale without sacrificing product quality. This kind of cooperation is central to making the final product desirable to the customers and commercially viable.

4.3.4.2 External Partnerships

Many companies collaborate with a third party, such as external partners, including food tech start-ups, academic institutions, and research organizations. It helps companies realize new technologies and expertise, and shorten the process of bringing new products to the market. For example, PepsiCo has collaborated with food tech start-ups to seek alternative routes to bring sugar content down in their beverages. Similarly, Unilever has collaborated with universities to seek suitable alternatives in packaging that conserve the environment. This alliance helps businesses to stay at the forefront of innovation and get new products to the market more quickly.

4.4 CHALLENGES IN FOOD PRODUCT DEVELOPMENT

Developing a new food product is a complex and multifaceted process that often encounters several challenges. From navigating regulatory environments to managing production costs, food companies must overcome numerous obstacles to bring a successful product to market. Understanding these challenges and finding solutions is crucial for companies to ensure their products meet consumer expectations and are viable in a competitive marketplace.

4.4.1 Regulatory and Compliance Issues

One of the most significant challenges in food product development is ensuring compliance with the complex web of regulations that govern the food industry. These regulations, which vary by region and product type, cover everything from ingredient sourcing and food safety to labelling and health claims. Failure to comply with regulatory requirements can result in product recalls, fines, and damage to a company's reputation.

4.4.1.1 Food Safety Regulations

Both companies and regulatory authorities ensure that the safety of food comes first. The government strictly regulates the level of contamination to ensure that imported foodstuffs reaching the nation do not harbour any danger for human consumption. In the United States, the regulatory agencies include the FDA and the USDA. Likewise, in Europe, the EFSA performs the function. New product development often becomes the time of safety evaluation and relevant data submitted to the regulatory authorities from the manufacturer's side. For example, if a company innovates a new ingredient, it is necessary to demonstrate that the new ingredient is safe for consumption. This would include toxicological studies, testing on allergens, and stability testing, ensuring the safety of the product during its shelf life.

Food manufacturing regulations also cover the manufacturing process. Manufacturers must, for instance, have a Hazard Analysis Critical Control Point (HACCP) system to detect and control potential hazards occurring in the production line. Such systems are designed to ensure that contamination cannot take place during manufacturing processes, such as cooking temperatures and sanitation practices.

4.4.1.2 Labelling and Health Claims

The other major challenge, to be faced by these companies, is labelling and health claims. Correct labelling is important, mentioning ingredients, allergens (if applicable), nutritional content, and expiration dates. This presents a challenge when the formulas are more complex as the company would have to ensure that every ingredient it contains is listed, in addition to all health claims made on the packaging falling within a scientific basis to support that individual claim.

Region-specific regulation can vary substantially from region to region. For example, in the European Union, health claims need preapproval from the EFSA while in the United States, the FDA recognizes "structure-function" or claims describing the function of a nutrient or dietary ingredient that contributes to normal structure or function in ways that are generally recognized among experts and allow provided these are not false or misleading. For instance, when a food processor wants to advertise a certain product as boosting immune health, it must prove that the exact ingredients used have been scientifically proven to enhance human immune functions. The amount of time and money it takes to gather such evidence to submit the claim for approval, ensuring that the claim is indeed done by local rules can be overwhelming.

4.4.1.3 Navigating International Regulations

Companies operating on a global scale might face discrepancies in regulatory requirements across all countries. Some may be permissible in one country but restricted in others, and companies need to ensure they can be compatible with the regulations existing in each market. For example, genetic organisms are regulated differently across regions. GMOs are generally accepted in the United States but, comparatively, are tightly controlled in the European Union, where they have to be identified; indeed, in specific instances, their cultivation and sales of genetically modified crops are prohibited. A company manufacturing a product with GMO must ensure that it meets the specific regulations of each country to sell its product.

4.4.2 Technical and Production Challenges

Scaling a new product from the development phase to full-scale production presents its own set of challenges. While a product may perform well in a small-scale test kitchen, replicating that success in a manufacturing facility can be difficult. Companies must navigate technical challenges related to production processes, ingredient sourcing, and quality control.

4.4.2.1 Scaling from Prototype to Mass Production

Probably the biggest challenge of creating a food product is scaling up. A recipe can work perfectly in small-batch cooking, but may not do well in large-scale manufacturing because the texture, taste, and consistency of the product may be affected by varying equipment, speeds, and batch sizes. Moreover, as the production scale of a fried or baked product increases, companies may have to change the cooking time and temperature for optimal results on the final product coming out cooked inside and retaining the appropriate flavour and texture. This is done after much trial and error, with multiple rounds of test productions to fine-tune the production process for consistency in high-batch outputs.

Equipment constraints may also be a source of difficulty. Some equipment is required to acquire large amounts of a new product. Specialized is expensive, and could require substantial capital investments in infrastructure. For example, an organization developing a carbonated beverage will need to invest in specifically designed bottling equipment that could handle the carbonation process without compromising the integrity of the final product.

4.4.2.2 Ingredient Availability and Quality

Another technical challenge faced is the consistent availability of good-quality ingredients. Most food products rely on specific ingredients that would give them their desired flavour, texture, or nutritional value. The challenge lies in sourcing these ingredients, especially in cases where they might be seasonal, rare, or where the supply and demand vary.

For example, demand for avocados has, in the past year or so, skyrocketed worldwide. The increased demand leads to price and supply shocks. A firm whose entire line depends on avocados should adjust as such by seeking a reliable supplier, negotiating longer-term contracts, or substituting other ingredients if avocado supplies are not at the required levels.

Similar challenges lie in controlling the ingredient quality since natural variations in crops, growing conditions, and processing can easily alter the final product. For example, variations in the fat content of coconut milk can affect the texture and flavour of ice cream or yogurt. Companies should have strict quality control measures to ensure their ingredients meet consistent standards even when sourced from different suppliers or regions.

4.4.2.3 *Ensuring Consistency and Quality Control*

Product consistency is crucial for establishing consumer trust. Consumers expect that the product they purchase will taste, look, and perform the same every time, regardless of when or where it was produced. That level of consistency is possible only with the implementation of robust quality control systems that monitor every aspect of the production process from ingredient sourcing to final packaging. For example, through sensory panels, the companies conduct regular testing of the products to ensure that they have attained the desired flavour profile. Using automatic systems, one can monitor variables of production like cooking temperature, ingredient ratios, and packaging integrity, which can be prevented before the products are shipped out of the facility.

Such consistency may sometimes require a recipe or process adjustment to produce the same. For example, a chocolate manufacturer needs to alter the formulation during the summer to avoid melting the chocolate during transportation and storage processes. Similarly, companies that produce products with a long shelf life must ensure that the product maintains its quality over time, which may require the use of preservatives or modified packaging to prevent spoilage.

4.4.3 Cost Management

Balancing innovation with cost management is one of the most critical challenges in food product development. While consumers may be willing to pay a premium for certain products—such as organic, gluten-free, or sustainably sourced—price sensitivity is still a major factor in purchasing decisions. Companies must find ways to keep costs down while maintaining quality and profitability.

4.4.3.1 *Ingredient Costs and Supply Chain Management*

Ingredient cost is one of the highest contributors to the cost of a food product, and premium ingredients such as organic or non-GMO tend to be more expensive than their conventional counterparts. It can create a ripple effect through commodity price fluctuations. For instance, the price of vanilla spiked by a huge margin in recent years due to crop failures and breaks in the supply chain, leading to increased costs for companies that use vanilla in their products. Controls of ingredient costs usually involve a range of approaches: holding long-term negotiations with suppliers, diversifying supply chains, or finding substitutes. Sometimes, a formula needs to be adjusted to reduce the cost of a part or modify the process to make the most product output while investing minimum for waste minimization. Natural disasters, political instability, and pandemics also impact the availability and supply chain, which may drive up prices and impact the timing of production runs. Companies should plan around these disruptions through alternative suppliers or altered production plans.

4.4.3.2 Production Costs and Economies of Scale

Another decisive factor in the overall profitability of a food product is the production cost. Economies of scale determine and decide such costs of production: more massive production runs in volumes subsequently ensure lower per-unit costs. However, heavy capital investment in equipment, infrastructure, and labour is one of the requirements for attaining economies of scale. Such an exercise might be costly for small companies or even a start-up. Production has to be outsourced to a co-packer, which is another manufacturer that produces and packages a product on behalf of another firm. By reducing the cost of production, co-packers pose other challenges in terms of quality control, consistency, and supply chain management. Companies try to ensure that such costs are minimized by the optimal use of ingredients and packaging materials for the automatic production process. For example, nesting packaging design that maximizes how containers nest snugly together is designed to cut shipping costs more by making the most of the interior space inside transport vehicles.

4.4.4 Competition and Market Saturation

The food industry is a highly competitive one, with new products being launched regularly and consumer preferences constantly shifting. Companies face significant challenges in differentiating their products from competitors and standing out in a crowded market. In some cases, even innovative products may struggle to gain traction if the market is saturated or if consumers are hesitant to try something new.

4.4.4.1 Differentiating Products in a Crowded Market

With so many options now coming at consumers, companies need to be able to differentiate their products from the competition. It could be possible through creative flavour profiles, innovative packaging, or creating brand messaging that will speak directly to consumers' core values. Such is the case with Oatly, the Swedish company behind the largest oat milk brand; the company differentiates itself within the plant-based milk category using bold packaging, quirky marketing campaigns, and a focus on sustainability. Some products in the firm can also be designed to look different from the rest of its competitors. It is the marketing of products that are said to be high quality and expensive by carrying high quality and prices. For example, Häagen-Dazs is positioned as premium ice cream. Its high-quality ingredients and sophisticated packaging ensure that products are costlier than those from its competitors in the market.

4.4.4.2 Market Saturation and Changing Consumer Preferences

Market saturation is yet another challenge to the success of a new product. When categories are crowded, like snack foods or beverages, companies have to exert more effort in grabbing consumer attention and market share. Without a clear point of differentiation, products will not succeed in gaining against established brands, even more innovative ones.

Shifting consumer preferences also presents challenges for such firms. A product that sells today might be unacceptable tomorrow, as there are shifts in consumer tastes. The low-carb diets became popular, and demand increased when low-carb lines were Atkins bars and keto-friendly snacks. With the declining interest in low-carb diets, most of the above products saw a decline in their sales. They have to monitor the changing trends in consumers' choices and are prepared to adapt their products accordingly based on changing preferences. It could range from reformulation, launching a new flavour, or adding several products into a category to lead the way.

4.5 CHAPTER SUMMARY

The development of food products is a dynamic, complex process, requiring coordination of input from different departments and innovative thinking for an in-depth understanding of consumer trends. Neither does technological advancement change the tastes and preferences of consumers overnight. However, these developments still require food producers to move ahead by developing products that meet not only the present demands but also potential trends.

This chapter demonstrated that the phases of food product development include idea generation and concept testing, prototype development, and final commercialization. Trends such as personalized nutrition, AI-driven innovation, and alternative proteins are shaping the future of food product development. As sustainability becomes an even more crucial aspect, companies should integrate ethical sourcing and sustainable practices into every stage of product development, from ingredient selection to packaging.

The food industry is shifting through an unprecedented change forced by the new technologies and behaviours of consumers, who are working against global challenges such as climate change. Only those companies that take up innovation seriously, favour sustainability, and focus on the type of products they build—keeping in mind the changing needs of the consumer—will turn out successfully and set the standards for the future food industry.

CASE STUDY

Chobani: Innovation in Yogurt and Product Diversification

Chobani transformed the yogurt industry in the United States by introducing Greek yogurt as a mainstream product. The brand's emphasis on quality ingredients, simple recipes, and health benefits resonated with consumers, allowing it to rapidly gain market share and become a leader in the yogurt category. Chobani's success is a testament to the power of innovation in product formulation and brand positioning.

Disrupting the Yogurt Category with Greek Yogurt

While Chobani launched in 2005, the U.S. yogurt market was represented by traditional yogurt products that happened to be rich in sugar and poor in protein. Sensing an opportunity, Chobani introduced Greek yogurt, which is thicker, creamier, and higher in protein than traditional yogurt. Practically unknown to American consumers, Greek yogurt started picking up speed rapidly due to the emphasis on its health benefits, the foremost of which is its high protein content.

This increased popularity of Greek yogurt occurred as an overarching consumer trend towards high-protein foods and clean-label products. Chobani followed this trend by positioning its yogurt as natural, nutritious food from simple ingredients. The fact that the brand uses non-GMO ingredients and promotes sustainable agriculture resonates well with health-conscious and environmentally-conscious consumers.

Product Diversification and Innovation

Along with its growth, Chobani has diversified the range of yogurt forms and flavours to cater to various consumer needs. Chobani Flip, for example, is a product that includes yogurt with mix-ins such as nuts, chocolate, and fruit; this line primarily targets indulgent, but health-minded consumers. Chobani has also launched a set of oat and coconut-based non-dairy products to meet the growing need of consumers to have a wider range of plant-based alternatives.

Apart from increasing yogurt products, Chobani has ventured into the oat milk and coffee creamer segments, benefiting from the brand equity in front of new consumer segments. Diversification has allowed the company to continue growing and succeeding in the unstable food landscape.

Social Responsibility and Brand Values

Chobani's success is due not only to its product innovation but also to its commitment to social responsibility. The company has created a strong brand image with all these values: supporting the cause of refugees, promoting diversity, and offering numerous employee benefits such as profit-sharing. These have enabled the company to create a loyal customer base, valuing the mission more than the products.

Chobani became one of the biggest brands in the U.S. yogurt market with innovations in new products, maintaining a very tight focus on core values, and setting trends in the food industry.

Discussion Questions

1. How did Chobani effectively position Greek yogurt as a healthier alternative to traditional yogurt, and what marketing strategies helped it resonate with consumers?
2. Considering the rise of plant-based alternatives, how has Chobani used product labelling and packaging to communicate the nutritional benefits of its non-dairy and oat-based products?

3. What role does food technology play in Chobani's ability to innovate and maintain product quality while scaling its operations?
4. What are the key technological challenges Chobani might face as it continues to diversify into new product categories such as oat milk and coffee creamers? How can they overcome them?
5. What labelling strategies should Chobani adopt to differentiate itself further in the competitive plant-based and health-food markets?

BIBLIOGRAPHY

Earle, M., Earle, R., & Anderson, A. (2001). *Steps in the Product Development Process.* Woodhead Publishing. https://www.sciencedirect.com/book/9781845697228/food-product-development#book-description

Fuller, G. W. (2016). *New Food Product Development: From Concept to Marketplace.* CRC Press. https://doi.org/9781439818657

CHAPTER 5

Branding and Positioning Strategies in Food Marketing

LEARNING OBJECTIVES

After completion of this chapter, you should be able to:

- Understand the role of branding in food marketing.
- Recognize key components such as brand identity, personality, story, promise, and positioning in the context of food marketing.
- Analyze positioning strategies in food marketing.
- Evaluate the role of packaging in branding.
- Understand brand loyalty and its components.
- Examine global vs. local branding strategies.
- Understand the impact of digital platforms such as social media, influencer marketing, and AR/VR technologies in creating effective brand experiences.
- Identify and discuss challenges related to food branding.

5.1 INTRODUCTION

Effective branding and positioning strategies play a crucial role in gaining market share and securing long-term success in the highly competitive food business. These methods are crucial for developing a distinct brand, distinguishing items from competitors, and building a formidable presence in the minds of consumers. This extensive examination of branding and positioning strategies in food marketing explores their significance, essential components, and practical implementations. The chapter will discuss the role of branding and positioning strategies in food marketing with real-world examples.

5.2 SIGNIFICANCE OF BRANDING IN FOOD MARKETING

Branding in the food sector extends beyond a mere logo or name, encompassing the complete consumer experience with a product. A robust brand has a number of advantages:

DOI: 10.1201/9781003473305-5

- Consumers are more inclined to make purchases from brands they have confidence in. Consistent excellence and favourable encounters cultivate allegiance, resulting in recurring acquisitions and brand endorsement.
- In a competitive market, having a unique brand allows products to be more noticeable and distinguishable. Efficient branding emphasizes distinctive selling points (USPs) that differentiate a product from its rivals.
- Established brands can frequently set higher pricing based on the perceived value they offer. Consumers are willing to incur a higher cost in exchange for reliable and superior items.
- A robust brand can facilitate the introduction of new items into the market by utilizing its established brand value to quickly acquire consumer acceptance.

5.3 ESSENTIAL COMPONENTS OF FOOD BRANDING

There are four primary essential components that should be considered in food branding:

- This encompasses the visual and linguistic components that symbolize the brand. Logos, colour schemes, packaging design, and brand names collectively contribute to the establishment and recognition of a company's identity.
- Food branding involves the human traits associated with a brand. A brand can be seen as enjoyable, refined, trustworthy, and so on. Ben & Jerry's is commonly perceived as an enjoyable and socially aware brand.
- The account conveys the brand's principles, purpose, and past. An engaging narrative about a brand has the power to deeply resonate with people, establishing a strong emotional bond.
- The worth is obtained from how consumers perceive the brand name rather than the actual product. Strong brand equity can result in client loyalty, higher prices, and competitive advantages.

5.4 SIGNIFICANCE OF POSITIONING IN FOOD MARKETING

Positioning refers to the process of determining how consumers view a product in relation to its competitors. Strategic placement is essential for a number of reasons:

- Positioning aids in identifying and appealing to particular consumer categories, guaranteeing that marketing endeavours are concentrated and efficient.
- Positioning emphasizes the primary advantages and characteristics of a product, clarifying the reasons why consumers should select it over alternative options.
- Through the identification and promotion of distinctive characteristics, positioning can generate a competitive edge.
- Effective positioning ensures that consumers see the brand as being in line with its intended image, thereby strengthening the company's identity and values.

5.5 FUNDAMENTAL COMPONENTS OF POSITIONING

- The process of determining the particular segment of consumers that the product is designed for takes into account their demographic characteristics, psychographic profiles, and purchasing patterns.
- Engaging in investigative activities to comprehend consumer requirements, inclinations, and perspectives. This also entails an examination of the strategic posture of competitors.
- Clearly expressing the distinct advantages and worth that the product provides to the intended audience.
- A succinct statement that delineates the target market, the brand's commitment, and the factors that set the product apart from competitors.

5.6 EFFECTIVE APPROACHES FOR BRANDING AND POSITIONING IN FOOD MARKETING

Among the significant approaches of branding and positing in food marketing are the following:

- Flourishing brands and positioning strategies commence with a profound comprehension of consumer demands and preferences. This entails conducting market research, administering customer surveys, and analyzing trends.
- The introduction of new and creative products can set a brand apart and meet the evolving needs of consumers, such as those seeking healthier options or environmentally-friendly packaging.
- Ensuring a consistent delivery of superior items establishes trust and strengthens customer loyalty to the brand.
- Digital platforms are effective tools for establishing and marketing a brand. Utilizing social media platforms, forming influencer collaborations, and implementing content marketing strategies can enhance brand communication and directly interact with consumers.
- Developing and conveying a captivating brand narrative may personalize the brand and establish an affective bond with consumers. Narratives pertaining to heritage, sustainability, or community engagement have the potential to be exceptionally impactful.
- Emotional branding aims to establish a deep emotional bond with consumers. This can be accomplished by implementing marketing strategies that elicit emotions of nostalgia, happiness, or a sense of belonging.
- In response to increasing consumer awareness of environmental and social issues, brands that prioritize sustainability and ethical practices can set themselves apart and appeal to a dedicated customer following.
- Attractive and practical packaging has the potential to impact consumer choices when making a purchase. The packaging should accurately convey the brand's distinctive characteristics and cater to the specific tastes of the target market.
- Collaborations with other businesses, influencers, or organizations can augment brand exposure and establish brand legitimacy.
- Actively interacting with customers and actively seeking their opinions can enhance the quality of products and cultivate a feeling of community surrounding the brand.

5.7 EXAMPLES OF EFFECTIVE BRANDING AND POSITIONING IN FOOD MARKETING

Several businesses have used a variety of successful branding and positioning strategies to maintain their competitive position in the market. The following section discusses some specific examples:

5.7.1 Coca-Cola

Coca-Cola's branding approach exemplifies the creation of an emotional bond with consumers. The brand has constantly established itself as a symbol of joy and unity, represented by its enduring tagline "Open Happiness." Coca-Cola's marketing campaigns frequently highlight communal experiences and nostalgia, strengthening its status as an enduring and universally adored brand.

5.7.2 Whole Foods Market

Whole Foods Market has successfully established itself as a high-end grocery store that prioritizes organic and natural items. The branding of the product highlights its commitment to excellence, well-being, and environmental responsibility. Whole Foods has successfully attracted health-conscious consumers and promoted ethical sourcing, resulting in a dedicated client base that is ready to pay higher costs.

5.7.3 Chobani

The success of Chobani in the highly competitive yogurt business can be ascribed to its adept branding and strategic positioning. Chobani established itself as a superior and more authentic alternative to conventional yogurts. The branding of the product emphasizes the use of high-quality ingredients and a strong dedication to non-GMO products. Chobani's brand identity has been further reinforced by its narrative of its founder's journey and its emphasis on community efforts.

5.7.4 Ben & Jerry's

Ben & Jerry's is notable for its distinctive brand personality and its dedication to social causes. The brand's idiosyncratic and light-hearted image is mirrored in its product names, packaging, and marketing activities. Ben & Jerry's presents itself as more than just an ice cream brand; rather, it is projected as a socially conscious corporation that actively promotes and supports a range of social and environmental initiatives. This strict adherence to consumer values has attracted a devoted following.

5.8 THE ANATOMY OF A FOOD BRAND

Establishing a prosperous food brand necessitates more than simply having a memorable logo or a catchy name; it demands a unified and carefully planned approach that incorporates various crucial components. These elements collaborate to develop a brand's distinctiveness, set it apart from its rivals, and establish a profound connection with consumers. In this discussion, we examine the fundamental components that make up the structure of a food brand.

5.8.1 Brand Identity

The brand identity encompasses all the components that a firm develops to effectively convey the desired image to its customers. The brand identity encompasses visual components such as the logo, colour scheme, typography, package design, and brand name. Consistency of these aspects across all touchpoints is essential for establishing a unified and easily identifiable brand image. For instance, the vivid red and white colours used in Coca-Cola's logo and packaging are immediately identifiable and have been closely associated with the company.

5.8.2 Brand Personality

Brand personality encompasses the anthropomorphic characteristics linked to a brand. The brand's communication and behaviour shape its persona, establishing a connection with consumers. For example, Ben & Jerry's is renowned for its humorous, unconventional, and socially aware character. The brand's personality is evident in its product names, packaging, and marketing efforts, resulting in an approachable and distinctive image.

5.8.3 Brand Story

An engaging brand narrative is essential for establishing an emotional bond with consumers. It conveys the brand's historical background, core principles, purpose, and future aspirations. An artfully constructed narrative has the power to distinguish a company and cultivate a devoted client following. Chobani's narrative of its founder's progression from immigrant to accomplished entrepreneur strikes a chord with customers and strengthens the brand's dedication to excellence and community.

5.8.4 Brand Promise

The brand promise is a declaration that guarantees buyers the advantages and worth they might anticipate from the goods. It is a pledge to provide a uniform and unvarying experience. Whole Foods Market guarantees the provision of natural and organic items that are devoid of artificial preservatives, colours, and flavors. This commitment establishes consumer awareness, fostering confidence and dependability.

5.8.5 Brand Positioning

Brand positioning refers to the way in which consumers view a brand in relation to its competitors. The process entails the identification of the target market and the effective communication of the distinct advantages and characteristics of the product. Kind Snacks differentiates itself in the snack industry by positioning itself as a healthier option, highlighting the use of natural ingredients and the nutritional advantages it offers.

5.8.6 Visual & Sensory Elements

The visual and sensory aspects, including packaging design, product presentation, and even the taste, texture, and perfume of the product, play a vital role in food branding. The pieces should embody the brand's identity and resonate with the target demographic. Efficient packaging design has the ability to allure consumers and effectively communicate crucial details regarding the product's excellence and advantages.

5.8.7 Emotional Connection

Establishing an emotional bond with consumers has the potential to foster robust brand loyalty. Brands that elicit favourable feelings, such as joy, sentimentality, or solace, have the potential to cultivate profound bonds. For instance, the emotional advertising campaigns of Campbell's Soup often invoke feelings of comfort and connection, strengthening the brand's emotional ties.

5.8.8 Brand Equity

Brand equity is the measure of the added value that a brand contributes to a product. Brand equity is developed gradually via consistent brand encounters, favourable consumer impressions, and emotional connections. Having a strong brand equity can lead to client loyalty, the capacity to charge higher rates, and a competitive edge in the market.

5.9 BRAND POSITIONING STRATEGIES

5.9.1 By Product Attributes

Brand positioning refers to the way in which consumers view a brand in relation to its competitors. One efficient method for establishing a brand's position is by emphasizing its product attributes, which are distinct traits or qualities that set it apart from competing products in the market. Here are some tactics that brands can employ to establish their position based on the characteristics of their products:

5.9.1.1 High Quality and Dependability

Positioning a brand on the basis of quality and reliability entails highlighting exceptional workmanship, longevity, and unwavering performance. Brands that demonstrate exceptional performance in this characteristic frequently demand a higher price and establish a reputation for reliability and client contentment. Some examples include:

- Toyota is renowned for its reputation of being reliable and durable. The company prides itself on providing dependable vehicles that people can trust to perform efficiently for an extended period of time.
- Apple strategically markets its products, including iPhones and MacBooks, as premium, well-crafted devices known for their dependable performance.

5.9.1.2 Advancement and Technological Progress

Brands that prioritize innovation and cutting-edge technology set themselves apart by consistently introducing novel features and developments. They target technologically proficient people seeking the newest and most cutting-edge products. Some instances include:

- Tesla has established itself as a frontrunner in electric vehicles and sustainable energy solutions, highlighting its commitment to pioneering advancements in automotive technology.
- Samsung is renowned for its pioneering approach in the field of consumer electronics. The company strategically presents its goods, such as smartphones and TVs, as being at the forefront of technology and offering a wide range of advanced features.

5.9.1.3 Visual and Artistic Aspects of the Design

Design-oriented positioning emphasizes the creation of visually attractive, fashionable, and skillfully made products. Brands emphasize the sophistication, minimalism, or distinctive aesthetic features that distinguish their products. Some examples include:

- IKEA strategically markets its furniture and home goods as fashionable, practical, and skillfully crafted, targeting cost-conscious consumers who desire reasonably priced, yet visually appealing products.
- Dyson is renowned for its stylish and inventive vacuum cleaners and other household products, with a strong focus on exceptional design and performance.

5.9.1.4 Cost and Worth

Certain companies establish their position in the market by emphasizing their capacity to provide the highest value in exchange for money, achieved either through competitive pricing or by emphasizing cost-effectiveness and affordability. This method is attractive to cost-conscious consumers seeking favourable bargains without sacrificing quality. Some examples are:

- Walmart strategically positions itself as a retailer that provides consistently cheap pricing, appealing to cost-conscious buyers by offering a diverse selection of affordable products.
- Aldi prioritizes affordability and value while maintaining high quality, promoting itself as a cost-effective grocery shop.

5.9.1.5 *The Concepts of Sustainability and Ethics*

Brands can distinguish themselves in the current environmentally-aware market by prioritizing sustainability, ethical practices, and social responsibility. Their focus lies in the promotion of the use of environmentally-friendly materials, implementing sustainable manufacturing methods, and ensuring ethical sourcing practices. Some examples include:

- Patagonia is a corporation that presents itself as a sustainable provider of outdoor apparel and gear. It places a strong emphasis on procuring its materials ethically and taking responsibility for the environment.
- Ben & Jerry's is a well-known brand. It is renowned for its dedication to social and environmental causes. The company presents its ice cream products as being sourced in an ethical manner and being environmentally conscious.

5.9.2 By User Profile

Brand positioning through user profiling entails customizing marketing strategies and communication to connect with distinct customer categories based on demographics, psychographics, or behaviours. By comprehending the distinct requirements, inclinations, and actions of various user segments, brands may strategically position themselves to attract their intended audiences. Below are a number of tactics for establishing brand positioning that are tailored to user profiles.

5.9.2.1 *Segmentation Based on Demographic Factors*

Demographic variables, including age, gender, income, education, and family status, exert a substantial influence on consumer preferences and behaviours. Brands can establish their stance by considering certain demographic profiles:

- Brands that aim to attract younger consumers typically prioritize trends, innovation, and youth culture. To interact with this group, they employ vivid images, active participation on social media, and collaborations with influencers. Illustrative instances encompass fashion labels such as H&M and technological enterprises like Snapchat.
- Brands that aim to attract wealthy consumers prioritize luxury, exclusivity, and high-quality products. To cater to this group, they employ premium materials, sophisticated designs, and customized consumer experiences. Illustrations encompass high-end automobile manufacturers such as Mercedes-Benz and prestigious fashion labels like Chanel.

5.9.2.2 *Segmentation Based on Psychographic Factors*

Psychographic segmentation refers to the process of dividing a population into distinct groups based on their psychological characteristics, such as their attitudes, values, interests, and lifestyles. Consumer behaviour and purchase decisions are influenced by psychographic elements, which include lifestyle, values, personality traits, and interests. Brands have the ability to establish their stance by utilizing these psychographic profiles:

- Brands targeting environmentally concerned consumers prioritize sustainability, organic ingredients, and ethical behaviour. They attract consumers who prioritize health and environmental sustainability. Among the examples of such companies are Whole Foods Market and the outdoor wear manufacturer Patagonia.
- Forward-thinking and technologically adept: Brands that cater to tech aficionados prioritize innovation, state-of-the-art technology, and advanced functionalities. They attract consumers who prioritize innovation and are quick to adopt new technology. Notable examples consist of prominent technology companies such as Apple and Tesla.

5.9.2.3 Segmentation Based on Consumer Behaviour

Behavioural segmentation refers to the process of dividing a target market based on the behaviour and actions of consumers. This approach focuses on understanding how consumers interact with a product or service, their purchasing patterns, brand loyalty, and other relevant aspects.

Behavioural elements, such as purchase habits, usage patterns, brand loyalty, and desired advantages, enable brands to comprehend consumer behaviours and preferences. Brands can establish their position by utilizing certain behavioural profiles:

- Brands that want to attract price-sensitive consumers prioritize affordability through providing good value for money, and frequently offering deals. They attract consumers who emphasize cost reductions and preferential pricing. Examples include low-cost retailers like Walmart and Aldi.
- Brand loyalty and premium experience are prioritized by brands that promote exceptional customer service, loyalty programmes, and exclusive perks. They attract consumers who prioritize customized experiences and have confidence in the company. Among examples of such firms are the subscription service Amazon Prime and high-end hotels such as Four Seasons.

5.9.2.4 Segmentation Based on Geographical Region

Consumer behaviour and brand impression are also influenced by geographic factors such as location, climate, and cultural preferences. Brands have the ability to establish their position by aligning with regional or cultural preferences:

- Brands that cater to local communities prioritize the use of locally sourced materials, actively engage with the community, and take into account regional tastes. They attract people who prioritize patronizing local companies and products. Illustrations encompass local coffeehouses and boutique stores.
- Brands that have a worldwide presence prioritize cultural diversity, flexibility, and broad appeal. They attract consumers from many locations and cultures while upholding a uniform brand image. Illustrations encompass transnational enterprises such as Coca-Cola and McDonald's.

5.9.3 By Occasion or Use

Brand positioning that is based on occasions or special uses entails customizing marketing methods to meet distinct customer requirements and situations. By comprehending the timing and manner in which consumers utilize their products or services, brands can strategically position themselves to fulfill these particular requirements. Here are some tactics for establishing a brand's position depending on certain occasions or usage:

5.9.3.1 *Daily Utilization*

Positioning a brand for everyday usage entails highlighting its convenience, dependability, and utility. Brands want to integrate themselves into consumers' everyday routines and behaviours. Some strategies that can be employed are:

- Fast-moving consumer goods (FMCG) include products such as Coca-Cola and Nestlé, which are marketed as necessary for daily use. These brands emphasize their widespread availability, reasonable prices, and reliable quality.
- Personal care: Brands like Dove and Colgate prioritize goods that cater to everyday hygiene and grooming requirements, emphasizing their dependability and efficacy.

5.9.3.2 *Events of Significance*

Brands can strategically position themselves for special occasions by highlighting opulence, festivity, and extravagance. These brands prioritize the creation of unforgettable experiences and the enhancement of festivities. Some strategies that can be employed are:

- Jewellery and watches: Brands such as Tiffany & Co. market their products as presents for significant events like weddings and anniversaries, highlighting their opulence and enduring grace.
- Champagne and spirits: Brands like Moët & Chandon and Grey Goose strategically market their products for celebratory occasions and exclusive parties, linking them with opulence and refinement.

5.9.3.3 *Utilization for Professional or Occupational Purposes*

Positioning for professional or work use entails emphasizing usefulness, efficiency, and performance. Brands address the requirements of customers in professional environments or for certain assignments. Among the strategies employed are the following:

- Office supplies: Brands such as Staples and HP market their products as dependable and effective instruments for office use, emphasizing productivity and excellence.
- Business services: LinkedIn and Slack promote their platforms as indispensable resources for professional networking and communication, highlighting their emphasis on connectedness and productivity.

5.9.3.4 *Use that is Dependent on the Seasons or Weather Conditions*

Brands have the ability to establish their place in the market by catering to the specific needs that arise from seasonal or weather-related factors. They can modify their products or services to align with the evolving preferences and conditions of consumers. Among the strategies that can be employed are:

- Winter apparel: Brands such as The North Face and Patagonia market their products as indispensable for frigid weather conditions, highlighting qualities such as insulation, resilience, and suitability for outdoor activities.
- Summer beverages: Brands like Coca-Cola and Pepsi market their drinks as invigorating choices for sweltering summer days, highlighting the benefits of hydration and pleasure.

5.9.3.5 *Tourism and Recreation*

Positioning for travel and leisure entails prioritizing the aspects of being easily transportable, providing comfort, and ensuring enjoyment. Brands respond to the demands of consumers when they are on vacation, travelling, or engaging in recreational activities. Among the strategies are:

- Luggage and travel accessories: Brands such as Samsonite and Tumi market their products as long-lasting, practical, and fashionable choices for travellers, with a strong emphasis on convenience and dependability.
- Outdoor equipment: Brands like REI and Patagonia market their items as indispensable for outdoor expeditions and leisure pursuits, highlighting their resilience, functionality, and eco-friendliness.

5.9.4 Against a Competitor

To help distinguish your brand in the eyes of consumers, it is essential to create a unique position in the market. This unique position should highlight your brand's advantages over competitors and resonate with your target audience. Below are a number of highly effective tactics for placing your brand concerning a competitor:

5.9.4.1 *Strategy of Differentiation*

A differentiation strategy emphasizes the distinct characteristics, advantages, or qualities that distinguish your brand from competitors. These factors may encompass product innovation, high quality, outstanding customer service, or distinctive brand values. Some strategies include:

- Prioritize ongoing innovation and product development to provide unique features or functionalities that surpass those of competitors. Apple's unwavering commitment to introducing new iPhone models and features distinguishes it from other competitors in the smartphone market.

- Emphasize the superiority of your brand by positioning it as providing superior quality or reliability in comparison to competitors. Volvo has a strong focus on safety and longevity in its vehicles, setting itself apart from other automobile manufacturers.

5.9.4.2 *Pricing Leadership*

A pricing leadership approach entails portraying your brand as providing superior value for money in comparison to competitors. This may entail providing discounted prices, promotional offers, or value-added services that justify a higher price. Among the strategies that can be employed are:

- Cost leadership: Engage in price competition by minimizing manufacturing costs and transferring the resulting savings to customers. Walmart and Aldi demonstrate exceptional proficiency in this approach by providing consistently low pricing on a diverse selection of products.
- Value proposition: Highlight the worth and advantages that clients obtain in exchange for the price they pay. Amazon Prime provides expedited shipping, streaming services, and special discounts for a membership fee, thereby offering additional value beyond the cost of products.

5.9.4.3 *Niche Marketing*

Niche marketing entails focusing on a particular subset of the market that competitors might neglect or inadequately cater to. This strategy is centred around catering to the distinct requirements or preferences of a smaller, more specialized audience. Some strategies that can be employed are:

- Segmentation: This involves the process of identifying a specific market segment that has unique and specific wants or preferences. For example, Lululemon caters to individuals who are passionate about yoga by offering top-notch athletic clothing that is designed to meet their unique needs, emphasizing both functionality and fashion.
- Personalization: This provides customized products or experiences that specifically target particular customer segments. Nike offers customers the ability to customize their shoes and gear, enabling them to create individualized goods that accurately represent their own interests.

5.9.4.4 *Customer Experience*

Customer experience-based positioning prioritizes the delivery of outstanding service, convenience, and satisfaction that exceeds that of competitors. The objective of this strategy is to establish a solid foundation of consumer loyalty and advocacy. Some of the strategies include:

- Service excellence: Deliver exceptional customer service by engaging in individualized contacts, promptly resolving issues, and offering proactive assistance. Zappos is widely recognized for its exceptional customer care, which includes complimentary returns and round-the-clock customer help.

- Convenience: Highlight the convenience factor by offering effortless online ordering, swift delivery, and smooth purchasing experiences. Amazon's Prime membership offers expedited shipping and supplementary services such as Prime Video, which enhances convenience for customers.

5.9.4.5 *The Importance of Brand Authenticity and Values*

Authenticity and values-based positioning involve aligning with consumer ideas, ethical standards, and social responsibilities. This approach fosters trust and loyalty by showcasing sincere dedication to significant issues or convictions. Among the strategies are the following:

- Corporate Social Responsibility (CSR): This refers to the practice of endorsing social or environmental initiatives that align with the values and interests of your intended customer base. Patagonia's dedication to environmental sustainability and ethical sourcing resonates with ecologically aware consumers.
- Transparency: Foster trust by openly sharing information about business methods, ingredients, and sourcing. Honest Tea presents itself as a company that is open and socially accountable, employing organic ingredients and fair trade methods.

5.9.5 By Price or Quality

Brand positioning, which is focused on either price or quality, is strategically utilizing these factors to establish a unique market position that attracts specific consumers. Brands can effectively distinguish themselves by either stressing affordability and value or by setting premium prices and offering higher quality. Brands can strategically position themselves by emphasizing either their price or their quality.

5.9.5.1 *The Strategy of Price Leadership*

Price leadership positioning is selling items or services at lower prices than competitors, targeting price-sensitive clients that emphasize affordability. Some important strategies include:

- Everyday low prices: Brands such as Walmart and Aldi are renowned for their dedication to providing consistently low prices on a diverse array of products. They utilize economies of scale, streamlined supply chains, and minimum extras to maintain low prices and transfer the resulting savings to customers.
- Brands can appeal to cost-conscious consumers by regularly providing discounts, promotions, or bundled offers. This approach promotes purchasing choices that are influenced by the perceived value and potential savings.
- Value proposition: Highlight the worth and advantages that clients obtain in exchange for the cost incurred. Amazon Prime provides expedited shipping, streaming services, and special discounts for a monthly fee, thereby offering supplementary benefits in addition to the price of the products.

5.9.5.2 The Strategy of Distinguishing Products Based on Their Quality

Quality differentiation positioning involves providing products or services that are considered superior in terms of their workmanship, longevity, effectiveness, or distinctive characteristics. Essential tactics comprise the following:

- High-end companies such as Rolex and Mercedes-Benz prioritize top-quality materials, precise craftsmanship, and rigorous attention to detail, attracting consumers who appreciate luxury and exclusivity.
- In the realm of innovation and technology, brands such as Apple and Tesla establish themselves as frontrunners, presenting "state-of-the-art" products equipped with advanced features and exceptional performance, thereby justifying their higher price tags.
- Prioritize delivering outstanding customer service, tailored interactions, and after-sales assistance to elevate the overall perception of the brand's quality. Brands such as Ritz-Carlton and Nordstrom are highly proficient at providing exceptional customer experiences that validate their higher prices.

5.9.5.3 Cost-Effective Approach

Value-based positioning achieves a harmonious equilibrium between cost and quality, providing items or services that offer the optimal blend of affordability and effectiveness. Essential tactics comprise:

- Mid-range pricing: Brands such as Samsung and Toyota target the mid-range market by providing products that have both good quality and features, all at competitive pricing. This strategy allows them to attract a wide spectrum of consumers.
- Stressing dependability: Highlight the dependability, longevity, and constant functionality of the product to justify its price and establish consumer confidence. Brands such as Honda and Sony prioritize the production of dependable products that have strong performance indicators.

5.9.5.4 Strategy for Setting Higher Prices

Premium pricing positioning is establishing higher prices to convey exclusivity, luxury, exceptional quality, or distinctive brand heritage. Some important strategies include:

- Luxury businesses such as Louis Vuitton and Gucci establish exclusivity by offering limited editions, personalized services, and exceptional client experiences, which justify their higher prices.
- Brands like Rolls-Royce and Chanel, known for their long-established lineage and reputation for excellence, utilize their brand legacy to demand higher prices and attract wealthy consumers who desire status and luxury.

5.10 ROLE OF PACKAGING IN BRANDING

Packaging is essential in branding as it serves as a physical reflection of a brand's identity, values, and products. It extends beyond the practical responsibilities of safeguarding and containing products to actively shape consumer perceptions, purchase choices, and overall brand encounters. This analysis delves into the pivotal functions that packaging fulfills in establishing a brand's identity, substantiated by illustrative instances:

5.10.1 Initial Perception and Visual Representation

Packaging functions as the initial interface with consumers, establishing a powerful first impression that can influence brand perception. The brand's visual identity is represented by its use of colours, typography, logos, and design elements. Some examples include:

- Apple is renowned for its minimalist and clean packaging designs, which effectively convey the brand's premium and sophisticated image. The pristine white containers adorned with delicate embossing exude sophistication and meticulousness, perfectly harmonizing with Apple's product design principles.
- The renowned Tiffany blue box, produced by Tiffany & Co., immediately brings to mind opulence, refinement, and a feeling of eager expectation. The unique hue and packaging layout enhance the brand's traditional roots and high-end reputation in the jewelry sector.

5.10.2 Distinguishing and Establishing the Unique Identity of a Brand

Packaging plays a crucial role in distinguishing a brand from its competitors in highly competitive marketplaces by effectively communicating its distinctive selling points and brand positioning. It conveys the brand's ideals, excellence, and desired consumer experience. Some examples include:

- Coca-Cola products can be easily recognized by their iconic contour bottle shape and vibrant red colour. The packaging enhances the brand's legacy, rejuvenates the flavor, and establishes an emotional bond with consumers globally.
- Lush is renowned for its environmentally-friendly and artisanal cosmetics. The company employs simple and recyclable packaging to uphold its dedication to sustainability and the use of natural ingredients. The packaging enhances the brand's ethical principles and attracts ecologically aware consumers.

5.10.3 Disseminating Product Information

Efficient packaging conveys vital product information, directions for use, and advantages to consumers. It provides education and information while improving the entire brand experience. For example:

- Amazon packages are typically labelled and accompanied by instructions that are straightforward and succinct, providing guidance on how to open, return, and handle them. The packaging guarantees a smooth customer experience and strengthens Amazon's dedication to convenience and consumer contentment.
- IKEA: The flat-packaging design employed by IKEA serves the purpose of minimizing transportation expenses and environmental consequences. Additionally, it

incorporates detailed assembly instructions in the form of illustrations. The packaging amplifies the perceived worth of IKEA products as convenient, functional resolutions for household decor.

5.10.4 Establishing Emotional Bonds

Packaging has the ability to elicit emotional reactions and generate lasting impressions that deeply connect with people. It boosts customer loyalty and promotes recurring sales. Illustrative instances encompass:

- Hershey's Kisses are instantly recognizable due to their classic silver foil wrapper, which evokes feelings of nostalgia and luxury. The minimalistic, yet unique packaging design amplifies the emotional allure of the brand's chocolate products.
- Oreo's packaging design incorporates the brand's emblem and distinctive colours, evoking a feeling of familiarity and dependability. The container design enhances Oreo's light-hearted and sentimental brand image, promoting emotional bonds with customers.

5.10.5 Advancement and Longevity

Introducing new packaging methods can strengthen a brand's dedication to sustainability, minimize ecological harm, and attract environmentally aware consumers. Examples include:

- Patagonia, renowned for its dedication to environmental sustainability, employs recycled and recyclable packaging materials for its outdoor clothing. The packaging mirrors the brand's principles and connects with consumers who are conscious about the environment.
- In addition to its basic appearance, Lush employs compostable packaging materials and promotes the practice of customers returning used containers for recycling. The packaging innovation is in line with Lush's ethical position on waste reduction and sustainability promotion.

Packaging serves as a potent instrument in branding, surpassing mere practical necessities to influence consumer perceptions, strengthen brand identification, and generate lasting impressions. Efficient packaging design effectively conveys brand principles, distinguishes products in competitive marketplaces, and improves consumer involvement. Brands can strategically utilize packaging to enhance their market position, cultivate brand loyalty, and achieve long-term success in ever-changing consumer environments by utilizing visual identity, differentiation, information communication, emotional connection, and sustainability practices.

5.11 BUILDING BRAND LOYALTY

Establishing brand loyalty is essential for organizations seeking to maintain a robust and enduring consumer base. It entails establishing favourable connections, confidence, and recurring acquisitions among consumers. Efficient tactics prioritize the improvement of client satisfaction, the cultivation of emotional bonds, and the provision of reliable value. This section provides an in-depth analysis of the strategies that brands employ to cultivate customer loyalty, accompanied by illustrative examples.

5.11.1 Unvarying and Dependable Customer Experience

Establishing a uniform and dependable client experience is crucial for cultivating brand loyalty. Brands that continuously fulfill or surpass client expectations establish trust and contentment over a period of time. Suitable examples include:

- Amazon is renowned for its customer-centric strategy, including dependable shipping, hassle-free returns, and tailored suggestions. These services improve convenience and foster confidence, thus promoting consumer loyalty and driving repeat transactions.
- Starbucks ensures uniformity in the excellence of its products, the level of customer service provided, and the atmosphere of its stores throughout its worldwide network. The constant experience enhances the perception of brand dependability and motivates customers to select Starbucks as their usual coffee provider.

5.11.2 Outstanding Customer Service

Providing exceptional customer service is crucial for building client loyalty by immediately and effectively resolving their requirements. Brands that place high importance on ensuring customer happiness and promptly resolving any issues are able to establish robust relationships with their customers:

- Zappos is well-known for its outstanding customer service and consistently goes the extra mile to guarantee client pleasure. Their highly responsive customer support service and exceptionally lenient return policies foster good experiences that bolster consumer loyalty.
- Ritz-Carlton: The renowned hotel chain is acclaimed for its bespoke client care and meticulous focus on details. Staff members are given the authority to predict and meet the needs of customers, delivering outstanding service that encourages loyalty and positive recommendations.

5.11.3 Customization and Customer Interaction

Customizing interactions and actively involving clients on an individual basis can greatly improve loyalty. Brands that customize their products and messages to suit individual interests and behaviours establish significant connections with customers. Examples include:

- Netflix utilizes tailored suggestions derived from users' viewing history and interests to optimize the user experience. Netflix enhances user pleasure and promotes ongoing subscriptions by recommending material according to individual preferences.

- Nike offers customers the opportunity to customize their sneakers with unique designs and colours through its NikeID platform. This customization cultivates a feeling of possession and emotional connection to the products, which enhances brand loyalty among sneaker fans.

5.11.4 Utilizing Emotional Branding and Community-Building Strategies

Establishing emotional bonds through narrative, mutual principles, and engagement with the community has the potential to generate robust brand allegiance. Brands that evoke strong emotional connections with their customers cultivate a feeling of inclusion and steadfast devotion. Examples include:

- The Dove brand's Real Beauty campaign aims to acknowledge and embrace the wide range of appearances and characteristics found among women, while also encouraging a positive sense of self-worth. Dove cultivates a devoted consumer base by promoting body positivity and inclusivity, attracting those who resonate with its principles.
- Harley-Davidson fosters a robust community of motorcycle aficionados through various events, clubs, and social gatherings. The brand's focus on liberty, uniqueness, and the expansive journey establishes a profound emotional bond with its clientele.

5.11.5 Loyalty Programmes and Motivational Initiatives

Integrating loyalty programmes and rewards can stimulate recurring purchases and enhance brand loyalty. Brands that provide awards, discounts, or exclusive benefits to loyal customers show gratitude and encourage ongoing involvement. Among examples are the following:

- Sephora operates a Beauty Insider programme that offers users points for their purchases, which may be exchanged for products or special experiences. The programme promotes regular buying and fosters loyalty among beauty enthusiasts.
- American Airlines offers the advantage programme, which enables frequent passengers to accumulate miles through flights, hotel accommodation, and partner transactions. The programme provides exclusive membership levels and incentives, cultivating loyalty among regular travelers.

Developing brand loyalty necessitates a methodical strategy that places emphasis on ensuring customer contentment, tailoring experiences to individual preferences, fostering emotional connections, and offering incentives. Through the consistent delivery of extraordinary experiences, a deep understanding of client preferences, and the establishment of meaningful connections, brands may create a loyal customer base that continually picks their products or services above those offered by competitors. These examples demonstrate how prosperous businesses utilize several techniques to establish and sustain robust relationships with their customers, ultimately leading to long-term success and profitability in competitive markets.

5.12 REBRANDING AND BRAND EXTENSIONS

Rebranding and brand extensions are deliberate actions taken by organizations to refresh their brand image, broaden their market presence, and use their existing brand value. Rebranding is the process of modifying aspects of a brand's identity to align with new positioning or market circumstances. On the other hand, brand extensions entail utilizing an established brand name to introduce new items or venture into new market categories.

5.12.1 Rebranding

Rebranding refers to the process of modifying various aspects of a brand's identity, including its name, logo, design, or messaging, to develop a new brand image that aligns more effectively with current market trends, consumer preferences, or corporate goals. The objective frequently involves distinguishing oneself from rivals, enticing fresh clientele, or rejuvenating the perception of the brand. Some illustrations include:

- Old Spice: Old Spice executed a triumphant rebranding initiative in the early 2010s with the aim of captivating a younger target audience. The brand altered its communication strategy from conventional masculinity to a focus on humour and disregard for authority, intending to appeal to a younger demographic. This was achieved through the implementation of memorable advertising campaigns featuring actor Isaiah Mustafa.
- Starbucks: Starbucks regularly modifies their logo and branding to remain current and align with changing consumer preferences. The rebranding initiatives of the coffee corporation prioritize sustainability, community involvement, and digital advancement, in line with current consumer ideals.

5.12.2 Brand Extension

Brand extensions refer to the strategic utilization of the reputation and familiarity of a well-established brand in order to introduce new products or penetrate new markets. This approach enables organizations to use established customer loyalty, minimize marketing expenses, and lessen the risks associated with introducing completely new brands:

- Apple is widely recognized for its effective brand expansions throughout its whole range of products. Apple has expanded its brand into other consumer electronics categories, including Mac computers, iPods, iPhones, and iPads. This expansion has been possible due to Apple's renowned reputation for innovation and high-quality products.
- Dove, first recognized for its beauty soap, effectively expanded its brand to encompass a wide range of personal care items, including body washes, shampoos, and skincare products. Dove's brand extension strategy prioritizes the aspects of moisturization, gentle washing, and the promotion of self-esteem, which are in line with its initial brand commitment.

5.12.3 Strategic Considerations

Brand equity utilization refers to the strategic use of a brand's existing reputation, trust, and recognition to introduce new products or enter new markets. This approach helps to expedite the acceptance and adoption of the brand in these new categories or areas.

- Incorporating brand extensions can be a more economical approach compared to creating completely new brands, as it minimizes the requirement for extensive marketing efforts to develop brand recognition.
- Brand dilution occurs when brand extensions are poorly implemented and fail to correspond with consumer perceptions or quality expectations, resulting in a decrease in the core brand's equity.
- Market cannibalization is a potential concern when introducing new items that directly compete with existing offerings within the brand portfolio, as it may lead to a decrease in sales of the current products.
- Conduct comprehensive market research to uncover opportunities, customer preferences, and any dangers related to rebranding or brand extensions.
- It is important to make sure that any rebranding or brand expansions are in line with the fundamental principles of the brand, meet the expectations of consumers, and retain the brand's position in the market. This is necessary to maintain a sense of coherence and integrity.
- Effectively convey changes or new offerings to consumers through comprehensive marketing efforts, highlighting advantages, distinctiveness, and value propositions.

Rebranding and brand extensions are strategic techniques that organizations employ to adjust to shifting market dynamics, broaden their range of products, and sustain their competitiveness. Revitalizing a brand's identity to attract new audiences or expanding a trusted brand name into different product categories necessitates meticulous planning, customer insights, and adherence to brand principles. Successful examples illustrate how organizations efficiently utilize their established brand value to stimulate growth, innovation, and long-term market significance.

5.13 GLOBAL VS. LOCAL BRANDING STRATEGIES

Global and local branding strategies are different ways to manage a brand, designed to take advantage of specific market conditions, customer habits, and competitive environments. Global strategies focus on maintaining a uniform brand image across different countries, while local strategies emphasis tailoring and adjusting the brand to cater to distinct regional preferences and cultural subtleties.

5.13.1 Global Branding Strategy

Global branding strategies are defined as approaches that prioritize consistency in brand identity, messaging, and positioning throughout many markets around the world. The objective is to establish a unified brand identity that surpasses local limitations and appeals to a worldwide customer demographic.

Notable attributes:

- Global brands maintain a consistent brand identity by using the same logos, colours, packaging, and messaging. This ensures that they are instantly recognizable and coherent throughout different markets.
- Economies of scale refer to the cost advantages that global businesses can gain through the standardization of processes and marketing campaigns. This allows them to streamline production, distribution, and marketing, resulting in cost efficiencies.

Illustrations:

- Coca-Cola demonstrates a triumphant worldwide branding strategy through the use of consistent branding features, including its renowned red and white logo, distinctive contour bottle design, and widely acknowledged advertising campaigns. The brand upholds a uniform portrayal of joy, rejuvenation, and pleasure on a global scale.
- Nike is widely known for its worldwide recognizable symbols of athletics, innovation, and empowerment, namely the "Just Do It" slogan and swoosh logo. Nike's worldwide branding strategy prioritizes overarching themes of performance and inspiration, effectively connecting with athletes and consumers from many cultural backgrounds.

5.13.2 Local Branding Strategy

Local branding strategies relate to the deliberate efforts made to tailor brand aspects, product offerings, and marketing methods to cater to the unique cultural, linguistic, and consumer preferences of distinct markets or areas.

Notable attributes:

- Cultural relevance refers to the practice of adapting products, marketing, and promotions to match the habits, traditions, and consumer behaviours of a specific region or culture.
- Market sensitivity: Local brands exhibit agility in adapting to market trends, regulatory landscapes, and competitive forces that are specific to each geographical area.

Illustrations:

- Unilever effectively executes local branding strategies by modifying its product formulas and marketing messages to align with regional preferences and consumer demands. Brands such as Dove and Lipton offer localized versions that specifically target a wide range of cultural preferences and beauty ideals.
- McDonald's tailors its menu options to suit the specific tastes and preferences of different countries. McDonald's caters to the preferences of different regions by providing vegetarian choices in India, adapting portion sizes in Asia, and including local flavours into its menu items.

5.13.3 Strategic Considerations

Strategic Considerations for Global Branding:

- Implementing a global branding strategy creates a cohesive brand image and enhances brand awareness on a global scale.
- Economies of scale refer to the cost advantages that arise from standardized procedures and marketing initiatives, resulting in improved cost efficiencies and streamlined operations.

Strategic Considerations for Local Branding:

- Cultural relevance is achieved through local branding, which fosters customer trust and acceptance by showing an understanding of local culture and tastes.
- Market adaptation refers to the ability to quickly and effectively adjust to the specific dynamics and trends of a local market, which can result in accelerated growth and increased market share.

5.13.4 Strategies for Putting Plans into Action

Global and local branding strategies are different methods of managing a brand, each with its own benefits and difficulties. Global strategies promote consistency and scalability across worldwide markets, while local strategies prioritize customization and cultural relevance to resonate with varied consumer segments. Effective global brands deliberately integrate global brand standards with localized modifications to efficiently navigate global markets while satisfying local consumer expectations. By comprehending the subtle distinctions between global and local branding strategies and utilizing pertinent illustrations, brands can strengthen brand value, consumer involvement, and market competitiveness in an ever-more interconnected global economy.

Some key strategies include:

- Harmonize: Achieve a balance between global brand standards and localized adjustments in order to preserve brand integrity while addressing regional variations.
- Market research: Perform thorough research to gain insights into regional consumer behaviors, preferences, and competitive environments.
- Coordination: Promote coordination between the global headquarters and local teams to ensure synchronization and efficient implementation of branding strategies.

5.14 BRANDING IN THE DIGITAL AGE

In the digital era, branding has undergone substantial transformation, utilizing technology and online platforms to creatively interact with consumers. Digital branding goes beyond traditional limits, providing organizations with exceptional chances to establish connections, enhance visibility, and cultivate loyalty.

Currently, marketers utilize the influence of social media, websites, and digital advertising to construct captivating stories and interact directly with their intended audience. Social media platforms such as Instagram, Facebook, and Twitter allow brands to engage in immediate communication, exchange genuine content, and foster communities based on common values and interests. By directly engaging with consumers, brands can personalize their identities, promptly address customer feedback, and customize messages for certain demographics.

Furthermore, digital branding facilitates accurate targeting by utilizing data analytics and individualized marketing methods. Brands have the ability to monitor and analyze consumer behaviour, preferences, and purchase habits to provide tailored content and offers, thereby improving the entire customer experience. The incorporation of SEO (Search Engine Optimization) and SEM (Search Engine Marketing) strategies enhances visibility and increases traffic to digital assets, thereby optimizing brand presence across various digital platforms.

Essentially, digital branding enables organizations to establish significant relationships, quickly adjust to market trends, and evaluate campaign efficacy in real time, creating a dynamic and responsive brand presence in the digital realm.

5.14.1 Social Media and Branding

Social media has transformed the process of establishing and promoting a brand by providing platforms that allow for direct engagement, narrative creation, and the formation of communities. This enables brands to connect with audiences on a worldwide level. Utilizing social media effectively not only increases the visibility of a company but also fosters relationships, encourages interaction, and impacts consumer opinions. This chapter examines how brands utilize social media to enhance their branding efforts, with the support of illustrative examples.

5.14.1.1 Active Involvement and Direct Communication with Consumers

Brands may use social media platforms like Facebook, Instagram, Twitter, and LinkedIn to directly participate in interactions with consumers. Brands have the ability to promptly answer queries, resolve complaints, and actively engage in conversations, thereby creating a perception of being easily approachable and accessible:

- Illustration: Wendy's
- Wendy's is renowned for its clever and captivating presence on Twitter. The fast-food franchise employs wit and rapid reactions to engage with its followers, establishing a distinctive persona that strikes a chord with its audience. This strategy not only increases the level of involvement, but also strengthens Wendy's brand as accessible and customer-oriented.

5.14.1.2 Brand Storytelling and Content Marketing

Social media empowers organizations to communicate captivating narratives, distribute content, and exhibit brand principles, fostering emotional bonds with audiences. Visual media, including movies, pictures, and infographics, may efficiently communicate brand stories and messages.

- Illustration: Airbnb
- Airbnb utilizes Instagram as a platform to narrate anecdotes of travel adventures that have been shared by its community members. Airbnb showcases genuine and varied travel experiences by utilizing user-generated content (UGC), effectively emphasizing its commitment to providing a sense of belonging in any location. This narrative technique adds a human touch to the brand and evokes a strong desire to travel among its followers.

5.14.1.3 Collaborations with Influential Individuals and Support for a Cause

Engaging in partnerships with influencers and brand supporters on social media can significantly enhance the extent of brand exposure, credibility, and interaction. Individuals who have a significant number of followers on social media platforms can promote items, convey genuine personal experiences, and sway the purchasing choices of their followers.

- Example: The brand name is Daniel Wellington.
- Daniel Wellington has utilized influencer marketing on platforms such as Instagram to endorse its fashionable timepieces. Daniel Wellington collaborates with fashion influencers and celebrities to produce captivating material that presents its products in various lifestyle settings, attracting fashion-savvy consumers and boosting sales.

5.14.1.4 Customer Feedback and Market Insights

Social media platforms function as great reservoirs of client input and market knowledge. Brands have the ability to observe discussions, analyze emotions, and collect information to comprehend consumer preferences, trends, and attitudes immediately, which aids in making informed strategic decisions and developing products.

- Illustration: Starbucks
- Starbucks utilizes social media networks to collect feedback regarding new products and customer experiences. Starbucks actively interacts with its audience through surveys, comments, and direct messaging to gain insights into their preferences and expectations. This helps Starbucks to consistently enhance and innovate its offerings.

5.14.1.5 Strategies for Handling Emergencies and Establishing a Positive Public Image

Social media is essential for managing crises and establishing a positive reputation. Brands have the ability to promptly react to crises, openly tackle problems, and control public opinion by engaging in timely communication and providing updates.

- Illustration: Domino's Pizza
- Domino's Pizza employed social media as a means of addressing a situation stemming from unfavourable consumer feedback. The brand acknowledged the problems,

publicly apologized and made improvements in response to customer criticism. This transparent methodology aided in the restoration of trust and credibility among consumers.

Social media has revolutionized the concept of branding by providing interactive channels for interaction, narrative development, collaborating with influencers, gaining customer insights, and handling crises. Brands that successfully utilize the influence of social media can establish robust connections with consumers, amplify brand exposure, and stimulate business expansion. Through smart utilization of platforms, development of engaging content, promotion of community involvement, and attentiveness to audience input, brands may generate a strong online presence that effectively connects with their desired audience in the competitive digital era.

5.14.2 Influencer Collaborations

Brands are using influencer partnerships as an effective method to connect with specific audiences by partnering with well-known individuals who have large followings on social media platforms. These collaborations utilize the credibility, authenticity, and reach of influencers to increase brand visibility, encourage engagement, and impact customer behaviour. This chapter delves into the realm of influencer collaborations in brand marketing, providing concrete instances to exemplify the concept.

5.14.2.1 Categories of Influencers

Influencers can be classified according to their extent of influence, specific area of expertise, and amount of interaction:

- Macro-influencers refer to celebrities or renowned individuals who have a substantial number of followers, usually exceeding 100,000.
- Micro-influencers are individuals who have smaller, yet highly engaged, specialty audiences, usually ranging from 10,000 to 100,000 followers.
- Nano-influencers refer to individuals who have a relatively small number of followers, typically less than 10,000. They are highly regarded for their strong engagement rates and genuine presence within specific interest groups.

5.14.2.1.1 Influencer Collaborations

One example is the partnership between Daniel Wellington and micro-influencers. Daniel Wellington, a watch brand, strategically partners with micro-influencers on Instagram. These influencers, typically lifestyle or fashion bloggers, frequently post pictures of themselves wearing Daniel Wellington watches, along with discount codes that their followers can use. This collaboration not only increases the exposure of the business among certain target audiences but also utilizes the influencers' trustworthiness and expertise in fashion to boost sales.

5.14.2.1.2 Nike's Collaboration with Athletes and Influencers

Nike regularly engages in partnerships with athletes and fitness influencers on several social media platforms. Nike engages in various partnerships that encompass sponsored postings, product evaluations, and motivational material that resonate with the company's core principles of athleticism, innovation, and empowerment. Nike enhances their brand image as a frontrunner in sports performance and fosters consumer loyalty and engagement by partnering with renowned athletes.

5.14.2.1.3 Sephora and Beauty Influencers

Sephora collaborates with beauty influencers on YouTube and Instagram to endorse and publicize new cosmetic goods, skincare regimens, and makeup instructional videos. These agreements involve influencers doing tests and evaluations of Sephora's goods, offering genuine recommendations to their followers. Sephora gains advantages from heightened product visibility, client confidence, and the ability to convert purchases within the beauty community.

5.14.2.2 Advantages of Influencer Collaborations

- Enhanced reach and visibility: Influencers provide brands with the opportunity to connect with certain target audiences and demographics which may be challenging to reach through conventional advertising methods.
- Authenticity and trustworthiness: The influencers' content and personal endorsements resonate with their followers, creating a sense of trust and legitimacy for the products they advocate.
- Engagement and conversion: The audiences of influencers who are actively involved are more inclined to interact with sponsored material, resulting in increased rates of engagement and a greater likelihood of converting into customers.
- Creativity: Influencers contribute to brand campaigns by bringing creativity and a range of opinions, resulting in compelling content that connects with their following and effectively amplifies brand messaging.

5.14.2.3 Obstacles and Factors to Consider

- Authenticity concerns arise when influencers face difficulties in maintaining genuine representation while endorsing several businesses or when they fail to disclose sponsored content.
- Relevance and alignment are essential factors in selecting influencers for a campaign. It is necessary to choose influencers whose beliefs, audience demographics, and content closely match those of the brand to achieve campaign objectives.
- Assessing the efficacy of influencer marketing necessitates the use of measures such as engagement rates, reach, conversions, and brand sentiment to determine the return on investment (ROI).

Influencer collaborations have transformed into a dynamic and efficient marketing tactic for brands aiming to genuinely engage with certain audiences. Brands may enhance their brand awareness, engagement, and positive customer behaviour by collaborating with influencers that align with their brand values and target audience characteristics. This is achieved by utilizing the influencers' credibility, inventiveness, and wide audience reach. As organizations adapt to the changing digital marketing environment, collaborating with influencers remains an effective strategy for increasing brand recognition, cultivating customer confidence, and accomplishing marketing goals in a competitive market.

5.14.3 Augmented Reality (AR) and Virtual Reality (VR) in Brand Experiences

Augmented Reality (AR) and Virtual Reality (VR) have revolutionized brand experiences by providing immersive and interactive platforms that captivate consumers in novel ways. These technologies combine digital components with the real world (AR) or generate completely artificial worlds (VR), improving customer engagement with businesses and products. This article examines the use of AR and VR in brand experiences, with the support of illustrative examples.

5.14.3.1 Augmented Reality (AR)

Augmented Reality (AR) refers to the integration of digital information or virtual elements into the real world, enhancing the user's perception and interaction with their surroundings. AR finds its applications in various brand experiences, allowing companies to create immersive and interactive experiences for their customers. This content is usually seen through smartphones or AR glasses. Brands utilize augmented reality (AR) to generate interactive experiences, amplify product demos, and include consumers in customized engagements.

Illustrations:

- IKEA Place is a mobile application: IKEA's augmented reality (AR) application, IKEA Place, enables customers to preview furniture in the context of their own homes before making a purchase. Through the utilization of the application, customers can visualize how IKEA merchandise fits and appears in their specific area, thus improving the overall shopping experience and minimizing any doubts regarding the appropriateness of the goods.
- Sephora Virtual Artist: Sephora's AR feature, Virtual Artist, allows customers to experiment with virtual beauty products by utilizing their smartphone camera. Customers have the opportunity to try out several hues of lipstick, eyeshadow, and other cosmetics, allowing them to make well-informed purchasing choices and interact with the brand in a fun and interactive way.

5.14.3.2 Virtual Reality (VR)

Virtual Reality (VR) refers to the use of computer technology to create a simulated environment that can be experienced through sensory stimuli. It has various applications in the field of brand experiences, which involve creating immersive and interactive experiences for consumers to engage with a brand. VR technology allows users to be fully immersed in virtual settings using headgear or VR devices. Brands utilize VR technology to transport consumers to simulated environments, provide virtual tours, and create impactful narrative experiences.

Illustrations:

- Virtual Test Drive by Mercedes-Benz: Mercedes-Benz provides a VR experience that enables prospective customers to partake in virtual test drives of their vehicles. Users get the opportunity to examine the car's interior, try out its features, and simulate different driving situations. This allows them to get a realistic sample of the driving experience and strengthens the brand's dedication to innovation.
- The North Face VR Expedition: The North Face developed a VR encounter that transports visitors on a simulated journey to Yosemite National Park. Participants get the opportunity to discover awe-inspiring landscapes, ascend mountains, and engage in outdoor escapades, which aligns with The North Face's brand story of exploration and adventure.

5.14.3.3 Benefits and Impact Increased Engagement

- Augmented Reality (AR) and Virtual Reality (VR) technologies capture consumers by providing immersive and interactive experiences, resulting in higher levels of engagement and better brand memory.
- Product visualization: Augmented Reality (AR) allows customers to view products in real-world settings, minimizing ambiguity and improving purchasing choices.
- Brand storytelling: Virtual Reality (VR) enables captivating storytelling by immersing people in virtual worlds that are in line with brand storylines and values.
- Customization: Both Augmented Reality (AR) and Virtual Reality (VR) present possibilities for personalized experiences, enabling marketers to customize content and interactions based on individual preferences and behaviors.

5.14.3.4 Obstacles and Factors to Consider

- The development of AR and VR experiences can incur significant costs, and the limited availability of high-quality VR equipment may hinder consumer adoption.
- Technological breakthroughs are happening quickly. Therefore, organizations need to keep up with the latest AR/VR capabilities and incorporate them successfully into their marketing campaigns.
- Privacy and security: The collection and management of user data in AR/VR experiences necessitate strong privacy protections to safeguard consumer information.

AR and VR technologies are continuously reshaping brand experiences by providing immersive, interactive, and customized interactions with consumers. Companies that adopt AR and VR technologies can set themselves apart from their competitors, improve client interaction, and cultivate stronger emotional bonds. The integration of these evolving technologies into marketing strategies will have a vital impact on moulding the future of consumer interaction and brand innovation in a digitally focused world.

5.15 CHALLENGES IN FOOD BRANDING

The food branding sector faces distinct issues as a result of intense competition, shifting customer preferences, regulatory obligations, and dynamic market trends. To effectively overcome these problems, brands must engage in innovation, adaptation, and effective differentiation. This section examines the primary obstacles in food branding, accompanied by illustrative instances:

5.15.1 Issues Related to Health and Nutrition

Consumers are becoming increasingly health-conscious and are demanding food manufacturers provide transparency and healthier choices. The change in consumer preferences presents difficulties for businesses to modify products, offer precise nutritional information, and effectively convey the advantages for health.

- Example: McDonald's and Health-Conscious Menu Choices
  McDonald's received backlash due to its menu options, which are rich in calories and fat. As a reaction, the company implemented more nutritious alternatives such as salads, fruit cups, and grilled chicken sandwiches. This effort sought to target health-conscious consumers and enhance the brand's reputation in terms of nutrition.

5.15.2 The Promotion of Long-Term Environmental and Social Responsibility

Consumers are increasingly prioritizing sustainability, ethical sourcing, and environmental effects in their selection of food goods. Brands are required to use sustainable practices, minimize their carbon footprints, and sustainably acquire ingredients in order to fulfill consumer expectations and comply with regulatory regulations.

- Example: Patagonia Provisions
  Patagonia Provisions prioritizes the use of sustainable and ethically sourced food items. This includes the incorporation of organic ingredients, the implementation of regenerative agricultural practices, and the establishment of Fair Trade relationships. The brand's dedication to environmental stewardship is following customer values and bolsters brand trust.

5.15.3 Ensuring the Genuineness and Openness of the Brand

Consumers desire genuineness and openness in food branding, anticipating that firms will honestly maintain their beliefs, ingredients, and production practices. Deceptive assertions or a lack of openness can gradually diminish the confidence consumers have in a company and harm the reputation of its brand.

- Illustration: Honest Tea
  Honest Tea prioritizes transparency by prominently displaying information about ingredients and sourcing processes on its package. Consumers who prioritize

authenticity and ethical honesty in their beverage selections are drawn to the brand's dedication to using organic ingredients and sustainable sourcing.

5.15.4 Adherence to Regulations

Food companies are required to comply with strict regulations that govern food safety, labelling, advertising, and health claims. Compliance guarantees the safety of consumers, adherence to legal requirements, and protection against regulatory penalties or negative reactions from consumers.

- Topic: FDA Regulations Regarding Labelling
  The Food and Drug Administration (FDA) in the United States requires precise labelling of food products, which includes providing the correct information about nutrition, disclosing allergens, and making health claims. Brands must adeptly negotiate these regulations in order to assure compliance and prevent any potential legal consequences.

5.15.5 Market Dynamics Characterized by Intense Competition

The food sector is characterized by intense competition, as multiple companies strive to capture consumer attention and foster loyalty. Brands need to distinguish themselves by employing product innovation, distinctive value propositions, and efficient marketing techniques in order to excel in a saturated marketplace.

- Example: Beyond Meat
  Beyond Meat revolutionized the food market by introducing plant-based meat substitutes that cater to health-conscious consumers, vegans, and ecologically conscientious individuals. The brand's pioneering efforts in developing plant-based protein products have established it as a frontrunner in the swiftly expanding market for alternative protein sources.

5.15.6 Conquering Obstacles and Strategic Factors to Consider

- Pursue innovation and adaptation to consistently address consumer preferences for healthier, environmentally-friendly, and convenient food choices.
- Enhance consumer trust by promoting clear communication, genuine storytelling, and active involvement on social and digital channels.
- Regulatory compliance: Stay informed about changing regulations and assure adherence to retain customer confidence and legal soundness.
- Differentiation: Distinguish your brand by producing unique products, adhering to ethical procedures, and crafting captivating brand narratives that strongly appeal to your target customers.

To successfully navigate the obstacles in food branding, firms must give priority to consumer health, sustainability, transparency, regulatory compliance, and differentiation. By proactively tackling these difficulties, businesses may establish trust, foster loyalty, and gain a competitive advantage in the ever-changing and dynamic food industry. McDonald's, Patagonia Provisions, Honest Tea, and Beyond Meat are examples of food brands that have successfully overcome hurdles and positioned themselves for long-term success in a competitive environment.

5.16 CHAPTER SUMMARY

In conclusion, branding and positioning strategies in food marketing play a pivotal role in shaping consumer perceptions, driving purchase decisions, and fostering brand loyalty within a competitive industry. Effective branding goes beyond product quality to encompass emotional connections, authenticity, and differentiation that resonate with target audiences. By leveraging storytelling, visual identity, and consumer engagement, brands can create compelling narratives that reflect their values and address consumer desires for health, sustainability, and transparency. Positioning strategies, whether based on product attributes, consumer profiles, occasions, or competitive landscapes, allow brands to carve out distinctive market positions and meet diverse consumer needs effectively. However, success hinges on consistent execution, adapting to evolving consumer trends and navigating regulatory landscapes to maintain trust and compliance. Ultimately, brands that excel in food marketing combine innovation with strategic insights, fostering strong brand equity and sustainable growth in an ever-changing marketplace.

CASE STUDIES

Case Study 1

An Analysis of Branding and Positioning Strategies in Food Marketing

In the highly competitive field of food marketing, it is crucial to have strong branding and positioning strategies in order to differentiate oneself, appeal to consumers, and cultivate brand loyalty. This case study examines the tactics utilized by Company Y, a fictional artisanal bakery, to build a unique brand identity and position itself in the nearby market.

Brand Strategy: Company Y distinguishes itself by prioritizing artisanal craftsmanship and utilizing high-quality products procured from local sources. The brand prioritizes conventional baking methods, organic components, and distinctive taste profiles that appeal to discriminating customers in search of genuine and luxurious bakery goods. Company Y effectively conveys its commitment to superior craftsmanship and high-quality products by employing a consistent branding strategy that emphasizes its legacy and artisanal expertise.

Positioning Strategy: Company Y establishes itself as a local bakery that provides high-end, artisanal baked products, emphasizing excellence, novelty, and the use of locally sourced ingredients. Company Y differentiates itself from mass-produced bakery chains by focusing on consumers who value artisanal products and are willing to pay a higher price for superior quality. The brand utilizes its local origins and active involvement in the community to cultivate a feeling of inclusion and confidence among customers, positioning itself as a reliable provider of high-quality handmade delicacies in the competitive bakery industry.

Effect on Consumer Perception: The branding and positioning initiatives implemented by Company Y have had a favourable impact on consumers. Customers view Company Y as a dependable supplier of premium, locally obtained bakery goods that exemplify skilled craftsmanship and genuine flavour. The brand's focus on freshness, artisanal methods, and community integration strengthens its reputation as a high-end bakery dedicated to culinary mastery. Consequently, Company Y benefits from a high level of customer loyalty and positive recommendations within its local market, which leads to continuous business expansion and a strong connection with its brand.

Company Y's strategic approach to branding and positioning in the artisanal bakery market demonstrates the significance of authenticity, quality, and community participation in establishing a robust brand identity and fostering consumer loyalty. Company Y achieved sustained growth and consumer trust in a competitive market by emphasizing artisanal craftsmanship, local sourcing, and community integration. This was accomplished through strong branding and positioning strategies.

Discussion Questions

1. What was the impact of Company Y's emphasis on artisanal craftsmanship on its brand differentiation?
2. How did local sourcing and community participation contribute to Company Y's positioning strategy?
3. By strategically utilizing its brand narrative and historical background, how did Company Y successfully cultivate and maintain customer loyalty?

Case Study 2

Augmented Reality (AR) and Virtual Reality (VR) in Brand Experiences

Augmented Reality (AR) and Virtual Reality (VR) have revolutionized brand experiences by offering immersive and interactive platforms that engage consumers in innovative ways. This case study examines how Company A, a fictional retail brand, leverages AR and VR technologies to enhance customer engagement and differentiate itself in the competitive retail landscape.

Implementation of AR and VR

Company A integrates AR and VR technologies into its retail experience to offer customers immersive product demonstrations, virtual try-ons, and interactive storytelling. Through its mobile app and in-store installations, customers can visualize products in their environment (AR) or experience simulated environments (VR), enhancing their shopping experience and decision-making process. Company A uses these technologies to create memorable brand interactions that resonate with tech-savvy consumers seeking personalized and engaging retail experiences.

Impact on Brand Perception

The adoption of AR and VR technologies has positively impacted Company A's brand perception among consumers. Customers perceive Company A as an innovative and forward-thinking brand that prioritizes customer experience and technological advancement. By offering interactive AR try-ons for clothing and VR tours of its product collections, Company A enhances customer satisfaction, drives engagement, and differentiates itself from traditional retail competitors.

Company A's strategic implementation of AR and VR technologies in its retail operations exemplifies the transformative impact of immersive technologies on brand experiences. By leveraging AR and VR to enhance engagement, differentiation, and brand perception, Company A successfully positioned itself as a leader in providing innovative and personalized retail experiences that resonate with modern consumers.

Discussion Questions

1. How did Company A effectively integrate AR and VR to enhance customer engagement?
2. What role did AR and VR play in differentiating Company A from its competitors in the retail sector?
3. How did the use of AR and VR technologies contribute to enhancing brand perception and customer loyalty for Company A?

CHAPTER 6

Distribution Channels and Logistics in Food Marketing

LEARNING OBJECTIVES

By the end of this chapter, you should be able to:

- Understand the structure of food distribution channels.
- Analyze the characteristics and roles of key food distribution channels.
- Explore logistics in food distribution.
- Assess the role of technology in distribution and logistics.
- Identify challenges in food distribution.
- Evaluate the strategies used for distributing food products globally and locally.
- Understand how e-commerce is reshaping food distribution.
- Analyze the impact of changing consumer behaviour on food distribution channels and logistics.

6.1 INTRODUCTION

Distribution channels and logistics are essential elements of food marketing, determining the movement of food products from producers to consumers in an intricate and ever-changing supply chain. The efficacy of distribution networks is a determining factor in the market presence, ease of access, and overall triumph of food goods.

Distribution channels refer to the routes and agents involved in the transportation of food goods, starting from the agricultural producers and ending with the final customers. These channels are crucial in guaranteeing the efficient and cost-effective delivery of food to its target market. The process encompasses multiple stages, including the acquisition of raw materials, manufacturing, packaging, distribution, and retailing.

DOI: 10.1201/9781003473305-6

The food marketing industry consists of two main types of distribution channels: Direct distribution refers to the process of selling products directly from manufacturers to customers, bypassing any intermediaries. This can be achieved through many channels such as farm stands, farmers' markets, community-supported agriculture (CSA), and increasingly through internet platforms that facilitate direct connections between farmers and consumers. Indirect distribution refers to the involvement of intermediaries such as wholesalers, distributors, retailers, and food service providers. These intermediates expedite the transportation of goods from manufacturers to end-users via various distribution channels such as grocery shops, supermarkets, restaurants, and food service providers.

Logistics include the supervision and coordination of the movement of products, data, and assets from their starting point to their final destination. Efficient logistics play a crucial role in food marketing by ensuring product quality, reducing waste, and satisfying consumer demand. This include the provision of transportation, storage, control of stock, and strategic planning for the efficient delivery of perishable food items, all customized to meet their unique requirements.

Distribution routes and logistics play a crucial role in food marketing, impacting the availability of products, the accessibility for consumers, and the competitiveness of the market. In order to maximize efficiency, minimize environmental effect, and satisfy the changing demands of the market, firms need to adjust their distribution strategy as customer tastes and technology capabilities advance. Gaining a comprehensive understanding of these dynamics is crucial for effectively navigating the intricacies of food distribution and attaining success in the fiercely competitive food sector.

6.2 OVERVIEW OF FOOD DISTRIBUTION CHANNELS

Food distribution channels are the complex network that food products go through, from producers to consumers. These channels include several routes, such as online retail platforms, farmers' markets, and community-supported agriculture (CSA) programmes. These approaches allow producers to directly reach consumers and focus on individualized relationships. Retail distribution channels, such as supermarkets, specialty food stores, convenience stores, and hypermarkets, offer widespread consumer accessibility with a wide range of product choices conveniently situated for everyday shopping. Food service channels, such as restaurants, cafés, catering services, and institutional providers, focus on creating and serving meals that are customized to meet the specific preferences of consumers, hence improving the overall dining experience in various environments. Furthermore, wholesale and distribution networks have a crucial function in overseeing large-scale purchases and effective distribution to retailers, food service operators, and institutions. This ensures that products are widely available and that the supply chain is managed efficiently. Collectively, these channels serve a wide range of consumer demands, tastes, and market trends, influencing the availability and consumption habits of food goods on a global scale.

6.2.1 The Path from Producer to Consumer

Food distribution channels refer to the several routes that food items take from producers to consumers, with the goal of ensuring that food is available, easily accessible, and of high quality in a variety of markets. Gaining a comprehensive understanding of these distribution channels is crucial for food businesses seeking to efficiently reach and fulfill consumer demand. Below is a summary of the primary food distribution channels:

Direct-to-Consumer Channels: Direct-to-consumer (DTC) channels refer to the method of selling products directly from manufacturers to customers, bypassing any intermediaries. This route has become prominent due to the increasing popularity of e-commerce platforms, enabling producers to circumvent traditional retail distribution. Some examples include:

- Online retail involves producers selling food products through their own websites or online marketplaces, providing convenience and direct engagement with consumers.
- Farmers' markets facilitate direct sales of fresh produce, baked goods, and artisanal products from farmers and small-scale producers to local consumers. This practice promotes community involvement and contributes to the growth of local economies.
- Community-supported agriculture (CSA) is a system in which consumers sign up to receive regular shipments of fresh, locally grown produce from nearby farms. This promotes sustainable farming methods and fosters direct connections between farmers and consumers.

Retail Distribution Channels: Retail channels encompass the process of selling food goods through intermediaries, including supermarkets, grocery stores, convenience stores, and specialized food shops. These channels offer a diverse selection of items and simple purchasing experiences for consumers. Important retail distribution routes include the following:

- Supermarkets and hypermarkets are expansive retail establishments that provide a wide variety of food products, encompassing fresh produce, packaged goods, and essential household items.
- Convenience stores are compact retail establishments that provide convenient and instant access to necessary food items, snacks, and beverages. They are designed to cater to consumers that prioritize convenience and immediate consumption.
- Specialty food stores are retail establishments that specialize in offering specialty products, including organic foods, gourmet items, ethnic foods, and health-focused products. These stores cater to unique consumer tastes and dietary demands.

Food Service Channels: Food service channels are specifically designed to serve consumers who choose to eat out or consume pre-prepared meals. The following channels are included:

- Restaurants and cafés are facilities where customers can enjoy a wide range of prepared meals, beverages, and desserts either on the premises or for takeout.

- Catering services: Companies that provide food preparation and delivery services for various occasions such as events, celebrations, and business meetings, guaranteeing convenience and customized menu choices.
- Institutional food services: This refers to the provision of meals and snacks in establishments such as schools, hospitals, and workplaces. These facilities cater to students, patients, employees, and visitors, with a particular emphasis on nutrition and adherence to dietary requirements.

Wholesale and Distribution Networks: Wholesale channels encompass the acquisition and dispersal of food goods in large quantities to retailers, food service providers, and institutional customers. Distribution networks facilitate the effective transportation of goods from manufacturers to consumers, streamlining the management of stock and logistical operations. Some examples include:

- Wholesalers and distributors are businesses that specialize in procuring, storing, and delivering food items to retail stores, restaurants, and other food service providers.
- Food brokers are intermediaries who facilitate transactions between food manufacturers and retail purchasers. They represent several brands and negotiate sales agreements in order to broaden the reach of the products.

Export and Import Channels: International trade plays a crucial role in the distribution of food, encompassing the shipment of domestically produced food to international markets and the importation of foreign foodstuffs to fulfill domestic demand. The export and import routes encompass:

- Exporters and importers: These are businesses which are engaged in the transportation of food goods across national borders. They must adhere to international trade norms and taxes in order to guarantee the safety of their products and get access to different markets.
- Global distribution networks: This refers to the complex systems of logistics that enable the efficient movement of food goods throughout the globe. These networks handle various tasks such as organizing transportation, managing customs clearance, and ensuring the distribution of products to foreign markets.

6.2.2 Differentiating between Direct and Indirect Channels

Food distribution routes are crucial for the transportation of food products from producers to consumers. These channels can be classified into two main categories: direct channels and indirect channels. Each category presents unique benefits and difficulties, influencing how manufacturers connect with their intended markets and how consumers obtain food items.

6.2.2.1 Direct Distribution Channels

Direct distribution channels refer to the process of distributing products directly from manufacturers to customers, bypassing any intermediaries. This technique has become increasingly popular as a result of technical improvements and evolving consumer preferences for transparency and customized buying experiences.

6.2.2.1.1 Examples

E-commerce:

- Companies such as Blue Apron and HelloFresh provide meal kits directly to clients' residences, enabling convenience and the ability to personalize meal planning.
- Benefits: Producers have the ability to manage and shape their brand identity, engage in direct communication with consumers, collect valuable customer data for personalized marketing efforts, and enhance pricing methods for maximum profitability.
- Challenges: The management of logistics, the assurance of product freshness, and the competition within a saturated online marketplace can pose significant difficulties.

Agricultural markets:

- Local farmers directly sell fresh vegetables, meats, dairy products, and handcrafted crafts to consumers at community gatherings or designated marketplaces.
- Benefits: Farmers can attain greater profit margins, establish strong connections with clients, enlighten consumers about their agricultural methods, and obtain prompt feedback.
- Challenges: The challenges include the fluctuating supply of products based on the seasons, reliance on weather conditions, and a narrower customer base compared to larger retail establishments.

Community-Supported Agriculture (CSA):

Community-Supported Agriculture (CSA) is a system in which individuals in a community financially support local farmers in exchange for a share of the farm's produce.

- Consumers opt to get periodic shipments of fresh, seasonal produce or other agricultural goods straight from nearby farmers.
- Benefits: This encourages the use of environmentally-friendly farming methods, cultivates community backing for nearby agriculture, and provides consumers with fresh, locally procured goods.
- Challenges: This endeavour necessitates consumers making a firm commitment in advance, coordinating logistics for deliveries, and dealing with fluctuations in produce supply due to farming conditions.

6.2.2.2 Indirect Distribution Channels

Indirect distribution channels utilize intermediaries, such as wholesalers, retailers, and food service providers, to enable the transportation of items from producers to consumers. These channels offer expanded market reach, logistical assistance, and enhanced product visibility.

6.2.2.2.1 Examples

Retail establishments known as supermarkets and hypermarkets:

- Chains such as Walmart and Kroger have a diverse selection of food goods obtained from different manufacturers, providing consumers with the convenience of shopping for everything in one place.

- Benefits: Extensive market coverage, substantial customer flow, cost advantages in procurement and distribution, and prospects for strategic product positioning and advertising.
- Challenges: Engaging in shelf space negotiations, fulfilling retailer demands for packaging and labelling, and contending with other brands for prominence.

Food distribution channels:

- Restaurants, cafés, catering services, and institutional food service providers procure food supplies from distributors or wholesalers in order to create meals for consumers.
- Benefits: Customized menu options that respond to individual preferences, consistent patronage from devoted clientele, and potential for large-scale sales through catering and institutional agreements.
- Challenges: Ensuring adherence to quality and consistency requirements, effectively handling perishable inventory, and adjusting to evolving consumer preferences and dietary trends.

Wholesale and distribution networks:

- For instance, distributors such as Sysco and US Foods provide food items to various establishments, including restaurants, hotels, schools, hospitals, and other food service facilities.
- Benefits: Streamlined supply chain, enhanced buying power, inventory optimization services, and the ability to distribute products regionally or nationally.
- Challenges: Overseeing intricate supply chains, guaranteeing product freshness and safety, and sustaining competitive pricing in a fiercely competitive sector.

Both direct and indirect distribution methods present distinct advantages and difficulties for food producers aiming to efficiently reach customers. Direct channels prioritize direct engagement with consumers, maintaining brand authority, and capitalizing on unique market prospects. On the other hand, indirect channels offer wider market coverage, logistical assistance, and potential for expansion. Food firms must comprehend the distinctions between these channels in order to create all-encompassing distribution strategies that are in line with their market goals, consumer preferences, and operational abilities in a constantly changing and competitive food industry environment.

6.3 TYPES OF DISTRIBUTION CHANNELS IN THE FOOD INDUSTRY

Distribution networks in the food sector are varied and essential for the transportation of food goods from producers to consumers. These channels include direct-to-consumer techniques such as online retail platforms, farmers' markets, and community-supported agriculture (CSA) programmes. They allow producers to sell their products directly to end-users, which helps build direct relationships and improve product transparency. Retail distribution channels encompass various establishments such as supermarkets, specialized food stores, convenience stores, and hypermarkets. These establishments provide consumers with a diverse range of products that are conveniently accessible for their everyday shopping needs. Food service channels, such as restaurants, cafés, catering services, and institutional providers, have expertise in the preparation and serving of food to clients, accommodating a wide range of dining preferences and situations. Wholesale and distribution networks are crucial in enabling the acquisition of large quantities of food items and their efficient distribution to retailers, food service operators, and institutions. They ensure that these products are widely available and delivered on time in regional or nationwide markets. Every distribution channel has its own specific purpose in meeting customer demands and enhancing the overall efficiency and availability of food goods in the market.

6.3.1 Direct-to-Consumer (DTC)

Direct-to-consumer (DTC) is a distribution technique in which manufacturers or producers sell products directly to consumers, without involving typical retail intermediaries. This technique has garnered considerable popularity in several sectors, such as the food and beverage industry, due to technology improvements, evolving consumer preferences, and the demand for more customized shopping experiences. This is an analysis of DTC practices within the food business.

6.3.1.1 The Development and Factors Influencing the Growth of Direct-to-Consumer

Several causes have contributed to the emergence of DTC in the food business. Advancements in e-commerce technologies and digital platforms have made online sales channels accessible to a wider range of people, including small-scale producers, allowing them to directly contact a worldwide audience. Consumers, who are becoming more familiar with the ease and openness of online purchasing, desire direct engagements with companies, placing importance on genuineness, details about products, and the ability to personalize their purchases. Furthermore, the COVID-19 pandemic expedited the transition to online shopping and DTC models. Lockdowns and social distancing measures compelled consumers to choose safer and more convenient methods of buying food products.

There are two categories of direct-to-consumer channels:

a. **Online Retail Platforms**: Food producers utilize their own websites or third-party e-commerce platforms (such as Amazon or Shopify) to directly sell their products to consumers. This channel provides brands with the ability to customize their product selection, pricing, and marketing approaches, enabling them to maintain control over their brand story and customer interactions.
b. **Subscription Models**: DTC firms frequently employ subscription-based business models, which involve offering services such as meal kits (e.g., Blue Apron, HelloFresh) or regular monthly delivery of specialized food products. Subscriptions provide consumers the advantages of ease, a wide range of options, and carefully tailored experiences, while also providing brands with reliable sources of revenue and chances to foster client loyalty and retention.

c. **Farmers' Markets and CSA Programmes**: Local farmers and producers engage in farmers' markets or run community-supported agriculture (CSA) programmes, where consumers buy seasonal food and artisanal goods directly from the origin. These channels facilitate community involvement, bolster local economies, and advocate for sustainable farming methods.

6.3.1.2 Advantages for Consumers

DTC arrangements provide numerous benefits for consumers. They offer access to a wider range of fresh and diversified products, often obtained from local or sustainable sources. DTC brands place a high importance on being transparent about where they source their materials and how their products are made. This enables consumers to make well-informed decisions that correspond with their personal beliefs. In addition, engaging directly with brands enables tailored customer care, feedback mechanisms, and direct contact channels, thereby strengthening consumer trust and brand loyalty.

6.3.1.3 Challenges and Considerations

Although DTC in the food business has significant benefits, it also presents certain difficulties. Managing logistics and fulfillment may be intricate, especially when dealing with perishable commodities that necessitate specialized packaging and delivery. Ensuring competitive pricing while covering transportation expenses can affect profitability, requiring effective supply chain management and strategic collaborations. Furthermore, DTC businesses face continuous obstacles due to regulatory compliance, food safety regulations, and customer expectations about prompt delivery and product quality.

6.3.1.4 Future Outlook and Business Trends

The future of DTC in the food business is optimistic, propelled by the continuous customer need for convenience, personalization, and openness. The growth of such models is expected to be influenced by technological advancements such as AI-driven personalization, augmented reality (AR) for virtual shopping experiences, and sustainable packaging solutions. DTC tactics will have a significant impact on the marketing, distribution, and consumption of food goods worldwide as brands adjust and introduce new ideas.

6.3.2 Retailers

Retailers are crucial in the food sector as they act as mediators connecting food producers and consumers. These firms cover a wide range of industries, including major supermarket chains and niche food stores, each serving specific consumer requirements, preferences, and shopping experiences. Below is an analysis of food merchants in the industry:

Supermarkets and Hypermarkets: Supermarkets are expansive commercial establishments that provide an extensive range of food products, vital home items, and frequently non-food merchandise in a single location. Some examples of these companies are Walmart, Kroger, Tesco, and Carrefour. These merchants utilize economies of scale to negotiate advantageous pricing with suppliers, offering consumers a wide range of products, convenience, and cheap prices. Supermarkets commonly have sections dedicated to fresh fruit, meat and fish, bread items, dairy products, frozen foods, and pantry essentials, in order to accommodate a wide range of culinary preferences and nutritional requirements.

Specialty Food Stores: Specialty food stores are establishments that specialize in particular product categories. They offer niche and gourmet food items that are attractive to discerning consumers who are looking for distinctive flavours, high-quality ingredients, and specialized dietary options. Among examples of stores that specialize in specific types of food are the following: Whole Foods Market, which focuses on natural and organic products; Trader Joe's, which offers unusual and cheap gourmet foods; and local health food stores. These merchants place a high importance on the quality of their products, their commitment to sustainability, and frequently provide support to local producers. This helps to build ties within the community and allows them to sell carefully selected products that set them apart from larger supermarket chains.

Convenience Stores: Convenience stores, alternatively referred to as corner shops or mini-marts, offer expedient access to necessary food products, beverages, snacks, and, occasionally, pre-made meals. Among examples of convenience stores are 7-Eleven, Circle K, and Wawa. Convenience stores are deliberately positioned in regions with heavy foot traffic, such as city centres, gas stations, and transit hubs. They provide consumers with convenience, longer opening hours, and rapid options for consumption. They offer a range of products and services that cater to spontaneous purchases, urgent need, and the increasing desire for convenient meal options.

E-commerce and online shops: The food retail scene has been altered by the emergence of e-commerce and the availability of online shops and supermarket delivery services. Among the examples of grocery delivery services are Amazon Fresh, Instacart, and Peapod. These systems enable users to purchase groceries online, choose between delivery or pickup options, and access a diverse selection of products from various stores and local suppliers. Online grocery shopping provides convenience, time efficiency, and adaptability, catering to many consumer choices for either home delivery or curbside pickup.

Wholesale clubs: Wholesale clubs function under a membership-based system, providing large amounts of food products and household goods at reduced prices. Among examples of wholesale clubs are Costco, Sam's Club, and BJ's Wholesale Club. These shops cater to frugal customers and businesses looking for affordable purchasing options, offering value through bulk discounts, exclusive member perks, and a diverse range of products, including fresh food, electronics, and fashion.

Ethnic grocery stores: Ethnic grocery stores are establishments that focus on selling food products and ingredients that are distinctive to certain cultural backgrounds or areas across the globe. Among examples of such establishments are Asian supermarkets, Hispanic markets, and Middle Eastern food stores. These stores specifically target immigrant groups, international students, and culinary lovers who are looking for genuine ingredients, spices, and traditional cuisines which are often not to be found in conventional supermarkets. Ethnic grocery stores foster cultural variety, commemorate culinary legacy, and provide distinctive shopping experiences that mirror global food trends and traditions.

6.3.3 Wholesalers and Distributors

Wholesalers and distributors play a crucial role as intermediates in the food sector, enabling the smooth flow of products from producers to retailers, food service providers, and institutional customers. Their responsibility involves overseeing inventory, logistics, and bulk procurement, in order to enhance the efficiency of supply chain operations and facilitate the wider distribution of food items. This section delves into the role and significance of wholesalers and distributors within the food business.

Role and Functions: Wholesalers and distributors serve as intermediaries between food producers and end-users, functioning at different points throughout the supply chain. Their main functions are as follows:

- Bulk purchasing involves wholesalers acquiring substantial quantities of food items directly from manufacturers or producers at reduced prices, taking advantage of economies of scale.
- Inventory management involves the storage, organization, and distribution of inventory in order to fulfill the needs of retailers, food service providers, and institutional consumers.
- Distributors manage the logistics and transportation of goods, guaranteeing prompt delivery to customers in regional or nationwide marketplaces.
- Marketing and sales support: Their role is to supply retailers and food service clients with marketing materials, promotional assistance, and product information. This helps increase product visibility and boosts sales.

6.3.3.1 Categories of Wholesalers and Distributors

a. **Broadline Distributors**: These distributors provide an extensive selection of food products and non-food items to various food service establishments such as restaurants, hotels, schools, hospitals, and others. Notable examples include Sysco and US Foods, both of which have vast distribution networks throughout North America. They provide a wide range of products, including fresh vegetables, meats, kitchen equipment, and cleaning supplies.
b. **Specialty Food Distributors**: These focus on specialized product categories such as gourmet foods, organic products, ethnic foods, or health-oriented goods. Examples of such companies in the food distribution industry are UNFI (United Natural Foods, Inc.), which specializes in distributing natural and organic foods, and KeHE Distributors, which is known for distributing specialty and artisanal food products.
c. **Regional and Local Distributors**: These distributors cater to smaller geographic areas or focus on serving local customers by offering products that are acquired locally. These distributors frequently prioritize establishing connections with area producers and retailers, advocating for the use of local resources and providing assistance to community enterprises.

6.3.3.2 *Instances of Wholesalers and Distributors*

Sysco Corporation is a global distributor of food products and services. It manages a large network that supplies a wide range of food items to restaurants, healthcare facilities, educational institutions, and hospitality enterprises.

UNFI, which is short for United Natural Foods, Inc., is a company that focuses on the distribution of natural, organic, and specialty foods to retail and food service channels in the United States and Canada. UNFI provides assistance to businesses such as Whole Foods Market and Sprouts Farmers Market, meeting the increasing need for health-conscious and sustainable food choices.

KeHE Distributors specializes in the distribution of natural and specialized food goods to independent retailers, chain stores, and food service operators around the country. KeHE prioritizes collaborations with up-and-coming businesses and provides them with logistical assistance, marketing, and merchandising services to enhance product visibility and market expansion.

6.3.3.3 *Challenges and Considerations*

Wholesalers and distributors encounter many obstacles in the food supply chain that require careful consideration:

- Logistical complexity arises from the need to effectively handle perishable items, ensure product quality, and meet delivery deadlines in various markets. This necessitates the presence of strong logistical capabilities and infrastructure.
- Market rivalry: Fierce rivalry among distributors requires them to distinguish themselves by their service offers, range of products, price methods, and customer support.
- Regulatory compliance involves following food safety rules, labelling requirements, and sustainability standards, which can create extra operational difficulties and expenses.

6.3.3.4 *Anticipated Developments and Adjustments*

Wholesalers and distributors are adjusting their operations to align with industry developments, including the integration of e-commerce, the growing demand for sustainable practices, and the advancements in supply chain management technology. Advancements in inventory tracking, predictive analytics, and digital platforms are improving operational efficiency and customer service, allowing wholesalers to meet changing consumer expectations and industry requirements.

6.3.4 Foodservice Channels

Foodservice channels are a vibrant segment of the food industry that focuses on the handling, delivery, and consumption of food outside of people's homes. This heterogeneous category includes a broad array of establishments and forms, each catering to certain consumer requirements, preferences, and dining experiences. Here is an analysis of different routes in the foodservice industry:

Restaurants: Restaurants are the most prominent sector of the foodservice industry, providing a range of dining options, including informal restaurants and high-end dining facilities. Some examples of restaurants include popular fast-food franchises such as McDonald's and Burger King, casual dining establishments like Olive Garden and Applebee's, and high-end restaurants renowned for their culinary mastery and atmosphere. Restaurants offer table service, a wide range of menu options, and often focus on certain cuisines or culinary concepts, catering to a variety of consumer preferences and dining situations.

Cafés and coffee shops: Cafés and coffee shops are establishments that focus on providing a variety of beverages such as coffee, tea, and specialty drinks, along with light meals, pastries, and snacks. Some examples of coffee shops are Starbucks, Dunkin', and local independent coffee houses. These facilities prioritize convenience and foster a sociable atmosphere, typically functioning as venues for mingling, conducting business meetings, or enjoying casual eating.

Institutional Foodservice: Institutional foodservice refers to the provision of catering services and foodservice to sizable establishments such as hospitals, schools, universities, corporate cafeterias, and government facilities. These operations prioritize the provision of nourishing meals, catering to specific dietary requirements, and ensuring adherence to food safety procedures. Notable examples of such companies are Aramark and Sodexo, who have a global presence and specialize in providing a wide range of services to institutional clients, including catering, on-site meals, and event management.

Mobile Food Services: Food trucks and street vendors provide on-the-go food options, usually offering pre-prepared meals, snacks, and drinks at outdoor venues, festivals, and public gatherings. These vendors offer convenience and novelty, and frequently focus on specialized cuisines or unique menu items that attract adventurous diners looking for fast and delectable meals on the move. Examples encompass gourmet food trucks that provide artisanal burgers, tacos, ethnic cuisine, and innovative culinary concoctions.

Catering Services: Catering services are experts in delivering food and drink services for exclusive occasions, such as private events, weddings, corporate gatherings, and social festivities. Caterers provide personalized menus, event coordination, and on-site cooking services that are specifically designed to meet the preferences and themes of their clients' events. Some examples include local catering firms, hotel banquet services, and specialist event planners who guarantee exceptional dining experiences and flawless event execution.

Airline and Railway Catering: Airline and railway catering services offer food and beverage services to passengers during their air or rail travel. These services prioritize the production, packaging, and transportation of meals to cater to the particular dietary requirements and preferences of travellers. LSG Sky Chefs and Gate Gourmet are just two examples of companies that provide airlines with pre-packaged meals, snacks, and beverages specifically tailored for consumption in small spaces that have high safety and hygiene requirements.

6.4 LOGISTICS IN FOOD DISTRIBUTION

Efficient and timely transfer of food items from producers to consumers, while maintaining quality, safety, and compliance with regulatory criteria, is of the utmost importance in logistics for food distribution. This intricate procedure encompasses various crucial components:

Transportation and Distribution: Transportation is a crucial aspect of food logistics. It involves many types of transportation, such as trucks, trains, ships, and airplanes. Refrigerated trucks are crucial for delivering perishable items such as fresh produce and dairy products over extended distances, ensuring that the temperature remains at an ideal level to avoid deterioration. Distribution centres strategically positioned along transportation routes assist the consolidation, sorting, and redistribution of goods to regional markets, assuring prompt delivery and minimizing travel durations.

Efficient Inventory Management: Efficient inventory management and storage are crucial in the area of food distribution in order to avoid excessive stock levels or shortages. Warehousing facilities with temperature-controlled settings and specific storage conditions, such as freezing or chilling, may suit a wide range of product needs. Cold storage warehouses ensure a constant temperature for frozen goods and seafood, thereby keeping their quality and prolonging their shelf life until they are distributed to retail outlets or food service providers.

Traceability Systems: Traceability systems allow for the monitoring and tracking of food products along the supply chain, assuring adherence to food safety rules and enabling quick response to recalls or quality problems. Radio-Frequency Identification (RFID) tags and blockchain technology improve traceability by offering up-to-date information on the origins of products, how they are handled, and the conditions they are stored in. This level of transparency fosters consumer confidence and provides protection against instances of food-borne diseases or contamination.

For instance, companies such as Amazon Fresh employ complex logistics networks to provide prompt delivery of food, making use of sophisticated routing algorithms and fulfillment facilities to enhance delivery efficiency. Global food manufacturers collaborate with third-party logistics providers (3PLs) such as DHL Supply Chain to optimize worldwide distribution, guaranteeing smooth customs clearance and adherence to import rules.

Logistics in food distribution encompasses a range of procedures and technology aimed at satisfying the requirements of worldwide supply chains, with a focus on ensuring food safety, preserving quality, and maximizing operating efficiency. Adapting to changing consumer expectations, regulatory regulations, and market dynamics in the food business requires continuous innovation and investment in logistical infrastructure.

6.4.1 Cold Chain Logistics

Cold chain logistics is a distinct component of the supply chain which guarantees the secure and effective transportation, storage, and management of products that are sensitive to temperature. This encompasses perishable goods, medications, and other products that necessitate regulated temperature conditions to uphold their quality, safety, and effectiveness throughout the entire process from manufacturing to consumption. Below is a comprehensive analysis of cold chain logistics, highlighting its significance and providing examples of its implementation.

Cold chain logistics plays a vital role in maintaining the quality and freshness of perishable items. Temperature variations can result in deterioration, decreased shelf life, and safety issues, especially when it comes to food products and medications. Cold chain logistics maintains a stable temperature range in the supply chain to reduce losses and ensure that consumers receive fresh, safe, and high-quality items. This is particularly vital in the food business, especially for products such as dairy, meat, seafood, and fresh vegetables, which are extremely perishable and susceptible to temperature fluctuations.

6.4.1.1 Elements of Cold Chain Logistics

The cold chain has multiple essential elements:

- Refrigerated transportation refers to the use of trucks, ships, airplanes, and trains that are equipped with refrigeration units in order to maintain the necessary temperature. Refrigerated trucks (also referred to as reefers) are extensively utilized for the transportation of dairy goods, frozen meals, and fresh vegetables over vast distances.
- Cold storage warehouses are specifically built to store commodities that are sensitive to temperature at different points along the supply chain. Cold storage warehouses can be categorized as chilled storage, which is used for preserving products at temperatures slightly above freezing, and frozen storage, which is used for preserving items at temperatures below freezing. Illustrative instances comprise storage facilities managed by corporations such as Americold and Lineage Logistics.
- Temperature monitoring and control systems: Sophisticated monitoring systems are used to track and regulate the temperature and humidity levels at every stage of the supply chain. IoT sensors and data recorders allow for the collection of real-time data, which facilitates proactive management of the cold chain and prompt response to temperature aberrations.

6.4.1.2 Difficulties in Cold Chain Logistics

Preserving the cold chain presents numerous difficulties:

- Infrastructure: The process of creating and upkeeping the essential infrastructure, such as transportation that can keep things cold and facilities for storing items at low temperatures, demands substantial financial commitment. In areas with underdeveloped logistics infrastructure, the maintenance of continuous temperature control can be very difficult.
- Regulatory compliance: Adhering to international and national standards on food safety and pharmaceutical storage increases the intricacy of cold chain logistics.

This encompasses compliance with regulations such as the Good Distribution Practice (GDP) for medications.
- Expense: Cold chain logistics incurs higher costs compared to regular logistics due to the necessity of specialist equipment, energy consumption for refrigeration, and additional handling requirements.

6.4.1.3 Instances of Cold Chain Logistics in Operation

The food industry, including companies such as Nestlé and Tyson Foods, depend heavily on cold chain logistics for the efficient distribution of its perishable products. Nestlé, for example, employs a system of refrigerated warehouses and trucks to guarantee that its ice cream products maintain their frozen state throughout the whole process, from production to retail distribution.

The pharmaceutical industry emphasizes the significance of cold chain logistics due to the need for precise temperature control in the distribution of vaccines. During the COVID-19 pandemic, for instance, Pfizer's COVID-19 vaccine necessitated storage at extremely low temperatures (–70°C), which mandated the utilization of specialist freezers and thermal shippers equipped with dry ice.

Online grocery businesses such as Amazon Fresh and Instacart have included cold chain operations to provide same-day delivery of fresh produce, dairy products, and frozen delicacies. These organizations employ insulated packaging, refrigerated delivery trucks, and real-time temperature monitoring to guarantee the quality of their products.

6.4.1.4 Innovations and Future Trends

The cold chain logistics industry is always progressing due to technology improvements. Technological advancements, such as the use of blockchain for enhanced traceability, the use of artificial intelligence for predictive analytics in temperature control, and the adoption of renewable energy sources for refrigeration units, are improving efficiency and promoting sustainability. Furthermore, the emergence of automated cold storage facilities and the utilization of drones for so-called 'last-mile' delivery are promising innovations that have the potential to completely transform cold chain logistics.

6.4.2 Transportation

Transportation plays a crucial role in the distribution of food, guaranteeing the efficient and secure movement of food goods from producers to consumers. Efficient transportation logistics are essential for preserving the quality, freshness, and safety of food goods during the whole supply chain. This encompasses a wide range of transportation methods, specifically designed vehicles, and sophisticated logistical approaches to effectively address the distinct requirements of the food business.

6.4.2.1 Transportation Modes

Various transportation modes are employed in the distribution of food, each tailored to certain product types and distances:

- Trucks are the predominant means of transportation for distributing food, especially across short to moderate distances. Reefers are necessary for the transportation of perishable commodities like dairy products, meat, and fresh fruit. Companies such as Tyson Foods employ such trucks to transport their products nationwide, guaranteeing that they arrive at merchants in the best possible state.
- Rail transport is a highly efficient method for transporting large quantities of non-perishable materials, such as grains and canned goods, over long distances. Rail transportation provides a financially efficient and ecologically sustainable alternative, particularly for moving substantial amounts of food across long distances. Companies such as Cargill utilize rail transportation to efficiently transfer substantial quantities of grain from production regions to processing plants and export terminals.
- Air transport is utilized for the swift delivery of high-value, perishable goods, such as fresh seafood, exotic fruits, and specialized cuisine. Air cargo expedites the transportation of these products to global markets, ensuring their prompt delivery and preserving their freshness and quality. As an illustration, freshly harvested Alaskan salmon is frequently transported via air to quickly reach markets in Asia and Europe within a 24-hour timeframe.
- Maritime transport is well-suited for transporting large quantities of both perishable and non-perishable foods over great distances. Cargo ships utilize refrigerated containers to deliver frozen groceries, fresh produce, and other perishable items across the globe. Maersk Line and similar companies provide comprehensive refrigerated container services, which enable the international commerce of perishable commodities such as bananas from Central America to Europe.

6.4.2.2 Difficulties in the Transportation of Food

The transportation of food poses various issues that need to be effectively addressed in order to guarantee the quality and safety of the products:

- Temperature regulation: Ensuring the accurate maintenance of temperature during transportation is vital for perishable goods. Temperature variations can result in rotting, decreased shelf life, and concerns regarding food safety. To mitigate these hazards, it is crucial to have advanced refrigeration equipment and real-time temperature monitoring.
- Promptness and effectiveness: Delivering goods on time is crucial, particularly for fresh produce and perishable items. Delays can result in substantial financial losses owing to the deterioration of goods. Logistics businesses utilize techniques such as improved routing and effective loading/unloading procedures to shorten the duration of transportation.
- Food transportation is required to adhere to multiple standards concerning the safety and handling of food in order to ensure regulatory compliance. This encompasses regulations established by regulatory bodies such as the Food and Drugs

Administration (FDA) in the United States and the European Food Safety Authority (EFSA) in Europe, which oversee aspects such as sanitation, temperature management, and traceability.

6.4.2.3 *Advancements in Food Transportation*

Ongoing technological progress is enhancing the efficiency of food transportation operations.

- IoT and real-Time monitoring: Internet of Things (IoT) devices and sensors offer immediate and continuous data on both the whereabouts and the state of food shipments. This enables prompt remedial measures in case of any problems, such as changes in temperature.
- Blockchain technology: This improves traceability and transparency in the food supply chain. Blockchain technology ensures food safety and authenticity by meticulously documenting every transaction and movement of food products.
- Sustainable strategies: The food business is progressively embracing sustainable transportation strategies in order to diminish its environmental footprint. This encompasses the utilization of electric or hybrid cars, optimizing routes to save fuel usage, and introducing more effective packing options.

6.4.3 Warehousing and Storage

Warehousing and storage play a vital role in the food distribution process, guaranteeing the safety, freshness, and adherence to regulatory criteria of food products from the moment they leave the producer right up to the point they are delivered to the customer. Efficient warehousing and storage methods are crucial for preserving the quality of food items, preventing decay, and guaranteeing a consistent provision to fulfill the needs of consumers.

6.4.3.1 *Warehousing Categories*

Food distribution relies on a range of warehouse designs to cater to diverse storage requirements:

- Dry storage: This refers to the use of warehouses to store items that are not perishable, such as grains, canned goods, and packaged foods. These warehouses ensure a consistent, low temperature and humidity level to prevent the deterioration of food quality. For instance, major distribution centres managed by corporations such as Walmart and Costco keep immense quantities of non-perishable commodities in dry storage sections, guaranteeing a continuous provision to their retail outlets.
- Refrigerated storage: Refrigerated or chilled warehouses are necessary for perishable goods that require special temperature control to preserve their freshness. These facilities keep dairy products, fresh produce, and specific beverages. One instance is the utilization of refrigerated storage by grocery chains such as Kroger, which guarantees that fresh fruits and vegetables maintain their optimal state until they are delivered to consumers.

- Cryogenic storage: Cryogenic warehouses are utilized for the preservation of perishable goods, such as meat, seafood, and frozen ready-to-eat meals, at extremely low temperatures below freezing point. These facilities employ cutting-edge freezing technology to maintain the nutritional integrity and overall quality of food goods. Companies such as Tyson Foods employ frozen storage to maintain the safety and freshness of their poultry products prior to distributing them to retailers and foodservice providers.

6.4.3.2 Cutting-Edge Storage Technologies

The efficiency and effectiveness of food storage have been greatly enhanced by technological improvements.

- Automated Storage and Retrieval Systems (ASRS): ASRS employ robotics and automation to efficiently handle goods in warehouses. These systems have the ability to rapidly and precisely store and retrieve items, hence minimizing human mistakes and enhancing productivity. Amazon's fulfillment centres employ ASRS to efficiently handle perishable and non-perishable food goods, guaranteeing prompt and precise order fulfillment.
- Temperature and Humidity Control: Contemporary warehouses are fitted out with sophisticated climate control systems that uphold accurate temperature and humidity levels. It is essential for preventing decay and prolonging the durability of perishable materials. Companies such as Americold offer temperature-controlled warehousing solutions that accommodate various food goods, guaranteeing their preservation during storage.
- The Internet of Things (IoT) with Real-Time Monitoring: IoT devices and sensors are employed to continuously monitor storage conditions in real time. These systems collect and analyze data on temperature, humidity, and other important parameters, enabling prompt remedial measures to be taken in case of any discrepancies. Food distribution company Sysco use IoT technology to monitor their storage facilities, guaranteeing that all food items are stored in ideal conditions.

There are multiple issues associated with managing food warehousing:

- Warehouses are required to comply with strict food safety requirements, including the Food Safety Modernization Act (FSMA) in the United States. These standards mandate the use of appropriate measures for cleanliness, pest management, and record-keeping to avoid any form of contamination.
- Inventory management: Effective inventory management is essential to avoid excessive stock or supply shortages. This entails monitoring the dates of expiration, overseeing the rotation of shares using the first-in, first-out (FIFO) method, and guaranteeing prompt distribution.
- Energy consumption: The act of maintaining temperature-controlled surroundings might require a significant amount of energy. Warehouses are implementing energy-efficient measures, such as the utilization of LED lighting, the enhancement of insulation, and the adoption of renewable energy sources, in order to decrease their carbon footprint.

Examples:

Nestlé runs a multitude of refrigerated and frozen warehouses worldwide to keep its wide array of food goods. These facilities are furnished with cutting-edge equipment to guarantee the utmost levels of food safety and quality.

Costco's distribution hubs feature huge dry storage spaces along with refrigerated and frozen sections to accommodate their diverse assortment of food products. They are able to regularly provide their retail locations with fresh and non-perishable supplies due to their effective warehousing operations.

6.5 THE ROLE OF TECHNOLOGY IN FOOD DISTRIBUTION AND LOGISTICS

Technology has a significant impact on food distribution and logistics by improving efficiency, guaranteeing food safety, and enhancing overall supply chain management. Advancements in this domain have empowered enterprises to satisfy the increasing expectations of consumers while upholding the excellence and authenticity of food items.

An important progress is the utilization of IoT devices and sensors. These technologies enable the continuous monitoring of temperature, humidity, and other essential factors in storage and transit situations. Logistics businesses such as Maersk utilize IoT sensors to monitor the state of perishable commodities as they are being transported. The use of real-time data allows for immediate action to be taken in response to any deviations from the necessary conditions, effectively preventing rotting and ensuring the safety of the food.

Blockchain is a crucial technology that improves the capacity to track and understand the flow of food in the supply chain. Blockchain technology maintains a comprehensive and immutable record of all transactions and movements of agricultural goods, starting from their origin on the farm and ending at the point of consumption. Walmart and other companies have adopted blockchain technology to trace the source of fresh fruit, significantly decreasing the time required to track a food item from several days to mere seconds. Not only does this enhance food safety, but it also fosters consumer confidence by guaranteeing transparency.

The implementation of Automated Storage and Retrieval Systems (ASRS) and robotics has brought about a significant transformation in the field of warehousing. These technologies, employed by corporations such as Amazon, mechanize the process of managing and storing goods, hence enhancing efficiency and minimizing human fallibility. Automated systems provide the capability to swiftly retrieve and dispatch commodities, thereby guaranteeing prompt delivery and optimal stock management.

Ultimately, technology plays a crucial role in contemporary food distribution and logistics by providing tools that optimize productivity, guarantee food security, and enhance the capacity to track and trace food. As the sector progresses, technological developments will enhance efficiency and fulfill the growing requirements of worldwide food supply chains.

6.5.1 Inventory Management Systems

Inventory management systems (IMS) are crucial instruments in the food distribution sector. They guarantee the availability of the correct products at the appropriate time, while also reducing waste and optimizing efficiency. These systems combine many technologies and processes to monitor inventory levels, oversee orders, and enhance stock management. Efficient inventory management is essential for preserving food quality, satisfying consumer expectations, and guaranteeing adherence to regulations.

6.5.1.1 The Significance of Inventory Management in Food Distribution

Inventory management is of the utmost importance in the food business, primarily because numerous products have a limited shelf life. Effective inventory management systems enable organizations to efficiently track expiration dates, ensure appropriate stock levels, and minimize food waste. For example, Tesco and other grocery chains utilize advanced IMS to monitor and manage stock levels of several products and multiple store locations. This ensures a consistent supply of fresh produce, dairy, and other perishable items while simultaneously reducing the amount of wasted or spoiled goods.

6.5.1.2 Main Characteristics of Inventory Management Systems

Contemporary IMS systems integrate numerous sophisticated characteristics that improve their functionality:

- IMS offers instantaneous monitoring of inventory levels, precise locations, and ongoing movement. This enables precise monitoring of stock and prompt reactions to fluctuations in inventory. Walmart utilizes real-time tracking to effectively oversee its extensive inventory, guaranteeing timely replenishment of products and minimizing the likelihood of stockouts.
- IMS has the capability to automate the process of reordering by using predetermined thresholds. This guarantees that inventory levels are sustained without the need for manual intervention. Amazon has an automatic reordering system for inventory management, which ensures that its warehouses are efficiently stocked, allowing for prompt fulfillment of consumer requests.
- Effective management of expiration dates is essential for perishable items. IMS monitors and manages these dates to enable stock rotation based on the first-in, first-out principle, ensuring that older items are sold before newer ones. This functionality is crucial for corporations such as Nestlé, which manage a diverse array of perishable food products.
- IMS frequently integrates with other systems, including Enterprise Resource Planning (ERP), Warehouse Management Systems (WMS), and Customer Relationship Management (CRM). This integration facilitates the smooth transfer of data and enhances collaboration among various business operations. For example, SAP (System Analysis Program) is a German multinational software company that provides ERP solutions to improve overall efficiency.

6.5.1.3 Advantages of Efficient Inventory Management

Establishing a proficient integrated management system provides a multitude of advantages to food distribution companies:

- Minimized waste: Through the precise monitoring of inventory and expiration dates, firms can effectively decrease food wastage. This not only reduces expenses but also promotes sustainability initiatives. For instance, Kroger has adopted inventory management strategies that effectively reduce food waste, thereby contributing to their objective of achieving zero waste.
- Enhanced customer satisfaction: By maintaining ideal inventory levels, consumers are guaranteed to find the products they require promptly and efficiently. Target and other retailers utilize sophisticated IMS to consistently maintain a well-stocked inventory of popular products. This practice enhances the overall shopping experience and increases consumer loyalty.
- Cost savings: These can be achieved by efficient inventory management, which helps to limit the amount of excess stock and therefore reduces the costs associated with holding that stock. Additionally, it reduces the likelihood of inventory shortages, which might result in missed revenue opportunities. Unilever and similar companies derive advantages from IMS by efficiently managing their inventory, minimizing storage expenses, and guaranteeing a steady supply of products.

6.5.1.4 Obstacles and Resolutions

While there are benefits, the process of installing and sustaining an IMS can be difficult. Typical obstacles involve ensuring the accuracy of data, integrating systems, and adjusting to dynamic market conditions. In order to tackle these difficulties, firms can:

- Allocate resources to training: It is imperative to provide comprehensive training to staff members in order to ensure accurate data and efficient system operations. Ongoing training programmes can assist personnel in staying current with system functionalities.
- Utilize sophisticated analytics techniques: IMS can be improved through the application of advanced analytics and machine learning, which can accurately forecast demand trends and optimize inventory levels. PepsiCo and similar companies employ predictive analytics to forecast consumer demand and make necessary inventory adjustments.
- Ensure scalability: As businesses expand, it is crucial for their information management systems to be able to scale proportionally. Opting for scalable solutions guarantees sustained efficiency and effectiveness in the long run, aligning with the company's evolving requirements.

Examples:

- Zara's inventory management techniques are frequently praised as exceptional, despite not being in the food industry. Their utilization of real-time data and swift replenishment cycles might serve as an exemplary model for food distributors striving to efficiently uphold a stock of fresh products.

- McDonald's employs an integrated inventory management system to effectively oversee its vast supply chain, guaranteeing the availability of fresh ingredients throughout its worldwide network of restaurants. This system enables them to uphold consistency and ensure high quality in their food offerings.

6.5.2 GPS and Route Optimization

The utilization of GPS (the Global Positioning System) and route optimization technology plays a crucial role in contemporary food distribution and logistics. These tools improve the effectiveness and dependability of delivery services by offering precise position tracking and appropriate routing solutions. By utilizing these technologies, food distribution corporations can guarantee prompt delivery, decrease operating expenses, and enhance overall consumer contentment.

6.5.2.1 Significance of GPS in Food Distribution

GPS technology enables the live tracking of vehicle locations, enabling logistics managers to oversee the transportation of goods from warehouses to delivery locations. This feature is essential for guaranteeing that perishable food items are transported within their designated time frames for optimal freshness. Companies such as Domino's Pizza utilize GPS tracking technology to offer consumers immediate and accurate information regarding the status of their pizza delivery. This improves transparency and customer satisfaction. In addition, GPS tracking aids in the management of extensive vehicle fleets, guaranteeing optimal use and minimizing the likelihood of delays.

6.5.2.2 Purpose of Route Optimization

Route optimization entails the utilization of algorithms and data analytics to ascertain the most optimal routes for delivery vehicles. This procedure considers multiple elements, including traffic conditions, delivery time windows, truck capacity, and road constraints. Through the process of route optimization, organizations can achieve substantial reductions in fuel consumption, decreased transportation expenses, and minimized delivery durations. UPS use a sophisticated route optimization software known as ORION (On-Road Integrated Optimization and Navigation) to minimize the number of miles driven and decrease fuel consumption. This not only leads to cost savings but also promotes environmental sustainability.

6.5.2.3 Advantages of GPS and Route Optimization

The incorporation of GPS technology and route optimization provides numerous advantages to food distribution companies:

- **Enhanced Efficiency**: Through the provision of real-time tracking and appropriate routing, these technologies contribute to a reduction in both the time and distance covered. This enhanced efficiency results in expedited delivery and the capacity to process a greater number of deliveries within a reduced timeframe. Amazon utilizes GPS and route optimization to enhance the efficiency of its delivery operations, guaranteeing prompt and dependable delivery of products to customers.

- **Cost Reduction**: By optimizing routes, fuel consumption and vehicle wear and tear are minimized, resulting in substantial cost savings. PepsiCo and similar companies utilize these technologies to enhance their delivery routes, leading to reduced transportation expenses and increased profitability.
- **Improved Customer Satisfaction**: Prompt delivery and up-to-date tracking notifications enhance the overall customer experience. Customers like having precise information about the current status of their deliveries, as this fosters trust and cultivates loyalty. For example, food delivery firms such as Instacart utilize GPS monitoring to offer clients precise delivery timings, hence improving the whole shopping experience.
- **Enhanced Environmental Sustainability**: Optimal routing decreases carbon emissions by decreasing both the distance covered and the amount of fuel consumed. This contributes to the company's sustainability objectives and aids in mitigating the environmental consequences of logistical operations. For instance, DHL has integrated GPS technology and route optimization strategies to improve their delivery efficiency and minimize their environmental impact.

6.5.2.4 Obstacles and Resolutions

Although there are advantages, the implementation of GPS and route optimization technology can sometimes provide difficulties:

- The charges associated with installing these technologies can be substantial, encompassing costs for hardware, software, and training. Nevertheless, the enduring advantages and financial savings frequently rationalize the original expenditure.
- Data management can be intricate when dealing with the substantial amounts of data produced by GPS tracking and route optimization systems. Robust data management solutions are necessary for companies to efficiently process and evaluate this data.
- Traffic circumstances and delivery demands have the ability to be altered rapidly. Systems should possess the ability to adjust and accommodate immediate modifications while ensuring ongoing optimization.

In order to tackle these difficulties, firms can:

- Invest in scalable solutions: Implementing scalable GPS and route optimization solutions allows for the firm to expand and adapt, offering flexibility and ensuring long-term viability of the investment.
- Employ cloud-based systems: Cloud-based solutions have the capability to efficiently process enormous amounts of data and offer real-time updates and analytics.
- Continuous improvement involves regularly updating and refining optimization algorithms in response to feedback and changing situations to maintain the system's effectiveness and relevance.

Examples:

- FedEx utilizes GPS and route optimization technology to effectively oversee its extensive delivery network, guaranteeing punctual and streamlined deliveries worldwide. Their method facilitates the reduction of delays and optimizes the utilization of their delivery fleet.
- Blue Apron is a company that delivers meal kits to customers. Blue Apron employs GPS monitoring to oversee its delivery vehicles and enhance routes, guaranteeing the quick and pristine delivery of fresh food to clients' doorsteps.

6.5.3 Blockchain in the Food Supply Chain

The food supply chain is being revolutionized by blockchain technology, which improves transparency, traceability, and efficiency. Blockchain guarantees the recording and verification of every stage in the food supply chain by establishing unchangeable records of transactions. This technological innovation tackles crucial concerns such as food safety, fraud, and inefficiency, ultimately benefiting customers, producers, and merchants.

6.5.3.1 Improving Clarity and the Capacity to Track and Verify Information

An essential advantage of blockchain technology in the food supply chain is its capacity to offer unmatched transparency and traceability. Every transaction or movement of food products is documented in a distributed ledger that can be accessed by all players in the supply chain. As an illustration, Walmart has employed blockchain technology to monitor the source of its fresh vegetables. Through the process of scanning a product, Walmart is able to quickly and efficiently identify its original farm of origin. This greatly decreases the amount of time needed to track food items and improves the company's capacity to address any potential food safety concerns.

6.5.3.2 Enhancing Food Safety

Food safety is an essential and significant topic in the food sector. Blockchain technology has the potential to mitigate food-borne infections by offering precise and prompt information regarding the entire process of food products, from the farm to the table. The Food Trust blockchain network developed by IBM enables stakeholders, including prominent companies such as Nestlé and Unilever, to obtain comprehensive data regarding the handling, processing, and transportation of food goods. This high level of visibility enables the rapid identification of sources of contamination, which in turn facilitates quicker recalls and reduces the likelihood of broad outbreaks.

6.5.3.3 Mitigating Fraudulent Activities and Guaranteeing the Genuineness of Information

Food fraud, which encompasses activities such as mislabelling or adulteration, presents substantial hazards to customers and erodes confidence in food companies.

Blockchain technology can effectively address the issue of food fraud by guaranteeing that all data recorded in the supply chain is resistant to alteration and authenticated by multiple entities. Carrefour, a prominent multinational retailer, use blockchain technology to authenticate the legitimacy of its organic agricultural products. Consumers have the ability to scan QR codes on product packaging in order to obtain information regarding the product's journey, such as specifics about the farm, production techniques, and certifications. This allows them to verify the legitimacy of organic claims.

6.5.3.4 Optimizing Supply Chain Operations

Blockchain technology optimizes supply chain operations by minimizing paperwork and human procedures. Smart contracts are contracts that are capable of completing themselves, with the provisions of the contract written directly into computer code. They have the ability to automate transactions and agreements. This minimizes the necessity for intermediaries and accelerates procedures such as payments and deliveries. Provenance, a business in the seafood market, use blockchain technology to monitor the whole journey of fish, from the moment they are caught to when they reach the plate. This automated system verifies the authenticity of the seafood and guarantees that it has been sourced responsibly.

6.5.3.5 Advocating for the Principles of Environmental, Social, and Economic Sustainability

Consumers and corporations are placing growing significance on sustainability. Blockchain can enhance sustainable practices by offering a transparent ledger of the ecological consequences of food production. Companies such as Bumble Bee Foods employ blockchain technology to monitor the sustainability of their tuna supply chain, guaranteeing that the fish are harvested using environmentally friendly techniques and that workers are treated with fairness. This level of transparency enables companies to achieve their sustainability objectives and empowers consumers to make well-informed purchase choices.

6.5.3.6 Obstacles and Prospects for the Future

Although blockchain has the capacity to revolutionize the food supply chain, its application encounters obstacles. These factors encompass the requirement for widespread implementation across the sector, incorporation with current systems, and the guaranteeing of the precision of data. Nevertheless, with the progression of technology and increasing awareness of its advantages, the utilization of blockchain is anticipated to expand. Effective collaboration among stakeholders, including industry consortia and alliances, will play a vital role in overcoming these hurdles and fully harnessing the promise of blockchain technology in the food supply chain.

Examples:

- Walmart used IBM's Food Trust blockchain to swiftly track the source of mangoes and other agricultural products, thereby improving food safety and promoting transparency.
- Nestlé's implementation of blockchain technology to trace the journey of milk, starting from farms in New Zealand and ending at factories and warehouses in the Middle East, showcases the effectiveness of this technology in guaranteeing the genuineness and safety of their products.
- Carrefour has implemented blockchain technology to provide consumers with comprehensive information about the whole path of its organic chickens, thereby enhancing transparency and fostering consumer trust.

To summarize, blockchain technology has the capacity to significantly change the food supply chain by increasing visibility, improved food safety, minimizing fraudulent activities, optimizing processes, and supporting sustainability. With the growing use and resolution of obstacles, blockchain is positioned to become a fundamental element of contemporary food supply chain management, providing advantages to all parties involved.

6.5.4 Drones and Autonomous Vehicles

Drones and autonomous vehicles are transforming the food distribution sector by increasing operational efficiency, lowering expenses, and raising delivery speed. These technologies are becoming more and more essential in the logistics industry, especially for 'last-mile' delivery, which is sometimes the most difficult and costly aspect of the supply chain.

6.5.4.1 The Use of Unmanned Aerial Vehicles (UAVs), Also Known as Drones, in the Delivery of Food

Drones provide a rapid and effective method for food delivery, particularly in metropolitan environments and remote areas that are difficult to access. By bypassing road congestion, they are able to greatly reduce delivery times. Companies such as Wing, a subsidiary of Alphabet, have been conducting trial programmes for drone delivery services in Australia and the United States. These services involve the delivery of various things, including coffee, groceries, and meals. Furthermore, Zipline, which initially concentrated on transporting medical supplies, has broadened its drone delivery services to encompass food and other vital items in Rwanda and Ghana. This demonstrates the adaptability and potential of drones in enhancing food accessibility.

6.5.4.2 Autonomous Vehicles in the Context of Food Distribution

Autonomous vehicles (AVs) are revolutionizing the logistics industry by providing dependable, 24/7 delivery services without the requirement for human drivers. These vehicles are especially advantageous for extensive distribution and have the potential to substantially decrease labour expenses. Companies such as Nuro have created compact, autonomous delivery vehicles that are expressly engineered for the purpose of conveying products. Nuro has formed strategic alliances with prominent merchants such as Kroger to provide transportation services for delivering groceries to clients in specific cities around the United States. This collaboration not only showcases the tangible implementation of autonomous vehicles but also underscores their capacity to optimize grocery delivery, enhancing efficiency and reducing costs.

Illustrations and Consequences:

- Wing's drone delivery service has successfully completed more than 100,000 deliveries across Australia, encompassing a wide range of items such as food and beverages. This achievement serves as a compelling demonstration of the practicality and effectiveness of drone technology in the context of urban food delivery.
- Nuro's autonomous vehicles have effectively executed grocery deliveries in Houston, Texas through its relationship with Kroger, serving as a blueprint for future development and acceptance in additional locations.

Ultimately, drones and autonomous cars are revolutionizing the food distribution sector by providing inventive measures that optimize supply effectiveness, decrease expenses, and enhance client contentment. As these technologies progress and legal frameworks adjust, their usage is expected to become more prevalent, leading to significant changes in the logistics and food delivery industry.

6.6 CHALLENGES IN FOOD DISTRIBUTION

Food distribution encounters numerous obstacles that might affect effectiveness, expenses, and the quality of products. Tackling these obstacles is essential for guaranteeing a seamless and dependable flow of goods from manufacturers to end-users.

6.6.1 Perishability

An important obstacle in the delivery of food is the fact that foodstuffs can easily spoil. Perishable items such as fresh produce, dairy, and meats have a short lifespan and need to be handled with care in order to preserve their freshness. For instance, the transportation of strawberries from farms to stores necessitates the implementation of precise temperature controls to avoid rotting. Insufficient refrigeration or delays might result in substantial product loss and wastage.

6.6.2 Supply Chain Management and the Movement of Goods

Effective transportation is crucial, yet it is frequently hindered by issues such as traffic congestion, extensive distances, and regulatory limitations. Transporting food across international borders necessitates adherence to diverse customs laws, which might result in delays. The COVID-19 pandemic exposed weaknesses in worldwide logistics, since interruptions in transportation routes resulted in substantial delays and scarcities.

6.6.3 Expense Control

Another problem is achieving cost optimization while simultaneously upholding quality and efficiency. Escalating fuel prices and labour costs have the potential to raise distribution expenses. Large corporations such as Amazon and Walmart allocate significant resources to their logistics operations in order to save expenses. However, smaller enterprises may find it challenging to rival these companies' advantages in terms of scale.

6.6.4 Integration of Technology

Although technology provides solutions, the process of incorporating sophisticated technologies such as GPS monitoring, inventory management, and blockchain into current infrastructures can be intricate and costly. Small and medium-sized firms (SMEs) may have financial constraints that hinder their ability to invest in these technologies, resulting in inefficiencies in their distribution processes.

6.6.5 Environmental and Social Sustainability Issues

Consumers are displaying a growing preference for sustainable techniques in the distribution of food. Minimizing the environmental impact of logistics operations is a complex task that necessitates the allocation of resources towards the adoption of more sustainable technology and practices. Companies such as PepsiCo are currently investigating the use of electric delivery vehicles as a means to decrease pollution.

Food distribution encounters various complicated challenges, such as the limited shelf life of food, intricate logistical operations, effective cost control, seamless technological incorporation, and the need for sustainability. To tackle these problems, it is necessary to make careful and calculated investments, come up with new and creative ideas, and be able to adjust to changing circumstances in order to guarantee a strong and effective system for producing and distributing food.

6.6.6 Regulatory and Compliance Issues

Regulatory and compliance difficulties pose significant obstacles in the food distribution sector, as they require strict adherence to food safety, quality, and ethical standards across the supply chain. Adherence to these laws is crucial in order to safeguard customers and uphold confidence in food items.

6.6.6.1 Regulations Pertaining to the Safety of Food

Ensuring food safety is of utmost importance, and regulatory authorities such as the FDA in the United States and EFSA in Europe establish rigorous protocols for the handling, storage, and transportation of food. As an illustration, the Food Safety Modernization Act (FSMA) mandated by the FDA necessitates that food facilities establish preventive measures and keep records to avert contamination. Failure to comply can lead to significant consequences, such as the recall of products and the closure of facilities.

6.6.6.2 *Legislation Regarding the Identification and Presentation of Products*

Precise labelling is crucial for providing consumers with information regarding ingredients, nutritional composition, and possible allergies. The EU's Food Information to Consumers (FIC) Regulation requires food products to include labelling that is both clear and accurate. Incorrect labelling might result in substantial penalties and legal consequences. In 2016, the US FDA sent a warning letter to KIND Snacks because they had incorrectly labelled their products as "healthy." This caused the company to make changes to how they label their products.

6.6.2.3 *Regulations on the Import and Export of Goods*

Global food distribution must adhere to diverse import and export restrictions, encompassing tariffs, quotas, and food safety standards. As an illustration, the US Department of Agriculture (USDA) mandates rigorous inspections for imported meat and poultry to guarantee compliance with US safety requirements. Adherence to these standards is crucial in order to prevent any delays or refusal of shipments at border crossings.

6.6.6.4 *Regulations Pertaining to the Environment*

Increasingly, there is a mounting interest in sustainability and the influence on the environment. Policies targeting the reduction of greenhouse gas emissions in the transportation of food, such as the European Union's Green Deal, incentivize corporations to embrace more environmentally friendly methods. Noncompliance with these environmental standards may lead to financial penalties and harm a company's reputation.

Illustrations and Consequences:

- Chipotle Mexican Grill encountered many food safety infractions in 2015, leading to the occurrence of *E. Coli* and norovirus outbreaks. As a result, there were temporary closures of stores, a substantial decline in sales, and a lasting effect on the brand's reputation.
- Nestlé has been subject to criticism and legal disputes surrounding its adherence to labelling regulations, particularly in relation to the promotion of its infant nutrition products.

Regulatory and compliance challenges in food distribution are complex and essential for guaranteeing food safety, precise labelling, lawful trading, and environmental sustainability. Compliance with these requirements safeguards customers, improves product authenticity, and upholds industry benchmarks.

6.6.7 Geographical and Infrastructural Challenges

Food distribution is greatly affected by geographical and infrastructural obstacles, especially in areas with varied terrains, insufficient infrastructure, and complex logistical issues. These issues hinder the achievement of efficient, timely, and cost-effective delivery of food goods to customers.

6.6.7.1 Geographically Isolated and Sparsely Populated Regions

Geographical isolation in remote and rural areas might pose difficulties for transportation and distribution. Restricted transportation infrastructure, challenging topographies such as mountains or deserts, and severe climatic conditions can impede the transportation of products. Delivering fresh produce to isolated settlements in the Himalayas, such as in this case, necessitates overcoming challenging topography and unpredictable road conditions, which have a direct impact on the quality and accessibility of food.

6.6.7.2 Urban Congestion

Urban areas, on the other hand, encounter infrastructure obstacles like traffic congestion and restricted parking, which have a negative effect on the effectiveness of food delivery. Delivery trucks manoeuvring through congested urban streets may encounter delays, leading to higher operational expenses and potentially impacting the quality of food. Metropolises such as New York and Tokyo enforce stringent regulations regarding delivery schedules and vehicle emissions in order to address these difficulties.

6.6.7.3 Development of Physical Structures and Systems

Investing in infrastructure is essential for enhancing the capacity of food distribution. Underdeveloped nations frequently face a shortage of sufficient storage infrastructure, efficient cold chain logistics, and well-established transportation networks. Sub-Saharan African countries encounter difficulties in preserving the quality of perishable goods because of an inconsistent electrical supply and insufficient refrigeration infrastructure.

6.6.7.4 Effects of Climate Change

Climate change worsens geographical issues by modifying weather patterns and intensifying the occurrence of extreme weather events. Floods, storms, and droughts have the potential to interrupt supply chains, inflict harm on infrastructure, and impact the availability of food. Coastal areas that are susceptible to rising sea levels may encounter the infiltration of saltwater, which can undermine agricultural output and jeopardize food security.

Illustrations and Consequences:

- Indigenous populations residing in the Amazon rainforest encounter logistical obstacles in obtaining food supplies as a result of their distant locations and inadequate transportation infrastructure.
- Mumbai, India: The presence of heavy traffic and limited street space in Mumbai presents difficulties for food delivery services, requiring the implementation of inventive strategies such as utilizing bicycle messengers and optimizing route planning.

6.6.8 Seasonal Variations

Seasonal oscillations pose substantial obstacles in the distribution of food, as they impact supply, demand, and logistical operations, which are influenced by shifting weather patterns and customer preferences.

6.6.8.1 Agricultural Production and Harvest Cycles

Fluctuations in agricultural production throughout the year have a direct impact on the supply chains of food. For instance, crops such as fruits and vegetables adhere to particular growing seasons determined by climate conditions. During periods of high harvest, there is frequently a surplus of agricultural products that must be transported and distributed efficiently to avoid deterioration. On the other hand, during off-seasons, there may be a lack of particular products and prices may rise, which can affect how easily consumers can acquire and afford them.

6.6.8.2 Consumer Demand and Preferences

Fluctuations in seasons also impact the choices consumers make when it comes to food products. During the summer months, there is generally a greater demand for fresh salads and fruits, whereas in the winter months, there is an uptick in demand for hearty soups and root vegetables. Food wholesalers and retailers should anticipate these changes in consumer behaviour and adapt their inventory and marketing tactics accordingly to fulfill customer expectations.

6.6.8.3 Management of the Flow of Goods and Services

Effectively handling fluctuations in seasons necessitates resilient logistics and supply chain management solutions. Companies frequently encounter difficulties in accurately predicting demand, optimizing storage capabilities, and preserving the freshness of perishable items throughout transportation. The preservation of temperature-sensitive items such as seafood and dairy in different locations and seasons is highly dependent on cold chain logistics.

6.6.8.4 Economic Consequences

Fluctuations in seasons can have an effect on the cost of food and the overall economic stability of areas that rely on agriculture. For instance, droughts or heavy rainfall can damage crop output, resulting in scarcity of food and fluctuation in prices. Seasonal variations in food supply in developing nations can worsen food insecurity and poverty rates, underscoring the significance of robust farming methods and efficient delivery systems.

Illustrations and Consequences:

- The agricultural sector in California undergoes periodic fluctuations, with the highest yield of almonds and citrus fruits occurring during specific months. This necessitates the need for effective distribution to international markets.
- Retailers face a surge in demand for festive foods and beverages throughout holiday seasons, requiring meticulous planning and inventory management to satisfy consumer expectations.

6.6.9 Sustainability Concerns

The prominence of sustainability concerns in food distribution has risen as worldwide awareness regarding environmental impact, resource conservation, and ethical behaviours throughout the supply chain has increased.

6.6.9.1 Ecological Consequences

Food distribution plays a substantial role in carbon emissions due to the shipping, refrigeration, and packaging involved. Sustainable practices seek to minimize these effects by streamlining logistics, embracing environmentally-friendly packaging materials, and investing in energy-efficient transportation systems. Companies such as Tesco in the United Kingdom have taken steps to decrease the distance that food travels and encourage sustainable procurement methods. This has resulted in a reduction in their total impact on the environment.

6.6.9.2 Minimization of Food Waste

One significant sustainability issue is the immense quantity of food waste produced throughout the whole supply chain. Efficient distribution networks, enhanced storage facilities, and novel packaging options have the potential to reduce waste. For instance, businesses such as Walmart have made substantial investments in technology to enhance their ability to monitor and control inventory, resulting in decreased waste and unsold merchandise.

6.6.9.3 Ensuring the Acquisition of Goods and Services from Ethical Sources and Promoting Fair Trade Practices

Consumers are placing greater importance on ethical factors when making selections about what food to buy. They are more likely to support firms that show a strong dedication to fair trade methods and ethical sourcing. Organizations such as Fair Trade USA certify products that comply with rigorous social, environmental, and economic criteria, guaranteeing equitable remuneration and sustainable agricultural methods.

6.6.9.4 *Adherence to Regulations and Standards*

Governments and international organizations enforce restrictions and establish standards to advance sustainability in the distribution of food. The Circular Economy Action Plan of the European Union seeks to reduce waste and optimize the use of resources in the food supply chain. Adhering to these standards not only reduces the negative effects on the environment but also improves the reputation of the company and fosters trust among consumers.

Illustrations and Consequences:

- IKEA has made a pledge to obtain all of its food items from sustainable and responsible sources. This includes purchasing MSC-certified fish and Fairtrade-certified coffee.
- Amazon has implemented packaging strategies aimed at minimizing waste, including the use of recyclable packaging materials and the introduction of "Frustration-Free Packaging" choices.

6.7 GLOBAL VS. LOCAL DISTRIBUTION STRATEGIES

Food firms must carefully assess the balance between global and local distribution systems in order to maximize efficiency, satisfy consumer preferences, and efficiently manage logistical challenges.

6.7.1 Strategies for Distributing Products Worldwide

Global distribution methods encompass the process of distributing items across several regions or countries, typically taking advantage of economies of scale and a wider market reach. This technique is appropriate for items that have a stable and predictable demand, as well as standardized production procedures. Companies such as Nestlé and Coca-Cola demonstrate effective global distribution strategies by establishing broad global networks to guarantee the consistent availability of their products worldwide. These organizations optimize their operations and reduce costs by consolidating production and distribution centres, allowing for efficient bulk transportation and centralized logistics.

6.7.2 Strategies for Distributing Products or Services within a Certain Geographic Area

Local distribution methods place a high importance on being able to quickly and effectively meet the needs and preferences of specific area markets and consumers. These tactics are especially advantageous for goods that are prone to spoilage, products that are exclusive to certain cultures, and objects that have a limited lifespan. Companies can rapidly respond to changes in local demand and regulatory requirements by utilizing local sourcing and production, together with decentralized distribution centres. For instance, farmers' markets and specialty food stores flourish by employing local distribution tactics, providing distinctive, handcrafted products customized to suit local preferences and interests.

6.7.3 Instances of Integration

- McDonald's employs a hybrid strategy by globally standardizing certain key products while customizing menus to suit local preferences. For example, providing the Maharaja Mac in India or the Teriyaki McBurger in Japan accommodates the tastes of the local population while still upholding a consistent global brand image.
- Unilever implements a dual approach by manufacturing internationally renowned brands such as Dove and Lipton, while also acquiring local brands to enhance its market dominance in certain countries. By adopting this strategy, Unilever can leverage local market knowledge and customer behaviour while also taking advantage of its global brand recognition and operational efficiencies.

6.7.4 Obstacles and Factors to Take Into Account

- Effectively managing both global and local distribution networks concurrently necessitates advanced logistical capabilities to guarantee prompt delivery, reduce disruptions in the supply chain, and comply with regulatory standards in many regions.
- Adapting products and marketing techniques to local cultures and regulatory environments is essential for achieving success due to cultural and regulatory differences. Companies must adeptly manage a wide range of consumer preferences, packaging requirements, and import/export constraints in order to be successful.
- Implementing global distribution techniques frequently involves transporting goods over longer distances, resulting in increased carbon emissions. It is becoming increasingly vital to balance global expansion with sustainability efforts, such as investing in carbon-neutral transportation and promoting local procurement, in order to reduce environmental harm. Ultimately, the decision between global and local distribution strategies hinges on variables such as product attributes, market needs, regulatory environment, and sustainability objectives. Global plans have the advantage of scale and efficiency, while local strategies allow for responsiveness and personalization. Successful organizations frequently incorporate these strategies strategically, utilizing their advantages to maximize their market share and enhance consumer contentment in a constantly changing global market.

6.8 THE IMPACT OF E-COMMERCE ON FOOD DISTRIBUTION

E-commerce has completely transformed the way customers buy groceries, meals, and specialty food items, leading to a revolution in food delivery. This transition has significant ramifications for businesses, consumers, and the wider food sector ecosystem.

6.8.1 Enhancing Consumer Convenience and Accessibility

The advent of e-commerce has significantly facilitated consumers' ability to conveniently obtain a diverse range of food products from the convenience of their residences or mobile devices. Online platforms such as Amazon Fresh, Instacart, and grocery store applications provide shoppers with the ability to browse, compare prices, and place orders for goods to be delivered to their homes or picked up at the store. This convenience is attractive to persons living in busy urban areas, older populations, and people looking for specialized dietary requirements or specialty items that are not accessible locally.

6.8.2 Expansion of the Market and the Improvement of Accessibility

The advent of e-commerce has facilitated market expansion for food producers and merchants, enabling them to extend their consumer base beyond their conventional geographic limits. Online marketplaces and direct-to-consumer websites have opened up opportunities for small and medium-sized food enterprises, artisanal producers, and specialized market players to compete globally. For instance, gourmet food companies have the ability to deliver perishable goods across the entire country, avoiding conventional retail distribution channels and reaching a wider range of customers.

6.8.3 Innovations in Logistics and Supply Chain Management

The incorporation of e-commerce into the food distribution industry has stimulated advancements in logistics and supply chain management. Companies allocate resources towards temperature-controlled packaging, 'last-mile' delivery solutions, and real-time tracking systems in order to guarantee the freshness and quality of perishable items during transportation. Companies such as HelloFresh and Blue Apron have optimized meal-kit delivery systems by sending pre-measured materials and recipes directly to customers' homes, reducing food waste and improving convenience.

6.8.4 Change in Consumer Behaviour and Expectations

The advent of e-commerce has fundamentally transformed customer behaviour, exerting a significant impact on expectations regarding ease, promptness of delivery, and the level of product information available. Modern consumers have high expectations for online purchasing, desiring smooth and uninterrupted experiences, tailored recommendations, and dependable delivery services. For the purpose of retaining client loyalty and happiness, it is imperative for companies to allocate resources towards digital marketing, customer care, and user-friendly interfaces.

6.8.5 Obstacles and Adjustments

Although e-commerce offers advantages, it also presents difficulties for the distribution of food. Factors such as adherence to food safety requirements, the proper handling of perishable commodities, and the management of delivery costs can have a significant impact on both profitability and operational efficiency. Companies must successfully negotiate these intricacies while simultaneously upholding product authenticity and adhering to regulatory benchmarks. Furthermore, the emergence of dominant online retail companies such as Amazon and Walmart in the food industry has increased rivalry, prompting conventional merchants to be more creative and adjust to changing consumer tastes.

Illustrations and Consequences:

- The acquisition of Whole Foods Market by Amazon in 2017 showcases the merging of online and offline retail methods, allowing Amazon to utilize Whole Foods' brick-and-mortar stores for grocery fulfillment and extending its grocery delivery service.
- The COVID-19 pandemic expedited the implementation of e-commerce in the food distribution sector. This was driven by consumers' desire for contactless purchasing and home delivery services to reduce health risks. As a result, existing supermarket chains swiftly improved their online capabilities.

Ultimately, e-commerce is transforming the way food is distributed by improving convenience, extending market access, and fostering advancements in logistics and consumer interaction. Food firms must embrace digital tactics and adapt to altering consumer tastes in order to succeed in an increasingly digitalized economy.

6.9 CHAPTER SUMMARY

Distribution channels and logistics are crucial factors in the achievement of food marketing, as they impact the movement of products from producers to consumers in a timely, efficient, and cost-effective manner. In order to keep up with changing customer needs, overcome logistical obstacles, and stay ahead of competitors, it is crucial for the food business to adopt creative distribution tactics.

Effective distribution networks are crucial for guaranteeing the availability and freshness of products while reducing waste and operational expenses. Businesses need to adapt their strategies to match market realities, regulatory demands, and technological improvements, whether they are using direct-to-consumer models, traditional retail networks, or foodservice channels.

Food marketing logistics involves a range of tasks, including storing and transporting goods, managing inventories, and fulfilling orders. The progress in technology, including GPS tracking, RFID systems, and route optimization software, has completely transformed the efficiency of supply chains. This has made it possible to monitor in real time, improve inventory accuracy, and expand delivery capacities.

Moreover, the incorporation of sustainability factors is progressively influencing distribution strategies, placing greater importance on the reduction of carbon footprints, the optimization of packaging materials, and the promotion of ethical sourcing practices across the whole supply chain.

Ultimately, the future of food distribution hinges on adopting digital advancements, maximizing the effectiveness of supply chains, and adapting distribution techniques to meet changing customer tastes and sustainability objectives. Through constant adaptation and innovation, food marketers can effectively address the difficulties of a rapidly changing marketplace while providing value to both consumers and stakeholders.

CASE STUDIES

Case Study 1

"Fresh Harvest," a Newly Established Business that Delivers Fresh, Locally-Sourced Produce Boxes Straight to Customers' Homes

Fresh Harvest is encountering tremendous expansion, yet has difficulties in upholding product quality and efficiency throughout its distribution routes. They presently employ two primary channels:

Fresh Harvest utilizes a direct delivery system by collaborating with local farms to acquire fresh food. Additionally, the company operates its own fleet of delivery cars to efficiently serve customers residing in adjacent suburbs.

Fresh Harvest employs a 3PL company to transport boxes to consumers residing in distant suburbs, in order to expand its customer base.

Problems

Ensuring Freshness: Direct deliveries guarantee freshness, while 3PL deliveries may encounter delays that affect the quality of the product.

Delivery efficiency: The cost of managing a fleet of vehicles might be high for Fresh Harvest. Depending exclusively on a 3PL could restrict the ability to manage and regulate delivery timetables.

Scalability: The existing paradigm may not be viable for future growth.

The hybrid distribution model proposed for Fresh Harvest involves maintaining direct delivery for adjacent locations and leveraging a 3PL provider with established expertise in cold chain management for larger distances.

Strategic Partnerships: Engage in collaborative efforts with nearby food stores to develop convenient pick-up locations for customers residing in distant suburbs, hence decreasing dependence on 3PL.

Delivery Optimization Software: Deploy software to strategize optimal routes for their delivery vehicles, hence lowering journey durations and fuel expenditures.

Real-Time Tracking: Employ advanced technology to monitor delivery in real time, enabling proactive communication with clients regarding possible delays.

Inventory Management Systems: Develop and implement efficient systems that analyze order history and seasonal patterns to minimize excess inventory and guarantee product availability.

Through the examination and utilization of these solutions, Fresh Harvest can establish a distribution network that is both more effective and capable of expansion. This will guarantee the freshness of products and the contentment of customers, while also allowing Fresh Harvest to keep a competitive advantage in the market.

Discussion Questions

1. What techniques may Fresh Harvest employ to enhance the efficiency of its distribution routes while guaranteeing the freshness of its products and the effectiveness of its deliveries?
2. How can Fresh Harvest utilize technology to enhance its distribution network and attain a competitive edge?

Case Study 2

Enhancing Distribution Channels for "Spice It Up!," a Worldwide Spice Brand

Company: Spice It Up!, a company providing distinctive and premium spice mixes acquired from various regions globally, distributed through online platforms and specific retail outlets.

Challenge: Spice It Up! aims to broaden its scope but encounters logistical obstacles in effectively managing its distribution channels for various product categories. The channels they currently have are:

Spice It Up! operates a Direct-to-Consumer (D2C) model, where they handle their online store and handle order fulfillment from their own warehouse.

Wholesale: They collaborate with wholesalers to provide their spice blends to grocery stores nationwide.

Problems

Inventory management can become intricate when handling distinct inventories for both direct-to-consumer (D2C) and wholesale channels, which can result in the risk of stockouts or overstocking.

Delivery Costs: Shipping individual spice mixes directly to consumers can incur higher expenses in comparison to bulk deliveries to wholesalers.

Spice It Up! may encounter difficulty in meeting the minimum order quantities (MOQs) set by grocery shops for less popular blends.

Potential Resolutions to Contemplate

Fulfillment by Amazon (FBA) allows Spice It Up! to utilize Amazon's extensive warehouse network for D2C fulfillment, resulting in reduced expenses for storage and order processing.

Dropshipping: To handle less popular mixes, Spice It Up! has the option to collaborate with a dropshipping provider who will store and deliver the goods directly to clients upon receiving an order.

Kitting involves the pre-assembling of commonly used spice blends in different quantities to suit the MOQ of grocery stores. This helps to reduce the chances of having excess inventory of individual spices.

Spice blend subscription boxes are available to provide customers with carefully selected assortments of spices. This helps to establish a dedicated customer base and ensures consistent demand.

Vendor-Managed Inventory (VMI) involves collaborating with distributors to develop a system in which they take charge of managing Spice It Up!'s inventory at their warehouses. This ensures that there is always enough stock available for grocery stores.

Spice It Up! can enhance its distribution network for both D2C and wholesale channels by implementing these techniques and investigating inventive options such as dropshipping or subscription boxes. These factors can result in enhanced inventory control, decreased shipping expenses, and a competitive edge in the worldwide spice industry.

Discussion Questions

1. What strategies can Spice It Up! implement to enhance its distribution networks in order to optimize inventory management, minimize delivery expenses, and effectively serve both retail customers and bulk purchasers?
2. Which novel logistical strategies could Spice It Up! investigate to get a competitive edge in the worldwide spice industry?

CHAPTER 7

Food Advertising and Promotion

LEARNING OBJECTIVES

By the completion of this chapter, you should be able to:

- Understand the fundamentals of food advertising.
- Differentiate between traditional and digital advertising methods.
- Evaluate the effectiveness of advertising and promotion.
- Understand different promotion methods, and their impact on consumer behaviour.
- Explore how advertising influences consumer preferences.
- Investigate the ethical and regulatory considerations in marketing practices.
- Identify emerging trends and technologies that are transforming food advertising.

7.1 INTRODUCTION

Food advertising and promotion play a crucial role in the modern food industry, influencing consumer choices and shaping market trends. As one of the most dynamic and competitive sectors, the food industry relies heavily on effective advertising strategies to differentiate products, build brand loyalty, and drive sales. This chapter delves into the multifaceted world of food advertising and promotion, exploring the various methods and techniques used by marketers to capture consumer attention and foster engagement.

The evolution of food advertising can be traced back to the early days of print media, where simple product descriptions and images in newspapers and magazines marked the beginning of commercial promotions. As technology advanced, so did advertising techniques, expanding into radio and television, and eventually embracing the digital revolution. Today, food advertising spans a wide array of platforms, from traditional media to social networks and search engines, reflecting the diverse ways consumers receive information and make purchasing decisions.

DOI: 10.1201/9781003473305-7

Promotional activities in the food industry are equally varied, encompassing everything from sales promotions and public relations to direct marketing and community engagement. These techniques aim not only to boost short-term sales but also to build long-term brand equity and customer loyalty. Effective promotions can create memorable brand experiences, encouraging repeat purchases and positive word-of-mouth.

Understanding consumer behaviour and psychology is fundamental to crafting successful advertising and promotion campaigns. By analyzing consumer preferences, habits, and motivations, marketers can tailor their messages to resonate with their target audience. This involves not only highlighting the unique benefits of their products but also addressing consumer concerns and values, such as health, sustainability, and ethical sourcing.

However, the power of food advertising comes with significant responsibility. Regulatory bodies oversee the industry to ensure that advertisements are truthful and not misleading. Ethical considerations also play a critical role, as marketers must navigate the fine line between persuasive promotion and manipulation. High-profile cases of misleading advertisements serve as reminders of the potential pitfalls and underscore the importance of transparency and integrity.

In this chapter, we will explore the strategies and techniques that define food advertising and promotion, supported by case studies and real-world examples. We will also examine the regulatory and ethical landscape, providing a comprehensive understanding of what it takes to succeed in this dynamic field. As we look ahead, we will consider emerging trends and technologies that are set to shape the future of food advertising, offering insights and recommendations for marketers striving to stay ahead in a competitive market.

7.2 ADVERTISING STRATEGIES IN THE FOOD INDUSTRY

In the competitive landscape of the food industry, effective advertising strategies are essential for capturing consumer attention and driving brand success. This section explores the various advertising methods employed by food companies, examining both traditional and digital approaches to reach and engage target audiences.

7.2.1 Traditional Advertising Methods

7.2.1.1 Print Media

Print media has long been a cornerstone of food advertising, with advertisements appearing in newspapers, magazines, and other printed publications. This medium allows for detailed descriptions, appealing visuals, and targeted placement in specific sections, such as food and lifestyle magazines. Print ads can effectively convey a brand's story and highlight product benefits, appealing to a broad demographic.

7.2.1.2 *Television and Radio*

Television and radio remain powerful tools for food advertising due to their extensive reach and ability to engage multiple senses. TV commercials combine visual and auditory elements to create compelling narratives, showcasing the product in various contexts. Radio ads, while limited to audio, use jingles, voiceovers, and sound effects to leave a lasting impression. Both mediums offer opportunities for time-specific advertising, such as during meal times or cooking shows, to maximize impact.

7.2.2 Digital Advertising Methods

7.2.2.1 *Social Media*

Social media platforms like Facebook, Instagram, Twitter, and TikTok have revolutionized food advertising by enabling direct interaction with consumers. Brands can use targeted ads to reach specific demographics, leverage influencers to endorse products, and create engaging content that encourages user participation. Social media campaigns often feature recipes, cooking tips, and behind-the-scenes looks at product development, fostering a sense of community and authenticity.

7.2.2.2 *Search Engine Marketing*

Search engine marketing (SEM) involves promoting food products through paid search ads and search engine optimization (SEO). Paid search ads appear at the top of search results for relevant keywords, driving traffic to brand websites. SEO strategies, on the other hand, focus on optimizing website content to rank higher in organic search results. Both approaches aim to capture consumers actively searching for food-related information, increasing visibility and driving conversions.

7.2.2.3 *Influencer Marketing*

Influencer marketing leverages the reach and credibility of social media influencers to promote food products. Influencers create content featuring the product, such as cooking demonstrations, taste tests, and unboxing videos, which they share with their followers. This form of advertising relies on the influencer's authenticity and connection with their audience, making it a powerful tool for building trust and driving engagement.

7.3 INTEGRATION OF TRADITIONAL AND DIGITAL STRATEGIES

Modern food advertising often involves a seamless integration of traditional and digital strategies to create cohesive and impactful campaigns. For example, a television commercial might be complemented by a social media campaign using the same theme and visuals. Print ads can include QR codes that direct readers to online content, bridging the gap between offline and online experiences.

7.4 MEASURING EFFECTIVENESS

To ensure the success of advertising strategies, food companies must continually measure and analyze their campaigns' effectiveness. Metrics such as reach, engagement, conversion rates, and return on investment (ROI) provide valuable insights into which approaches resonate most with consumers. By leveraging data analytics tools, brands can refine their strategies, optimize their spending, and achieve better results.

7.5 PROMOTION TECHNIQUES IN THE FOOD INDUSTRY

Promotion techniques in the food industry are vital for stimulating consumer interest, increasing sales, and enhancing brand loyalty. This section delves into the various promotion methods used by food companies, highlighting their effectiveness and providing real-world examples.

7.5.1 Sales Promotions

7.5.1.1 Discounts and Coupons

Discounts and coupons are widely used promotional tools that offer immediate price reductions, encouraging consumers to try new products or purchase more of their favourites. Coupons can be distributed through various channels, including print media, online platforms, and in-store displays. Discounts and coupons create a sense of urgency, motivating consumers to take advantage of the limited-time offers.

7.5.1.2 Contests and Giveaways

Contests and giveaways engage consumers by offering the chance to win prizes, ranging from free products to larger rewards such as kitchen appliances or vacation packages. These promotions generate excitement and buzz around a brand, encouraging participation and sharing on social media. Contests and giveaways can also collect valuable consumer data, such as contact information and preferences, which can be used for future marketing efforts.

7.5.2 Public Relations

7.5.2.1 Event Sponsorships

Sponsoring events, such as food festivals, cooking competitions, or community gatherings, allows food companies to showcase their products in a positive and engaging environment. Event sponsorships provide opportunities for direct interaction with consumers, product sampling, and brand visibility. By aligning with events that resonate with their target audience, brands can enhance their reputation and foster goodwill.

7.5.2.2 *Community Engagement*

Community engagement initiatives, such as partnering with local charities, supporting food drives, or hosting educational workshops, demonstrate a brand's commitment to social responsibility. These activities build strong connections with the community, enhancing brand image and loyalty. Community engagement also offers positive media coverage and word-of-mouth promotion.

7.5.3 Direct Marketing

7.5.3.1 *Email Campaigns*

Email campaigns are a cost-effective way to reach consumers directly with personalized messages, product updates, and special offers. By segmenting their email lists based on consumer behaviour and preferences, food companies can tailor their communications to specific audiences. Effective email campaigns can drive repeat purchases, increase website traffic, and promote new product launches.

7.5.3.2 *Loyalty Programmes*

Loyalty programmes reward consumers for their repeat business, encouraging long-term brand loyalty. These programmes often use a point-based system, where customers earn points for each purchase that can be redeemed for discounts, free products, or exclusive experiences. Loyalty programmes provide valuable insights into consumer behaviour and preferences, enabling brands to tailor their marketing efforts and foster stronger customer relationships.

7.6 INTEGRATION OF PROMOTIONAL TECHNIQUES

Successful promotional strategies often integrate multiple techniques to create a cohesive and impactful campaign. For example, a new product launch might involve a combination of in-store discounts, a social media contest, and an email campaign announcing the launch and promoting a limited-time offer. This multi-channel approach ensures broad reach and maximizes consumer engagement.

7.7 MEASURING EFFECTIVENESS

To assess the success of promotional techniques, food companies must track key performance indicators (KPIs) such as sales lift, customer acquisition cost, redemption rates for coupons and discounts, and engagement levels in contests and giveaways. Analyzing these metrics helps brands refine their promotional strategies, ensuring they resonate with consumers and deliver a strong return on investment.

7.8 CONSUMER BEHAVIOUR AND PSYCHOLOGY

Understanding consumer behaviour and psychology is crucial for developing effective food advertising and promotion strategies. This section explores how consumers make food-related decisions, the psychological factors influencing these choices, and the role of branding in shaping consumer perception.

7.8.1 Understanding Consumer Preferences

7.8.1.1 Factors Influencing Food Choices

Consumers' food choices are influenced by a variety of factors, including taste, price, convenience, health considerations, and cultural preferences. Taste is often the primary driver, but other factors, such as dietary restrictions, lifestyle choices, and social influences, also play significant roles. Marketers must understand these preferences to create products and advertising messages that resonate with their target audience.

7.8.1.2 Demographic and Psychographic Segmentation

Segmenting the market based on demographic factors (age, gender, income, education) and psychographic factors (lifestyle, values, personality) helps marketers tailor their strategies to specific consumer groups. For example, health-conscious consumers might be more receptive to advertisements highlighting nutritional benefits, while convenience-focused consumers may prioritize quick and easy meal solutions.

7.8.2 The Impact of Advertising on Consumer Choices

7.8.2.1 Emotional Appeal

Emotional appeal is a powerful tool in food advertising, as it can create a strong connection between the brand and the consumer. Advertisements that evoke positive emotions, such as happiness, nostalgia, or excitement, are more likely to be remembered and shared. Emotional branding can also build long-term loyalty by associating the brand with positive experiences and memories.

7.8.2.2 Social Proof and Influencer Endorsements

Social proof, or the influence of others' actions and opinions, significantly impacts consumer behaviour. Positive reviews, testimonials, and influencer endorsements can boost a product's credibility and attractiveness. Influencers, in particular, can sway their followers' purchasing decisions by showcasing the product in authentic and relatable contexts.

7.8.2.3 *Cognitive Dissonance and Post-Purchase Behaviour*

Cognitive dissonance occurs when consumers experience doubt or regret after making a purchase. Effective advertising can mitigate this by reinforcing the positive aspects of the product and reassuring consumers that they made the right choice. Post-purchase communication, such as follow-up emails and satisfaction surveys, can further enhance the consumer experience and reduce dissonance.

7.8.3 The Role of Branding in Consumer Perception

7.8.3.1 *Brand Identity and Positioning*

A strong brand identity, encompassing elements such as logo, packaging, and messaging, helps consumers recognize and remember the brand. Effective brand positioning differentiates the brand from competitors by highlighting its unique value proposition. Consistent branding across all touchpoints ensures a cohesive and memorable consumer experience.

7.8.3.2 *Perceived Quality and Value*

Perceived quality and value are critical determinants of consumer preference. High-quality products and positive brand associations can justify premium pricing and foster brand loyalty. Conversely, brands that emphasize affordability and value for money can attract price-sensitive consumers. Marketing communications should align with the brand's positioning to reinforce these perceptions.

7.8.3.3 *Trust and Credibility*

Trust and credibility are essential for building long-term consumer relationships. Brands can establish trust through transparency, ethical practices, and reliable product performance. Third-party endorsements, certifications, and awards can also enhance credibility. Addressing consumer concerns and providing accurate information further solidifies trust.

7.9 MEASURING CONSUMER BEHAVIOUR

To gain insights into consumer behaviour, food companies use various research methods, including surveys, focus groups, and observational studies. Data analytics tools can track consumer interactions across digital platforms, providing valuable information on preferences, purchasing patterns, and engagement levels. Analyzing this data helps marketers refine their strategies and better cater to consumer needs.

Understanding consumer behaviour and psychology is fundamental for creating effective food advertising and promotion strategies. By recognizing the factors that influence food choices and leveraging psychological principles, marketers can craft compelling messages that resonate with consumers, drive sales, and build lasting brand loyalty.

7.10 REGULATIONS AND ETHICAL CONSIDERATIONS

Food advertising and promotion are subject to numerous regulations and ethical guidelines to ensure that marketing practices are fair, truthful, and not misleading. This section examines the regulatory landscape governing food advertising, explores common ethical issues, and provides case studies illustrating the importance of adhering to these standards.

7.10.1 Advertising Regulations in the Food Industry

7.10.1.1 Government Regulations

Various government agencies oversee food advertising to protect consumers from false or deceptive claims. In the United States, the Food and Drug Administration (FDA) and the Federal Trade Commission (FTC) are the primary regulators. The FDA focuses on ensuring that food labelling is accurate and informative, while the FTC monitors advertising practices to prevent misleading advertisements.

7.10.1.2 Labelling Requirements

Regulations mandate that food labels provide essential information, such as nutritional content, ingredient lists, and potential allergens. Claims about health benefits, such as "low fat" or "high fibre," must meet specific criteria established by regulatory authorities. Marketers must ensure that their advertising aligns with these labelling requirements to avoid legal repercussions.

7.10.1.3 Advertising to Children

Special regulations exist to protect children, a vulnerable demographic, from deceptive food advertising. For example, in the United States, the Children's Food and Beverage Advertising Initiative (CFBAI) sets guidelines for marketing food to children, promoting healthier dietary choices and limiting the advertising of unhealthy foods during children's programming.

7.10.2 Ethical Issues in Food Advertising

7.10.2.1 Truthfulness and Transparency

Ethical advertising requires truthfulness and transparency. Marketers must avoid exaggerating product benefits or making unsupported health claims. Transparency involves providing clear and accurate information about the product, including its ingredients, nutritional value, and potential health effects. Misleading consumers not only violates ethical standards but can also damage brand reputation and lead to legal consequences.

7.10.2.2 Targeting Vulnerable Populations

Ethical considerations extend to the targeting of vulnerable populations, such as children, the elderly, and low-income individuals. Advertisers must be cautious not to exploit these groups with manipulative or aggressive marketing tactics. Ethical marketing practices prioritize the well-being of consumers and aim to provide value without exploitation.

7.10.2.3 Sustainability and Ethical Sourcing

Increasingly, consumers demand that the food they purchase is produced sustainably and sourced ethically. Ethical advertising should reflect a commitment to environmental stewardship, fair labour practices, and animal welfare. Brands that misrepresent their sustainability efforts or engage in "greenwashing"—falsely portraying products as environmentally friendly—risk losing consumer trust and facing public backlash.

7.11 MEASURING THE EFFECTIVENESS OF ADVERTISING AND PROMOTION

Measuring the effectiveness of advertising and promotion is critical for understanding the impact of marketing strategies and optimizing future campaigns. This section explores KPIs, tools and techniques for measurement, and methods for analyzing ROI in food advertising.

7.11.1 Key Performance Indicators (KPIs)

7.11.1.1 Reach and Impressions

Reach refers to the total number of unique individuals who see an advertisement, while impressions represent the total number of times an ad is displayed, regardless of whether it is clicked or not. These metrics are fundamental for gauging the overall exposure of an advertising campaign. High reach and impressions indicate broad visibility, essential for brand awareness.

7.11.1.2 Engagement Metrics

Engagement metrics measure how consumers interact with advertisements and promotional content. Key engagement metrics include:

- **Clicks**: The number of times consumers click on an ad.
- **Likes, Shares, and Comments**: Social media interactions indicating consumer interest and engagement.
- **Video Views and Watch Time**: Metrics for video ads, showing how many times a video is viewed and for how long.

7.11.1.3 Conversion Rates

Conversion rates measure the percentage of consumers who take a desired action after engaging with an advertisement, such as making a purchase, signing up for a newsletter, or downloading a coupon. High conversion rates indicate that the ad effectively drives consumer action and achieves campaign objectives.

7.11.1.4 Customer Acquisition Cost (CAC)

Customer acquisition cost (CAC) represents the total cost of acquiring a new customer through advertising and promotional efforts. It is calculated by dividing the total marketing expenses by the number of new customers acquired. Lower CAC indicates more efficient and cost-effective marketing strategies.

7.11.1.5 Return on Investment (ROI)

Return on investment (ROI) measures the profitability of advertising and promotional activities by comparing the revenue generated to the costs incurred. It is calculated using the formula:

$$\text{ROI} = (\text{Revenue} - \text{Cost}) / \text{Cost} \times 100$$

High ROI signifies that the marketing efforts are yielding substantial returns relative to the investment.

7.11.2 Tools and Techniques for Measurement

7.11.2.1 Web Analytics Tools

Web analytics tools, such as Google Analytics, provide detailed insights into website traffic, user behaviour, and conversion rates. These tools track how consumers interact with online advertisements, offering valuable data for optimizing digital marketing strategies.

7.11.2.2 Social Media Analytics

Social media platforms offer built-in analytics tools that track engagement metrics, audience demographics, and the performance of social media campaigns. Tools like

Facebook Insights, Twitter Analytics, and Instagram Insights help marketers understand how their content resonates with audiences and identify areas for improvement.

7.11.2.3 Customer Surveys and Feedback

Surveys and feedback forms collect direct input from consumers about their experiences with advertisements and promotions. This qualitative data provides insights into consumer perceptions, preferences, and satisfaction levels, complementing quantitative metrics.

7.11.2.4 A/B Testing

A/B testing involves comparing two versions of an advertisement or promotional content to determine which performs better. By testing variables such as headlines, images, and call-to-action buttons, marketers can identify the most effective elements and refine their strategies accordingly.

7.11.3 Analyzing ROI in Food Advertising

7.11.3.1 Attribution Models

Attribution models help marketers understand the contribution of different marketing channels and touchpoints to the overall conversion. Common models include:

- **First-Touch Attribution**: Credits the first interaction as the most influential.
- **Last-Touch Attribution**: Credits the final interaction before conversion.
- **Multi-Touch Attribution**: Distributes credit across multiple interactions, providing a comprehensive view of the customer journey.

7.11.3.2 Cost–Benefit Analysis

A cost–benefit analysis evaluates the financial benefits of advertising and promotion relative to their costs. This analysis helps marketers determine which campaigns are most profitable and allocate resources more effectively.

7.11.3.3 Lifetime Value (LTV) Analysis

LTV analysis estimates the total revenue a customer will generate over their lifetime. By comparing LTV to CAC, marketers can assess the long-term value of their customer acquisition efforts and prioritize strategies that maximize LTV.

7.12 FUTURE TRENDS IN FOOD ADVERTISING AND PROMOTION

The food advertising landscape is continually evolving, influenced by technological advancements, changing consumer behaviours, and globalization. This section explores the future trends in food advertising and promotion, focusing on the role of technology and innovation, predicting future consumer trends, and the impact of globalization on food advertising.

7.12.1 The Role of Technology and Innovation

7.12.1.1 Artificial Intelligence and Machine Learning

Artificial intelligence (AI) and machine learning (ML) will continue to revolutionize food advertising by enabling hyperpersonalized marketing strategies. AI can analyze vast amounts of consumer data to predict preferences, optimize ad placements, and create highly tailored advertising experiences. For instance, AI-driven algorithms can recommend products based on individual browsing and purchasing history, enhancing the relevance and effectiveness of marketing campaigns.

7.12.1.2 Augmented Reality (AR) and Virtual Reality (VR)

AR and VR technologies are set to transform how consumers interact with food brands. AR can allow consumers to visualize products in their own environments or provide interactive packaging experiences, while VR can offer immersive brand experiences, such as virtual tours of production facilities or cooking demonstrations. These technologies can create engaging and memorable interactions, driving consumer engagement and brand loyalty.

7.12.1.3 Blockchain for Transparency and Traceability

Blockchain technology offers a solution for increasing consumer demand for transparency and traceability in the food supply chain. By providing an immutable record of a product's journey from farm to table, blockchain can ensure the authenticity of claims about sustainability, ethical sourcing, and product quality. This transparency can be leveraged in advertising to build trust and differentiate brands in the marketplace.

7.12.1.4 Voice Search and Smart Speakers

The growing popularity of voice search and smart speakers, such as Amazon's Alexa and Google Home, presents new opportunities for food advertising. Brands can optimize their content for voice search to enhance discoverability and create voice-activated ads or interactive experiences that engage consumers through their smart devices. As voice technology continues to improve, it will become a critical component of digital marketing strategies.

7.12.2 Predicting Future Consumer Trends

7.12.2.1 Health and Wellness Focus

Consumers are increasingly prioritizing health and wellness, seeking out products that offer nutritional benefits, natural ingredients, and functional health properties. Food brands will need to emphasize these attributes in their advertising, highlighting health benefits and transparent ingredient sourcing. Trends such as plant-based diets, organic products, and functional foods are likely to gain further traction.

7.12.2.2 Sustainability and Ethical Consumption

Sustainability and ethical consumption will continue to be significant drivers of consumer behaviour. Brands that prioritize eco-friendly practices, such as reducing plastic use, minimizing carbon footprints, and supporting fair trade, will resonate with environmentally conscious consumers. Advertising that communicates these values effectively can build strong brand loyalty and attract a growing segment of ethical consumers.

7.12.2.3 Convenience and Digital Integration

The demand for convenience will shape future food advertising strategies. As consumers increasingly rely on online shopping and delivery services, brands must integrate digital solutions into their marketing efforts. This includes developing user-friendly e-commerce platforms, leveraging mobile apps, and using digital advertising to promote seamless purchasing experiences.

7.12.2.4 Personalized and Experiential Marketing

Personalized marketing, driven by data analytics and AI, will become even more sophisticated. Consumers will expect highly tailored experiences that reflect their individual preferences and lifestyles. Additionally, experiential marketing – creating memorable experiences through events, pop-ups, and interactive content – will become crucial for building emotional connections and brand loyalty.

7.12.3 The Impact of Globalization on Food Advertising

7.12.3.1 Cross-Cultural Marketing

Globalization has led to increased cultural exchange and diversity, requiring brands to adapt their advertising strategies to resonate with diverse audiences. Cross-cultural marketing involves understanding and respecting cultural differences, values, and preferences. Brands must tailor their messages to be culturally relevant and sensitive, ensuring they appeal to a global audience while maintaining authenticity.

7.12.3.2 Global Brand Consistency

While adapting to local markets, brands must also maintain global consistency to reinforce their identity and values. This balance can be achieved by developing core brand messages that are universally relevant and adaptable to local nuances. Consistent branding across different markets builds a strong global presence and enhances brand recognition.

7.12.3.3 Digital and Social Media Expansion

The rise of digital and social media platforms has facilitated the global reach of food advertising. Brands can leverage these platforms to connect with international audiences, engage with consumers in real time, and gather insights on global trends. Social media influencers and digital content creators play a crucial role in amplifying brand messages across borders, making digital strategy an essential component of global marketing.

7.12.3.4 Regulatory and Ethical Challenges

Globalization introduces regulatory and ethical challenges in food advertising. Brands must navigate varying regulations and standards across different countries, ensuring compliance with local laws on advertising, labelling, and consumer protection. Ethical considerations, such as avoiding cultural appropriation and respecting local customs, are also critical for maintaining brand integrity and consumer trust.

7.13 CHAPTER SUMMARY

The dynamic field of food advertising and promotion is continually evolving, driven by technological advancements, shifting consumer behaviours, and the growing influence of globalization. As explored in this chapter, effective advertising and promotion strategies are critical for brands to connect with their target audiences, drive consumer engagement, and achieve their business objectives.

Technology and innovation play pivotal roles in shaping the future of food advertising. From AI and machine learning enabling hyperpersonalized marketing to AR and VR creating immersive brand experiences, technological advancements offer exciting opportunities for innovation. Brands that embrace these technologies can enhance their advertising strategies, improve relevance, and engage consumers in novel ways.

Predicting future consumer trends is essential for staying ahead in the competitive food market. As consumers increasingly prioritize health and wellness, sustainability, and convenience, brands must adapt their advertising messages to align with these evolving preferences. Personalized marketing and experiential approaches will become increasingly important for fostering deeper connections and driving consumer loyalty.

Globalization has introduced new challenges and opportunities in food advertising. Cross-cultural marketing requires a nuanced understanding of diverse audiences, while maintaining global brand consistency ensures a cohesive brand identity. The expansion of digital and social media platforms enables brands to reach international markets and engage with consumers globally, but it also necessitates careful navigation of regulatory and ethical considerations.

The landscape of food advertising and promotion is marked by rapid change and innovation. By leveraging emerging technologies, understanding and anticipating consumer trends, and adapting to the global marketplace, food brands can craft effective advertising strategies that resonate with audiences and drive success. As the industry continues to evolve, staying informed and adaptable will be key to maintaining a competitive edge and achieving long-term growth in the ever-changing world of food marketing.

CASE STUDY

"FreshBite" – The Challenge of Launching a New Snack Product in a Competitive Market

FreshBite, a well-established company in the healthy food sector, is preparing to launch a new line of organic snack bars. The company aims to differentiate its product from competitors through a unique blend of marketing strategies, including a focus on nutritional benefits, innovative packaging, and a comprehensive promotional campaign.

Campaign Strategy

***Nutritional Benefits Highlight*:** FreshBite's advertising emphasizes the health benefits of its snack bars, including high protein content, low sugar, and natural ingredients. The marketing materials feature detailed nutritional information and testimonials from nutritionists and fitness experts.

***Innovative Packaging*:** The snack bars are packaged in eco-friendly, resalable pouches that are designed for convenience and freshness. The packaging also includes QR codes that link to a mobile app with additional product information, recipes, and user reviews.

***Integrated Marketing Campaign*:** FreshBite's campaign includes a mix of digital advertising, influencer partnerships, and in-store promotions. The digital ads target health-conscious consumers through social media platforms and search engines, while influencer partnerships help amplify the message through reviews and lifestyle content.

***Sampling and Demos*:** FreshBite organizes in-store sampling events at grocery chains and health food stores. These events include product demos and special promotions to encourage trial and purchase.

***Customer Feedback and Adaptation*:** FreshBite actively seeks customer feedback through surveys and online reviews. The company plans to use this feedback to make product adjustments and improve marketing strategies as needed.

Challenges

FreshBite faces several challenges in launching the new snack product:

- **Market Saturation**: The snack food market is highly competitive, with numerous established brands and new entrants. Standing out in this crowded space requires a distinctive value proposition and effective marketing.
- **Consumer Trust**: Gaining consumer trust for a new product can be difficult,

especially when competing against well-known brands. FreshBite needs to demonstrate the unique benefits of its snack bars convincingly.

- **Cost Management**: The costs associated with innovative packaging and extensive promotional activities may impact the product's pricing and profitability.

Discussion Questions

1. How does FreshBite's emphasis on nutritional benefits contribute to its marketing strategy?
2. What are the advantages and potential drawbacks of using innovative packaging with QR codes in FreshBite's marketing strategy?
3. Discuss the role of sampling and in-store demos in FreshBite's promotional campaign. How do these activities benefit the brand?
4. What strategies can FreshBite use to overcome the challenge of market saturation and stand out from competitors?
5. How should FreshBite manage the costs associated with innovative packaging and extensive promotional activities to maintain profitability?

CHAPTER 8

Packaging Design and Its Impact on Consumer Perception

LEARNING OBJECTIVES

By the end of this chapter, you should be able to:

- Understand the strategic importance of packaging in shaping consumer perception and driving purchasing decisions.
- Identify the key elements of packaging design that influence consumer behaviour, including visual design, material choices, shape, size, and labelling.
- Evaluate the role of packaging in building brand loyalty.
- Understanding what makes certain designs effective and how packaging mistakes can impact a brand's market position.
- Explore future trends in packaging design, including the rise of smart packaging, personalized packaging, and sustainable solutions.
- Understand the growing importance of sustainability in packaging and how eco-friendly innovations are influencing consumer behaviour.

8.1 INTRODUCTION

Packaging plays a pivotal role in the food industry, serving as both a protective barrier for products and a powerful marketing tool. As consumers navigate through a sea of choices on crowded supermarket shelves or online platforms, packaging often becomes the first point of contact between a product and its potential buyer. It is through packaging that a product can capture attention, communicate its value, and, most importantly, influence consumer perception.

At its core, packaging is not merely about encasing a product; it is about creating an experience that appeals to the consumer's senses, emotions, and rationality. The design of a product's packing can make or break its success. It could be the simple, clean lines of minimalist packaging or the bright, bold colours of a fun brand. Beyond aesthetics, packaging also reflects the brand's identity and values, conveying important messages about sustainability, quality, and trustworthiness.

DOI: 10.1201/9781003473305-8

Moreover, packaging is essential in shaping how consumers perceive a product's quality, freshness, and even its health benefits. Consumers often associate sleek, well-designed packaging with premium products, while poorly executed designs may lead to negative assumptions about the quality or reliability of the item inside. In a marketplace where consumer expectations are constantly evolving, packaging becomes an indispensable tool in standing out from the competition and fostering brand loyalty.

As we dive deeper into this chapter, we will explore how various elements of packaging – such as visual design, material choices, and labelling – affect consumer perceptions and influence purchasing decisions. We will also examine how sustainability trends and technological advancements in packaging are reshaping the future of consumer interactions with products. Understanding the multifaceted role of packaging design is crucial for businesses looking to build strong, lasting connections with their target audience.

8.2 THE ROLE OF PACKAGING IN BRANDING

Packaging is one of the most effective tools a brand can use to communicate its identity and values to consumers. It serves as the face of the product, often making a lasting impression on consumers even before they experience the product itself. Strong, cohesive packaging design not only captures attention but also conveys the essence of the brand, influencing how consumers perceive and engage with it.

For many companies, packaging is a direct reflection of their brand's positioning in the market. Whether it's luxury, sustainability, health-consciousness, or fun, packaging allows a brand to project these qualities immediately and visually. Take Coca-Cola, for example, whose red-and-white colour scheme and iconic script have remained largely unchanged for decades, instantly evoking feelings of nostalgia, familiarity, and refreshment. Similarly, brands like Ben & Jerry's have used playful, colourful designs to reflect their unique, fun, and socially conscious brand image, aligning their packaging with their core values of creativity and activism.

8.2.1 Packaging as a Differentiator

In crowded markets, packaging plays a crucial role in setting a product apart from its competitors. The unique design elements of packaging – such as the shape of the container, the typography, and the overall presentation – help consumers quickly identify a product on the shelf. This is particularly important in industries like food, where multiple products often serve similar functions. A visually distinct packaging design can become a key selling point, helping consumers make a quick decision at the point of sale.

For example, in the premium chocolate market, brands such as Lindt and Godiva utilize elegant, sophisticated packaging that reflects the high-quality nature of their products. The luxurious design of their packaging communicates to consumers that these chocolates are indulgent, a premium product, and worth the higher price point.

In contrast, brands that target more budget-conscious consumers often use simpler, functional packaging designs that emphasize value and practicality over luxury.

8.2.2 Building Brand Recognition

Consistency in packaging design also helps build brand recognition. When consumers see the same logos, colors, or packaging shapes repeatedly, they start to connect those visuals with your brand. Over time, this creates brand loyalty, as consumers grow to trust that the product inside the familiar packaging will meet their expectations.

Take Tropicana as an example. The brand's original packaging featured a prominent image of an orange with a straw, symbolizing freshness and purity. However, when the brand rebranded with a more modern, minimalist design in 2009, removing the orange image, it led to a drastic decline in sales. Consumers no longer recognized the product, highlighting how critical consistent packaging is in maintaining brand recognition and trust.

8.2.3 Packaging and Emotional Connection

Brand packaging also plays a significant role in establishing an emotional connection with consumers. Well-designed packaging can evoke feelings of excitement, trust, or even nostalgia, which enhances the overall product experience. In the food industry, this emotional connection can drive purchasing decisions, as consumers are often drawn to products that align with their values or create positive associations.

For instance, brands focused on sustainability often incorporate eco-friendly materials in their packaging, appealing to environmentally conscious consumers. The use of recycled materials or minimalist designs can signal to consumers that the brand is committed to reducing its environmental impact, fostering a sense of alignment between the brand's values and the consumer's.

In the digital age, packaging has also become an extension of the brand's storytelling. Brands increasingly use packaging as a platform to communicate their values, mission, and narrative. This storytelling can be conveyed through the text on the packaging, the design elements, or even the materials used, all contributing to how consumers perceive and connect with the brand.

8.3 PACKAGING ELEMENTS THAT SHAPE CONSUMER PERCEPTION

Packaging is a multidimensional element that goes far beyond its functional role of protecting the product. In today's competitive marketplace, it is a powerful marketing tool that communicates the product's value, builds brand identity, and influences consumer decisions. To fully understand how packaging impacts consumer perception, it is essential to dissect its core elements: visual design, material choice, shape and size, and labelling. Each of these elements plays a crucial role in shaping how consumers perceive not only the product but also the brand behind it.

8.3.1 Visual Elements: Colours, Typography, Imagery, and Layout

The visual design of packaging is the most immediate and influential factor in consumer perception. Humans are inherently visual creatures, and packaging that appeals to the senses can capture attention and generate interest. The combination of colours, typography, imagery, and layout creates a visual story that consumers interpret, often within seconds.

8.3.1.1 Colours

Colour psychology plays a significant role in terms of how consumers perceive a product. Each colour has several meanings and emotions set off. For instance, red often indicates excitement or passion/energy and often occurs frequently in impulse-buying packaging such as snacks or drinks. On the other hand, green describes health, nature, or even the environment and due to this fact, it is often applied in brands that produce organic or natural products.

Red, for example, is more or less identical to the vivacious and cool Coca-Cola logo, whereas entire brands, like Whole Foods, trend more towards the use of greenish colorations in their efforts to associate themselves more closely with health and the environment. A study states that as much as 85% of customers make purchasing decisions based on colours. In short, it is indispensable when it comes to perception.

8.3.1.2 Typography

Typography, which also has much to do with the overall look of text on packaging, has a profound effect on consumer perception. An elegant flowing font conveys luxury, whereas a bold and clear-looking font speaks for simplicity and reliability. There is a specific type of typography that goes with the personality of a brand and the message it would print on packaging. For example, if a product is something meant for children, a playful, rounded font may well do the job. For a premium brand like tea, however, a serif font will do justice.

Typography also takes a big share in terms of readability and accessibility. This is essential when communicating information about the product. The legible and clear text builds trust with the client, and if the text is out of focus or too complex and hard to read, it may frustrate an interested buyer and cause them to turn away from the product.

8.3.1.3 Imagery

The use of imagery on packaging can be a very emotive tool for creating a feeling or ideation of the value proposition that the product brings across. Images of the product, where the product quality is high, especially for food items, allow a vision to the consumer about what to expect. Packaging images help tell the brand's story. They can use appealing photos to show how to use the product or lifestyle images that connect emotionally with consumers.

For example, mouth-watering images of products on packaging can even command an instant buy by evoking the taste buds and satisfaction of the consumer. On the other hand, minimalist brands such as KIND Bars rely on really simple, transparent packaging to ensure that it is its product that will speak for itself with all purity and honesty.

8.3.1.4 Layout

The layout of packaging involves the arrangement of text, imagery, and branding elements on the surface. A properly structured layout makes sure that a buyer can easily identify essential product information from the appearance, including its name, benefits, and other qualifications, like organic or non-GMO. On the other hand, a disorganized layout will make it difficult for the consumer to understand any of these points and, therefore, is likely to degrade a commodity's attractiveness.

The design also determines whether or not a product would be perceived as premium or budget. A premium brand uses more "white space" with a simpler layout to convey sophistication, and a budget brand has a more hectic layout conveying value through real discounts and promotional offers that are visible in the design.

8.3.2 Material Choice: Texture, Durability, and Sustainability

The materials used in packaging are not just functional but also contribute significantly to how consumers perceive a product. The texture, durability, and sustainability of packaging materials can all influence whether or not a consumer sees a product as high-quality, reliable, or environmentally responsible.

8.3.2.1 Texture

The feeling of the package, the tactile experience of holding it in the hands of a consumer, elicits very powerful responses. It is a rather common characteristic of most luxury goods, but it grants exclusivity and refinement when touching such goods. The touch experience can be an addition to the unpacking experience, which makes it memorable and brings an even deeper emotional connection to the product.

However, at the same time, low-quality or crumpled papers may also convey a lack of quality or not properly cared for in consumers' perception. For example, high-quality chocolates covered with a nice, glossy wrapper could give an impression of indulgence. On the other hand, chocolates covered with thin and flimsy wrappers would be perceived as poor quality, no matter how great is their taste.

8.3.2.2 Durability

Other crucial aspects related to packaging include how durable it is, especially if kept fresh for some time or if the transportation time is relatively long. Packaging should not only serve its purpose but also be aesthetically pleasing. A strong and sealed package will ensure that the product inside will not get damaged, thereby safe to consume, and hence the consumer will rely on the brand more.

Durable packaging also speaks well of the commitment to quality of the brand. Some premium wines are packaged in heavier glass bottles, further relating to bulkier glasswork that speaks well of notions of quality and luxury. Fragile packaging can then speak to frustration on behalf of the consumer if it proves to break products and does not speak well to a sense of reliability for the brand.

8.3.2.3 Sustainability

There is considerable pressure on packaging materials due to increasing concerns about sustainability. As concerns over the environment build up among consumers, brands that believe in sustainable packaging materials tend to gain preference. Biodegradable plastics, recycled paper, and containers that can be used many times are fast becoming a favourite when brands want to establish themselves as eco-friendly.

For example, Lush Cosmetics offers little packaging with biodegradable materials to attract their audience. Although the perception of sustainability also attracts a consumer who is concerned by green issues, it also provides an overall boost in the brand's image because of its relation to globalized worldwide patterns towards waste reduction and sustainability.

8.3.3 Shape and Size: Ergonomics and Distinctiveness

The shape and size of packaging play a crucial role in both the functionality and attractiveness of a product. Ergonomically designed packaging which is easy to handle, store, and its use is more likely to be perceived positively by consumers. At the same time, distinctive packaging shapes can create a sense of uniqueness and enhance brand recall.

8.3.3.1 Ergonomics

Packaging that is easy to open, hold, and store is more user-friendly and this develops a positive attitude towards the brand. For example, resealable snack bags would be convenient and functional when they allow a consumer to enjoy the product over time without having to worry about its freshness. Packaging that is hard to open or use frustrates consumers, and frustration translates into a negative perception of the product.

In addition, ergonomic packaging sometimes takes on sustainability aspects. Lessened, compact, and space-efficient packaging waste or packaging that can be reused and recycled is often deemed more innovative and socially accountable.

8.3.3.2 Distinctive Shapes

Unique and unconventional packaging can be the source of really great visual interest and differentiation in your case from the competing companies. For example, a cylindrical can for Pringles is instantly recognized, and has become synonymous with the brand itself. Such unique shapes can help differentiate products on the crowded retail shelf, making them more likely to be noticed and bought.

There are also unique packaging shapes that allow customers to have memorable and enjoyable unboxing experiences to enhance brand loyalty. When brands offer something unique like an interactive package or, say, one that is reusable as a container, this makes a lasting impression on the customer.

8.3.4 Labels and Information Clarity: Transparency and Trust

Labels are an essential part of packaging that provides consumers with critical information about the product, such as ingredients, nutritional values, certifications, and usage instructions. Clear, transparent, and truthful labelling builds trust and plays a crucial role in terms of how consumers perceive both the product and the brand.

8.3.4.1 Transparency and Trust

Consumers are increasingly seeking transparency from the brands they purchase, particularly in the food industry. The clear labelling and ingredient list, nutritional facts, and even certifications such as organic or non-GMO, for instance, will help the consumer purchase in the right manner. Brands with clear labelling and truthful claims are very likely to win the trust of their target customers. For instance, brands such as KIND Snacks and RXBAR feature key ingredients on their packaging as a form of transparency. It connects with health-conscious customers who are seeking to know exactly what they are putting into their bodies.

8.3.4.2 Legal and Ethical Considerations

Proper labelling is more than an aspect of trust; it is also a legal requirement. Companies are therefore compelled by local and international requirements on food labelling, which should include declaring allergens, and nutritional information, as well as proofing health claims. Not complying with such measures can add damage to a brand's reputation and might even result in legal actions. As much as people may be interested in what is in the food, they might also want to know the story behind the brand, the mission, and the values of the company. For example, a sustainable brand, concerned about the indication of its eco-friendly practices, will attract even more environmentally conscious consumers.

8.4 EMOTIONAL AND PSYCHOLOGICAL IMPACT OF PACKAGING

Beyond its aesthetic and functional appeal, packaging holds a powerful psychological influence over consumers. The emotional relationships that packaging engenders can substantially influence the way the product is perceived, remembered, and, ultimately, bought. Colours, textures, shapes, and symbols that are used on the packaging are not arbitrary; they serve certain purposes of triggering some kind of emotion within a consumer. Packaging may have a psychological impact on the customer, such as feelings of trust, excitement, nostalgia, or even luxury, depending on the goals of the brand and the psychological profile of the target group.

8.4.1 The Power of Colours and Shapes in Evoking Emotion

More directly and clearly than almost any other factor, packaging speaks directly to the consumer through its use of colour. As discussed above, colours hold inherent psychological powers that brands can play on whenever they wish an emotional response or action to be triggered. However, the emotional influence of colour is more complex than simple associations alone can do to influence mood, enhance appetite, and stimulate action.

Warm colours are always so inviting, warm, comfortable, and full of excitement, such as red, orange, and yellow. They thus can be used to gain attention and impulse purchasing. For instance, McDonald's uses red and yellow on its branding and packaging materials to create a feeling of happiness, comfort, and energy. These colours are also more likely to stimulate appetite, making them good choices for fast-food brands.

Cool colours, such as blue and green, are generally associated with calmness, trust, and health. Green is often used in packaging for healthy and environmentally-friendly products. Companies like Whole Foods use green packaging, communicating to consumers their commitment to organic and natural foods, thus creating an emotional bond with their audience, those interested in sustainability and health.

Neutral colours such as black, white, and gray will create a sophisticated feel, simplicity, or luxury, depending on the usage. Black and white packaging may then make up a highly premium and minimalist look, mainly for premium brands to give an idea of elegance and exclusivity. Examples of such usage are in the area of luxury chocolate brands or fine spirits such as Johnnie Walker Black Label, where black mixed with gold gives a sophisticated timeless idea.

Shapes also influence how consumers instinctively connect with packaging. Soft shapes can move the customer emotionally since they can transfer feelings of warmth, safety, and accessibility. It becomes more approachable and accessible as rounded edges are paired with comforting feelings in general. However, angular shapes associated with sharp lines provoke a feeling of stability, strength, or even sophistication. This tends to possess a premium or masculine target that creates a feeling of boldness and reliability.

8.4.2 Minimalism vs. Elaborate Design: The Psychology of Simplicity

Minimalism has become the dominating trend in packaging design, deeply rooted in thinking that less is more. Simple designs look sophisticated and premium, not to mention generating a sense of trustworthiness and transparency for the consumer. More and more consumers associate minimalism with modernity, quality, and honesty in this world saturated by overly complex or flashy marketing tactics.

Consider Apple's iconic packaging. The minimalist design, dominated by clean lines, white space, and a simple product image, speaks to the brand's ethos of simplicity, innovation, and high quality. Clear packaging from clutter gives a consumer a sense of peacefulness, which is highly going to instill confidence – reinforcing the premium nature of that product.

In addition, minimalist packaging appeals to an increasing consumer desire for transparency, particularly in the food industry. Honest and uncluttered labelling—as in the case of RXBAR, where their few simple ingredients are displayed on the front of the package—is clear in its communication, reflecting the person who buys it. This resonates well with health-conscious consumers who want to see a product that is "clean" and free from unnecessary additives or artificial ingredients.

While intricate designs may trigger different psychological factors, such as nostalgia, excitement, or indulgence in the product, brands that wish to create an experience of luxury or being individualistic have more intricate designs on packaging for the sake of exclusivity. For instance, confectionery brands with premium products can be ornamented with intricate ornaments to create an experience of indulgence and appeal to be given as a gift.

The bottom line for brands will be to define simplicity about how much necessary product information can be communicated and how one should emotionally respond. Minimalism suits products targeting consumers who respect clarity, modernity, and health; more elaborative designs suit consumers looking for indulgence, tradition, or luxury.

8.4.3 Creating Perceptions of Quality and Value

Packaging tells consumers what they think about the quality of it and the value for money that they will get, and also does not just influence what consumers feel about a product. For example, many studies have proven that consumers make use of packaging as a proxy to figure out the quality of the product even before getting to taste it. Luxurious, careful designs of packaging often make consumers perceive the product to have higher quality even if the product inside is similar to that of competitors.

Superior products spend money on premium packaging materials and design elements, such as embossing, metallic accents, or textured surfaces, which communicate to the consumer that the product is worth it. Consumers see the packaging features and subconsciously attribute their loyalty to the superior product. As an example, a heavy, dark-glass wine bottle with an embossed label is an expensive and luxurious bottle, while a standard wine bottle containing similar-quality wine is less expensive.

On the other hand, simple and low-cost packaging can also communicate value and value for money. Functionally oriented consumers, who will pay for value but are sensitive to price, may prefer packaging that addresses perceived attributes of the product and efficiency. Store brands are in simple but uninspiring packaging that signals value through plain design and clear prices.

The weight and sturdiness of packaging also become factors in impressions of quality. Heavier, sturdier materials for packaging often feel more solid and heavier, which can be perceived as speaking to the robust nature of the product contained within. This is especially the case with premium products, where expectations of quality go not only for the product itself but for the entire experience surrounding handling, opening, and using the item.

8.4.4 Packaging's Role in Building Emotional Connections

Emotional connections between consumers and brands are critical in building long-term loyalty and brand advocacy. Packaging can play a key role in fostering these emotional connections by creating memorable experiences and reinforcing brand values that resonate with consumers' lifestyles and beliefs.

One of the most significant emotional 'buy' triggers for a brand would be nostalgia and the leverage that can be derived from the packaging of products. Packaging designs will create an emotional linkage concerning the consumer's memories by applying a retro design element, a traditional logo, or a classic color scheme. Coca-Cola often applies its vintage branding to evoke in consumers the memory of times passed by, hence creating an emotional association with consumers.

Exclusivity and luxury become emotional stimuli. With the use of limited-edition packaging designs, especially on high-end items like cosmetics, watches, or luxury spirits, an aura of exclusivity is created. Consumers, thereby, become special or indulged because they associate a product as rare or premium, hence increasing their emotional attachment to the brand.

Brands that appeal to the environmentally sensitive consumer draw on a connection emotionally linked to shared values concerning environmental stewardship by using sustainable packaging materials. Consumers who feel deeply about sustainability attach themselves to companies that are highly visible in their commitment to waste reduction, recycling, or reducing the amount of packaging. To illustrate this, Patagonia and Method brands have effectively used "green" packaging as a way to create great emotional ties to their consumers, who take environmental ethics seriously.

8.5 SUSTAINABILITY AND CONSUMER PERCEPTION

The food industry, over the years, has included enhanced levels of eco-friendly packaging in product development and branding. Indeed, in recent times, it is becoming increasingly inclined to more sustainable practices. Packaging aspects play a major role in the decisions of modern consumers, especially conscious consumers who care about environmental issues. Sustainability in packaging fulfills a functional need while at the same time indicating a brand's alignment with global environmental movements, reinforcing positive consumer perceptions about the brand.

8.5.1 The Rise of Eco-Friendly Packaging

Sustainable packaging, from being a niche concept, has entered the mainstream expectation of industries from food and beverages. More aware consumers seek a greener impact through their purchases; hence, companies have had no option but to embrace packaging solutions that limit waste, reduce carbon footprints, and encourage recycling. The ideals of using biodegradable material, eliminating extra plastic, or even innovating into reusable packages are increasingly adopted by brands for the sake of sustainability as well as actual consumer demands and regulatory requirements.

The transition towards becoming eco-friendly is being driven by several factors. The number of government policies relating to the use of green practices, and increased consciousness about climate change, plastic pollution, and their ghastly effects, and many more aspects, have made several countries come forward with stringent legislation governing the usage of single-use plastics, compelling companies to look for sustainable alternatives. Brands taking proactive steps towards sustainable packaging avoid any impending legal repercussions and move ahead to brand themselves as responsible and thinking companies.

8.5.2 Consumer Demand for Sustainable Products

The demand for packaging sustainability is more pronounced among the two younger generations, millennials and Gen Z, who are known for preferring purchase based on environmental concerns. According to reports, it has been observed that over 70% of consumers are willing to pay more for products with eco-friendly packaging, as they identify brands with sustainability as those of higher ethics. Sustainability can no longer be considered as an add-on by these particular consumers; it has now become a prerequisite.

Brands that align their packaging to eco-friendly values build greater emotional connections with green-conscious customers. Patagonia and Method are two great examples of brands with dedicated customer followings due to environmental responsibility, most notably via practices in packaging. Brands like these come to stand for social good, which deeply resonates with consumers concerned with the environmental or ethical implications.

8.5.3 Key Sustainable Packaging Trends

Several emerging trends are shaping the future of sustainable packaging. By understanding and implementing these trends, brands can not only reduce their environmental impact but also enhance their appeal to increasingly eco-conscious consumers.

8.5.3.1 Biodegradable and Compostable Packaging

Biodegradable and compostable packaging is another phenomenon that has gained ground under sustainable packaging. These materials normally decompose in time, reducing their volume to fill the volumes of landfills or contribute to the pollution of oceans. Biodegradable packaging, made from natural materials like corn starch, sugarcane, or bamboo, is less harmful to the environment than traditional plastics.

A good example is Pangea Organics, which uses biodegradable packaging that can be planted to grow herbs; this would not only make it biodegradable but also ensure that it fits well with eco-friendly values. For the food sector, various brands are nowadays switching towards compostable options to replace items like cutlery and takeaway containers, as well as wrappers, which degrade even faster in composting conditions.

8.5.3.2 Reduced Use of Plastic

Once the 'go-to' material for packaging, plastic is now undergoing increasing scrutiny over its environmental impacts. A lot of companies have already phased out single-use plastics, to be replaced by an alternative material like paper, glass, and metal. For instance, Evian aims to become fully circular through the use of 100% recycled plastic for bottles by 2025 which aspect in reduce virgin plastic production.

Many companies have also reduced the number of plastics used in their packaging in response to consumer pressure. Known as lightweight, this trend entails thinning out plastics used for packaging while maintaining strength and functionality. Brands reduce carbon footprint through lesser plastics while appealing to environmentally-conscious consumers without sacrificing product protection.

8.5.3.3 Recyclable and Reusable Packaging

Added importance is also being given to recyclable or reusable packaging. Brands are reconsidering the lifecycle of their packaging by making it easier for consumers to recycle or even reuse the material. Most food and beverage companies can determine whether their packaging can be recycled or not, thereby increasing recycling among the people.

Another area that is increasingly used is reusable packaging, related to the larger circular economy movement. Brands like Loop started to pioneer systems where customers can take their packages back for cleaning and reuse, taking in fewer new package productions and waste productions. This will minimize waste, but it will also make consumers feel responsible and engaged.

8.5.3.4 Edible Packaging

Unlike this, edible packaging is emerging as a solution to the problem of waste. Edible packaging is going to be eaten with the product rather than being trashed. Packaging made from food-grade materials like seaweed, rice paper, or starch is common.

Though edible packaging is not yet widely used, the huge potential of single-serve products like condiments, snack bars, or even coffee cups will convert the concept into a reality. Brands can depict their commitment to unprecedented sustainability standards in the approach towards edible packing products and could attract highly environmentally conscious consumers who await novelties.

8.5.4 The Impact of Sustainable Packaging on Brand Perception

Consumers increasingly view sustainable packaging as a reflection of a brand's values and ethics. Brands that adopt eco-friendly packaging solutions can build trust, enhance their reputation, and create a positive emotional connection with consumers who prioritize sustainability.

8.5.4.1 Building Brand Loyalty Through Sustainability

The closer the relationship between brands and consumers, the more serious brands are about sustainability. An enthusiastic consumer will defend a company that does care about environmental issues. Research indicates that nearly 80% of customers have a better perception of companies that use environmentally-friendly packaging. The likelihood of buying the same brand again is one thing, and then, of course, there are chances to recommend the same brand, thus increasing its reputation.

8.5.4.2 Sustainability as a Differentiator

Sustainability can serve as a strong differentiator in an overly saturated market. Brands that differentiate themselves with more sustainable packaging stand out from their competition, especially when today's consumers have become particularly finicky in the products they purchase. Suppliers who can offer biodegradable or recyclable packaging can target customers who consciously look for methods to reduce their impact on the environment.

Sustainability also opens up an avenue for brands to expand into new markets. With the new awareness of the environment, most consumers will seek products from companies that share the values of the consumer. In case a brand advocates for sustainable packaging, it is most likely to catch the eye of such a consumer and warm itself up as distinct from the rest in any cross-boundary competition in a world now aware of environmental stewardship.

8.5.5 Challenges and Considerations in Sustainable Packaging

While the benefits of sustainable packaging are clear, there are challenges that brands must navigate when adopting eco-friendly solutions. These challenges include cost, functionality, and supply chain limitations, all of which can impact the feasibility of implementing sustainable packaging.

8.5.5.1 Higher Costs

Sustainable packaging materials like biodegradable plastics or compostable ones are also pricey compared to alternatives such as plastic or aluminium. This can be a big turn-off for smaller companies because of the high cost. However, green packaging is probably going to drop in the long run because consumer demand increases and technology advances with each passing day.

8.5.5.2 Functionality and Durability

Eco-friendly packaging must be not only sustainable but also functional. Material that biodegrades or is compostable may not be as durable, protective, or shelf-stable as its traditional competitors the case of the food industry, the packaging has to protect the content from becoming spoiled. Brands have got to find the sweet spot between sustainability and practicality such that what they do doesn't compromise the integrity of what they're creating.

8.5.5.3 Supply Chain Considerations

Adjustments to sustainable packaging need to sometimes occur across the supply chain from raw material to methods of production. This is often complex for large companies because they have very complex global supply chains, and therefore lots of logistical inconveniences. Moreover, just as with recycling or composting, not all regions have the infrastructure in place, which can limit the sustainability of packaging in certain markets.

However, despite all this, most brands remain aware that the long-run benefit for their brand and to the environment outweighs short-term costs. More and more consumers will require sustainable solutions, and, therefore, companies implementing sustainable packaging today will be more adapted to fulfilling future market expectations.

8.6 THE INTERACTION BETWEEN PACKAGING AND CONSUMER BEHAVIOUR

Packaging does more than just protect a product—it plays a critical role in influencing consumer behaviour, both consciously and subconsciously. The way a product is presented through its packaging can determine whether it grabs attention, encourages interaction, or drives a purchase decision. By analyzing how packaging interacts with consumer behaviour, brands can fine-tune their designs to maximize appeal, foster trust, and enhance the overall consumer experience.

8.6.1 How Packaging Influences Purchasing Decisions

Packaging has a direct and measurable impact on purchasing decisions. In a typical retail environment, consumers are often faced with a vast array of choices, and they make rapid judgements about products based on visual cues like colour, shape, and branding. Research indicates that consumers can form a perception about a product within just a few seconds of encountering it, and packaging plays a central role in this snap decision-making process.

8.6.1.1 The Initial Attention Grab

One of the major responsibilities of packaging is to grab consumer attention in the sea of competing products. Packaging that has some form of eye-catching appeal, perhaps its bright colours, an unusual shape, or a creative design, has much chance of catching a consumer's eye. This is particularly useful in highly competitive categories like beverages, snack foods, or personal care, wherein brands have got to compete to stand out.

Packaging that features high visual contrast, such as complementary color schemes or unique graphics, is more likely to capture a consumer's attention. For example, the slender, silver-and-blue can of Red Bull immediately springs to a store shelf mind when there are many larger, more conventional soft-drink bottles waiting to be seen by shoppers looking for energy. The unique packing design will not only attract attention but also incorporate the concept of Red Bull as a product for high-performance and energy-driven consumers.

8.6.1.2 First Impressions and Brand Perception

After attracting attention, packaging becomes what will give the first impression that the consumer may have about the brand and the product. Well-designed packaging can communicate a whole range of qualities, such as the premium status of the product, its environmentally-friendly nature, or its health benefits. First impressions are often critical because they tend to make consumers associate some of the attributes one may consider quality, reliability, and innovation with the brand.

For instance, the minimalist and sleek packaging design adopted by Apple reflects its brand identity as a premium high-end innovative technology company; the way the product is unboxed creates a feeling of excitement and anticipation, thereby reaffirming the consumer's perception that Apple products are premium, high-end cutting-edge devices.

On the other hand, low-quality or overly complex packaging may discourage consumers from buying a particular product. Packaging that looks cheap or flimsy can give the impression to the consumer that the quality of the product it contains may be of a lower standard, whereas the properties of the product may be all right. Thus, it is relatively important to consider how these types of packaging influence the primary impression as well as the ultimate perception of the consumer.

8.6.1.3 Communicating Product Benefits

Packaging not only makes a first impression but also communicates essential information about the product. It tells consumers about the product's features, benefits, and unique selling points. Packaging can be used to emphasize attributes, such as organic ingredients, health claims, or eco-friendly materials that may also play a part in a consumer's decision to purchase the product.

For instance, packaging that prominently display terms such as "low fat," "high protein," or "organic" appeals to health-conscious consumers who look for such items. The KIND Bars brand communicates its health benefits through simple, clear messages on the packaging shown upfront, appealing to their consumers' requirements for wholesome and natural ingredients.

In addition to the overt messages, packaging can communicate benefit promises on a more subtle, emotion-evoking level through the deployment of texture, colour, or imagery. For example, a product using earthy tones and natural landscape images will tell a consumer that it's wholesome and pure, even though those two words may never be written on that packaging. That technique gives control to the brand in the behaviour of the consumer, thereby raising the likelihood of purchase.

8.6.2 The Impact of Packaging on Consumer Retention and Brand Loyalty

Packaging not only influences first-time buyers but also plays a crucial role in retaining customers and building long-term brand loyalty. Positive experiences with a product's packaging – whether through ease of use, functionality, or aesthetic appeal – can create lasting impressions that drive repeat purchases and foster emotional connections with the brand.

8.6.2.1 Ease of Use and Convenience

Packaging that is easy and convenient for the consumer to use ensures a much more satisfactory experience from a consumer's point of view and can therefore increase their loyalty to a particular brand. Features such as resealable pouches, ergonomic designs, and clear instructions enhance a product's usability and increase the likelihood of consumers returning to the brand for future purchases. Convenience is a particularly salient consideration in basic product categories such as food and beverage, where packaging itself needs to fulfill functionalities such as freshness or transportability for the consumer to be satisfied with the overall product.

For instance, convenient and practical resealable food packaging, such as Ziplock bags, caters to busy consumers with products that make life easier. Convenience through packaging is a good reason to repeat the experience because it promises a positive experience for the brand.

8.6.2.2 Brand Consistency and Recognition

The more recognizable and trusted the brand, the higher the probability that customers will repurchase the product. Consistent packaging design for a series of products improves recognition and recall of the brand, making the consumer more likely to locate what they are looking for either on store shelves or while shopping on the internet. Trust and loyalty will ultimately emerge from this strength of brand recognition based on consistent packaging.

For instance, the Coca-Cola Corporation has maintained a design language which does not seem to have changed since its decades-old red colour scheme, its curvy bottle shape, or its logo were first produced decades ago. This could happen because consistency reinforces brand recognition and trust. That is why consumers easily identify and choose Coca-Cola products without hesitation.

8.6.2.3 Packaging and Emotional Connections

Packaging that provides an emotional connection can lead to very loyal customers as they feel personally connected to the brand. Today, the unboxing experience is one of the most important things that makes for memorable and emotionally impactful consumer experiences. Whether it's opening up an otherwise well-designed package with excitement or discovering thoughtful details inside, this emotional response can make consumers and brands more strongly bonded.

Luxury goods, electronics, and much more build the unboxing experience to a multimillion-dollar level by adding perceived value: Therefore, in the case of luxury goods and electronics, for example, packaging can become a tool for restrengthening the perceived value of the product at the time of unboxing. However, if there is anything that something this method does for Apple, it is the ultimate definition of such techniques. This is in terms of quality material, sleek design and detailed finishing touches all coming together to create excitement and thereby reinforcing a premium status.

In addition to this, packaging that shares a consumer's values – whether it is sustainability, health consciousness, or social responsibility in its production – ensures higher levels of an emotional connection with this product. More loyalty from consumers who share such values will be found in brands working on natural, sustainable materials or whose packaging contributes to a social cause.

8.6.3 Packaging's Role in Online Shopping: Unboxing Experience and E-Commerce

As e-commerce continues to grow, packaging has now taken on a new role in the online shopping experience. In the absence of physical store interactions, packaging becomes a crucial touchpoint for brands to engage with consumers and differentiate themselves in the digital marketplace.

8.6.3.1 The Unboxing Experience in E-Commerce

In e-commerce, the unboxing experience has become an important factor that has become one aspect of how consumers perceive a brand. Comparatively, if a customer happens to go shopping in brick-and-mortar stores, they would probably get a better view and feel of the products before buying them. However, online shopping involves less exploration with senses. As a result, the first physical exposure of the consumer to this product is through its packaging.

Many online-only brands focus on creating visually appealing customized inserts and personalized messages to entice the user to feel like excitement when the order arrives. A custom insert, personalized message, and attention to detail make for an enjoyable unboxing experience and place the brand more effectively in the consumer's eye.

This is even more critical in the case of luxury or premium brands. It is possible to multiply perceived value once a client experiences a great unboxing of his product.

It may help increase that sense of exclusiveness that has become increasingly important for product differentiation. Finally, high-quality or personalized packaging reinforces in the consumer's mind that they have, indeed, made a premium purchase, increasing both satisfaction and loyalty.

8.6.3.2 Sustainable Packaging in E-Commerce

Like e-commerce growth, concerns over packaging waste have also been on the rise. Online customers are becoming increasingly environmentally savvy through their awareness and exposure to the excess packaging that continues to mar many items, especially those shrouded in extra packaging for the purposes of shipping protection. Sustainable, minimalistic packaging for their e-commerce orders will give these brands an edge with environmentally conscious consumers.

As an offset to this increased demand, many e-commerce brands now use recyclable or compostable materials for their shipping boxes, mailers, and packaging inserts. Companies such as Amazon and Loop have taken a step forward with sustainable packaging initiatives that reduce waste and encourage recycling. Prioritizing eco-friendly packaging in e-commerce will enhance a brand's reputation and attract consumers who are concerned with sustainability.

8.6.3.3 Personalization in Packaging for Online Consumers

Since e-commerce allows brands to personalize their package, there can be a special experience for the consumer. Such personalized experiences include bespoke packaging with the customer's name, special deals based on customer preferences, or special packaging for limited-edition specials. In this manner, personalization increases the attachment of the consumer and the feeling of belonging to that brand, and they feel appreciated.

That is why a company such as Birchbox, proposing monthly subscription boxes with personalized samples of beauty, uses custom packaging in order to be exciting and surprising with each delivery. In addition to strengthening the emotional bond that exists between the consumer and the brand, it encourages social sharing, and many consumers post photos or videos of their unboxing of products on Instagram or YouTube platforms.

8.7 FUTURE TRENDS IN PACKAGING DESIGN

As consumer expectations evolve and new technologies emerge, the future of packaging design is poised to undergo significant transformations. Brands are continuously seeking ways to innovate, not only to capture consumer attention but also to respond to global concerns such as sustainability, personalization, and digital integration. This section will explore the emerging trends shaping the future of packaging design, along with their potential impact on the food and consumer goods industries.

8.7.1 Smart Packaging: Integration of Technology

Smart packaging is one of the most exciting developments in the packaging industry. By integrating digital technology into packaging, brands can offer consumers enhanced functionality, product information, and interactivity. Smart packaging includes technologies such as QR codes, Near Field Communication (NFC), and augmented reality (AR), transforming packaging from a passive container to an active engagement tool.

8.7.1.1 QR Codes and NFC

QR codes and NFC tags on packaging open up quick, accessible information on a product, proof of provenance, and even advertising information from a consumer's smart mobile device. Wine brands are increasingly using QR codes on bottles, meaning that the consumer can access tasting notes, vineyard info, and pairing advice. This sort of interactive feature enhances the consumer experience by adding value without cluttering up the packaging with too many words.

Packaging will also be equipped with NFC capabilities to provide product authentication, so that consumers can rest assured of the authenticity of their luxury goods or gourmet foods. It offers help in their fight against fake products and also helps to build brand integrity.

8.7.1.2 Augmented Reality (AR)

Brands looking to get consumers to interact and engage with packaging are finding that augmented reality (AR) is a very powerful tool. Through their smartphone or AR glasses, AR allows interaction with the physical product on which virtual elements are layered. For instance, through AR, brands can use virtual animations or 3D models that appear once the packaging is scanned, thereby culminating in a much more interactive and engaging experience.

One successful example is the Australian wine brand 19 Crimes, which has used AR to bring to life the historical figures which are featured on its wine labels. Buyers can use an app to scan the label and then watch an animated version of the character telling their life story. This unique and engaging approach not only grabs attention but also prolongs the story of the brand.

8.7.1.3 Sustainability Monitoring

Another significant application of smart packaging lies in tracking the freshness and quality of perishables. The packaging could be incorporated with sensors that can detect environmental changes such as temperature and humidity, thereby giving instant information regarding the state of the product. This would come in very handy in reducing food waste, as the consumer would know the freshness of the produce at every point of consumption.

8.7.2 Personalized Packaging: Customization at Scale

Personalization has become a significant trend across industries, and packaging is no exception. Consumers increasingly value products that feel tailor-made for them, and advances in digital printing technology have made it easier for brands to offer customizable packaging at scale.

8.7.2.1 Personalized Labelling

Brands can now design their custom packaging for individual customers. This may be in the form of adding their names, special messages, or even designs that customers want on the packaging. One good example of how the personalization of packaging enables consumers to feel a sense of belonging or an emotional connection with the product is Coca-Cola's "Share a Coke" campaign, in which they printed people's names on the bottles. This marketing campaign led to an amazing surge in sales as people rushed over the need to catch their names or those of their loved ones written in the product. The "Share a Coke" campaign generated over 1.3 million mentions on social media in its first year. Coca-Cola's Facebook page gained 6.8% more followers. The company also reported a 2% increase in sales in the US, ending a ten-year decline in sales.

8.7.2.2 Customizable Product Experiences

Personalized packaging is not limited to labelling. For example, the manufacturers that produce Nutella and M&M's each allow consumers to have control in personalizing their product packaging with custom colours, images, or text for unique, personalizing features of the product.

8.7.3 Sustainable and Zero-Waste Packaging: Eco-Friendly Innovations

Sustainability continues to be a dominant force in shaping the future of packaging design. As consumers grow more environmentally conscious, they are demanding packaging that reduces waste, minimizes environmental impact, and uses fewer resources. In response, brands are adopting innovative approaches to sustainable packaging, from using alternative materials to rethinking packaging's role in the product lifecycle.

8.7.3.1 Biodegradable and Compostable Packaging

As the regulatory pressure on single-use plastics continues to increase, biodegradable and compostable packaging may surface as an alternative. Formulated from renewable sources such as plant-based plastics, corn starch, or cellulose, these are broken down naturally, thereby reducing volumes in landfills and oceans.

For example, Alter Eco bars of chocolate use compostable packaging sourced from sustainable origins, such as the peelings from their chocolate bar wrappers. Such an action will not only appease consumers' expectations regarding responsible packaging but also position it favourably among organic and fair trade chocolate markets.

8.7.3.2 Edible Packaging

Edible packaging is an innovation that has the potential to revolutionize the food industry completely by going through with eliminating packaging waste. Edible films are pieces of paper made from materials such as seaweed rice paper or starch. They are used for single-serve items and, most especially, for items such as sauces, snacks, or coffee cups.

Still, in its nascent stages, edible packaging can be a significant pusher in minimizing plastic waste, especially in the food and beverage sectors. Brands that embrace this technology will not only cater to the needs of a waste-reducing market but also place themselves in pioneering positions concerning ecological innovation.

8.7.3.3 Reusable and Refillable Packaging

There is an emerging trend in reusable and refillable packaging. Brands are no longer using disposable packaging and instead are introducing durable, reusable containers that consumers can either refill in-store or return to the manufacturer. For example, Loop provides consumers with the opportunity to buy products in reusable containers, which will then be returned to the brand for cleaning and reuse, thereby eliminating single-use packaging.

Therefore, this trend has matched with the growing consumer interest in the reduction of waste as well as support for the circular economy. Brands following the use of reusable packaging not only decrease their impact on the environment but also make a more sustainable relationship with customers and have a good chance of recurrence for buying in addition to enhancing brand loyalty.

8.7.4 Minimalist Packaging: Less is More

Minimalism has been a dominant design trend in recent years, and it is becoming increasingly important in packaging design. Minimalist packaging focuses on simplicity, clarity, and elegance, often using fewer materials, reducing visual clutter, and emphasizing the product itself.

8.7.4.1 Clean Aesthetic

Minimalist packaging designs are characterized by clean lines, simple typography, and an uncluttered layout. These appeal best to both transparency and authenticity in branding. For instance, in packaging design related to food items, designs that emphasize clean, simple wholesome ingredients could attract the interest of the health-conscious consumer in bringing clean, natural products to the fore.

For example, RXBAR is famous with regard to the minimalist design of packaging, indicating the basic contents of the product on the front in big, bold letters. One of the key expressions of brand simplicity, transparency, and commitment to such values in a complicated market.

8.7.4.2 Eco-Friendly Alignment

Minimalist packaging is not necessarily in opposition to eco-friendly values. Rather, minimalism incorporates more use of fewer materials in design, instead of inconsequential aspects like redundant packaging elements. Such minimalist designs can help reduce waste, cut down on costs of production, and suit consumers' demands for sustainable products without compromising aesthetics and functionality.

8.7.5 Transparency and Ethical Packaging: Building Trust with Consumers

As consumers become more informed and demand greater transparency from brands, ethical packaging is emerging as a critical trend. Brands are increasingly using packaging to communicate their commitment to ethical practices, such as fair trade, sustainable sourcing, and animal welfare.

8.7.5.1 Clear Labelling and Honest Communication

Today, consumers need to be seen as knowing where their products come from, how they are developed, and what type of impact leave on the environment. Packaging that talks about these facts or details through a fair trade certification, ingredient sourcing description, or the carbon footprint will reassure consumers and end up being loyal in the long run.

For instance, Tony's Chocolonely uses the outside of its packaging to tell part of the history of what they set out to do to eradicate the existence of child labour in the chocolate industry. In essence, transparency and coming through in their crusade for ethical practices help such brands appeal to their socially conscious target audience, who would not mind being responsible consumers.

8.7.5.2 Ethical Materials

Another important aspect of responsible packaging is the source of responsibly sourced raw materials. Brands are increasingly opting for packaging made from materials that are harvested sustainably, such as Forest Stewardship Council (FSC)-certified paper or bio-based plastics. These help reduce the level of environmental damage and bring consumer values of ethical consumption to the fore.

8.8 CHAPTER SUMMARY

Packaging, which was historically a merely functional element, has evolved into one of the most sophisticated tools in the business world, since it can shape consumer perception, drive purchasing decisions, and encourage brand loyalty. In this chapter, we have explored the multidimensional role packaging plays within the food and consumer goods industries, and the aesthetic and practical ways in which products are perceived and experienced by consumers.

It now does far more than the physical properties of the product. Packaging is a storyteller, an emotive connection builder, and a trust builder. It is important for the total brand experience. Established brands support recognition, but new or evolving brands with their innovative packaging might outshine their competitors, capture consumer attention, and unlock new avenues of engagement.

As we look to the future of packaging design, innovation, and sustainability become more driven in terms of how the brand will now interact with the consumer. Leading through these channels we have smart packaging, personalization, and eco-friendly materials, which are going to be at the heart of the new methods with which brands have come to understand the dynamics of creating functionality, reducing environmental impact, and delivering experiences tailored to the needs of the modern consumer.

Sustainability is no longer a trend. It is now an expectation. Demand for changes in how packaging is done is now coming from today's consumers. They now want eco-friendly solutions that will reduce waste and have minimal impacts on their environment, looking at responsible corporate practices from brands. From biodegradable materials to reusable containers, the brands that prioritize sustainability as it relates to packaging are not just going to meet the consumers' high demands but also position them to be leaders in taking the world toward a more sustainable future.

Packaging is an area that now plays a great role in altering consumer behaviour, given the fact that it brings in the first sale and also builds long-term loyalty. In a populous and competitive retail market, packaging that captures value and convenience can turn a potential buyer into repeat business. Brands continue to create unforgettable and enjoyable packaging experiences as the strength of the connection built with their consumers strengthens loyalty.

As consumer preferences shift towards the triple focus on sustainability, transparency, and personalization, brands need to show these values in their packaging strategies. Where once packaging was the most invisible form of customer contact, packaging that communicates a brand's values with the consumer at all touchpoints, from when the consumer lifts a product off the shelf to when they open it, will have a huge impact on customer satisfaction and brand loyalty.

Once upon a time, packaging was simply a box or a can in which to put products; it is now a marketing tool, an indispensable bridge between the product and its consumer. Packaging shapes perceptions, stories communicate the core values of the brand, and, therefore, it is the essential ingredient of a successful brand strategy when the marketplace becomes more crowded and consumers more analytic. Packaging will increasingly become a crucial touchpoint to meaningfully connect with customers and promote the organization's growth.

CASE STUDY

Procter & Gamble's Old Spice – A Packaging Revamp Success

Procter & Gamble's Old Spice offers a lovely example of how the design of packaging can, from a great bigger scheme in marketing, reinvent a brand and, in turn, change how it is seen by customers. Established in 1938, Old Spice had long been recognized as a traditional men's grooming brand, catering mostly to an older demographic. By 2009, however, this was becoming their biggest problem: staying relevant in a sea of fresh young competition. The brand needed a facelift to appeal to the younger generation, and packaging design played a vital role in that process.

The Challenge

Old Spice's original packaging design was simple and low-key, with non-offensive visuals that suited its elderly target markets. In the men's grooming market, however, there were increasing numbers of younger consumers who were seeking modern, sexy, and edgy brands. Old Spice was becoming old and irrelevant in such a changing marketplace.

The Packaging Redesign

In 2010, Proctor & Gamble (P&G) made a bold decision, restyling the image of Old Spice from head to toe, starting with its packaging design. Bright colours, edgy graphics, and bold typography appealed to modernity instead of revealing a relic from the olden days. Such vivid imagination appeared to address a new kind of young, tasteful, and dangerous customer group. It also chimed in with Old Spice's light-hearted, energetic advertisement campaign, which highlighted "The Man Your Man Could Smell Like."

The pack itself was not just eye-catching; it also conveyed the attitude of the brand, which was now being characterized as bold, funny, and confident. Using not-so-traditional or quirky fonts, Old Spice was sure to stand out on retail shelves and immediately draw the consumer's attention. The new design reflected the brand's shift towards appealing to younger men who valued both style and substance in their grooming choices.

Consumer Perception and Market Response

As a result of the redesign, consumer perception changed overnight. Punchy new packaging told the world that Old Spice was no longer the brand of aged men; rather, it was now a brand that spoke to relevance, youth, and coolness. Along with viral advertising, packaging helped Old Spice launch an entirely new identity appealing to millennials, who had previously utterly ignored the brand.

Sales ballooned after the redesign. According to reports by P&G, Old Spice saw sales rise 125% in the first six months following the introduction of the new packaging and the launch of the associated campaign. The brand regained market share, becoming one of the leaders within the category of men's grooming. This study proved how a well-thought-through packaging design, executed by an overarching brand strategy, can yield extraordinary business results.

The Old Spice case study shows clearly how innovative design can change consumer attitudes and increase sales. In other areas, this practice has taught food marketing companies not to relax but to continue to assess and evolve their packaging in adapting to changing consumer desires and tastes. Successful packaging goes beyond aesthetics; it has to create the essence of the brand and connect at a visceral level. In our case study that is how Old Spice, with its bold, youth-centred revamp, could achieve results.

Discussion Questions

1. How do you think packaging design influences consumer decisions, especially for products that have many similar competitors on the market?
2. In the case of Old Spice, how did the brand use packaging design to shift consumer perception and target a new demographic?
3. What risks are associated with redesigning the packaging of an established brand? Can you think of any examples in which a packaging redesign has failed?
4. Old Spice incorporated bold and playful designs to appeal to a younger audience. How can companies balance modernize their packaging while retaining their brand identity?
5. Packaging serves both functional and aesthetic purposes. In what ways can the packaging of a product impact consumer perception beyond just its visual appeal (e.g., eco-friendliness, convenience, etc.)?

CHAPTER 9

Pricing Strategies in the Food Industry

LEARNING OBJECTIVES

By the end of this chapter, you should be able to:

- Understand the core principles of cost-based, value-based, and competition-based pricing.
- Identify the advantages and disadvantages of each pricing approach.
- Analyze various pricing strategies.
- Examine factors influencing pricing decisions.
- Compare and analyze how different pricing models are employed in the food industry to adapt to varying customer needs and market conditions.
- Understand consumer behaviour and its impact on pricing.
- Recognize pricing for different market segments.
- Identify common pricing challenges and potential solutions to overcome these challenges.
- Analyze the impact of technological advancements, changing consumer behaviours, global market shifts, and regulatory factors on future pricing strategies.

9.1 INTRODUCTION

Pricing is one of the most critical components of marketing strategy, particularly in the food industry, where margins can be thin and competition fierce. The price of a product not only determines its affordability and profitability but also influences consumer perceptions and purchasing decisions. In a market characterized by diverse consumer preferences, fluctuating costs, and regulatory constraints, the development of an effective pricing strategy is essential for success. In this chapter, we will delve into the fundamentals of pricing within the food industry, exploring various strategies that businesses can employ to optimize their pricing models. We will examine the key factors that influence pricing decisions, from production costs to market demand, and discuss how different pricing approaches can be tailored to meet the needs of various market segments. Understanding the dynamics of consumer behaviour is paramount

DOI: 10.1201/9781003473305-9

in setting the right price. Consumers' perceptions of value, their sensitivity to price changes, and their loyalty to brands play a significant role in how they respond to different pricing strategies. In this chapter we will explore these behavioural aspects and how they can be leveraged to enhance pricing effectiveness.

By the end of this survey, you will have a comprehensive understanding of the various pricing strategies available to food businesses and the tools needed to implement them effectively. Whether you are launching a new product, entering a new market, or looking to refine your current pricing approach, this chapter will equip you with the knowledge and strategies necessary to navigate the complex landscape of food pricing.

9.2 FUNDAMENTALS OF PRICING

Pricing is more than just a number; it is a strategic tool that communicates value, influences demand, and drives profitability. To develop a robust pricing strategy, it is essential to understand the foundational principles that underpin pricing decisions. This section explores the three primary pricing approaches: cost-based, value-based, and competition-based pricing.

9.2.1 Cost-Based Pricing

Cost-based pricing is the simplest and most straightforward method of setting prices. It involves calculating the total cost of producing a product, including raw materials, labour, overheads, and other expenses, and then adding a mark-up to ensure profitability. The key components of cost-based pricing include:

- **Direct Costs**: These are costs directly attributed to the production of goods, such as raw materials and direct labour.
- **Indirect Costs**: These are overheads and administrative expenses that are not directly tied to production but are necessary for operations.
- **Mark-up**: A percentage added to the total cost to achieve desired profit margins.

Cost-based pricing ensures that all expenses are covered and provides a straightforward way to determine pricing. However, it does not consider market demand or competitor prices, which can lead to over-pricing or under-pricing in a competitive market.

9.2.2 Value-Based Pricing

Value-based pricing focuses on the perceived value of the product to the consumer rather than the cost of production. This approach involves understanding the benefits that the product delivers and setting a price that reflects its value to the target market. The steps to implement value-based pricing include:

- **Customer Perception**: Understanding how customers perceive the product's value in terms of quality, functionality, and brand reputation.
- **Benefit Analysis**: Assessing the tangible and intangible benefits that the product offers compared to competitors.
- **Willingness to Pay**: Determining the maximum price that customers are willing to pay for the perceived value.

Value-based pricing can lead to higher profit margins and stronger customer loyalty, as it aligns with consumer expectations and willingness to pay. However, it requires a deep understanding of the target market and continuous monitoring of consumer preferences and perceptions.

9.2.3 Competition-Based Pricing

Competition-based pricing, also known as market-oriented pricing, sets prices based on the prices of similar products offered by competitors. This approach involves:

- **Market Analysis**: Identifying key competitors and analyzing their pricing strategies.
- **Price Positioning**: Deciding whether to price the product higher, lower, or on par with competitors based on the desired market position.
- **Dynamic Adjustments**: Continuously monitoring competitor prices and market conditions to make necessary adjustments.

Competition-based pricing helps businesses stay competitive and respond to market changes quickly. However, if not managed carefully it may lead to price wars and reduced profit margins.

9.3 PRICING STRATEGIES

In the competitive landscape of the food industry, selecting the right pricing strategy is essential to attract customers, enhance profitability, and achieve sustainable growth. Each pricing strategy is tailored to different business objectives and market conditions. By understanding and implementing these strategies, food businesses can effectively position their products and respond to market dynamics.

9.3.1 Penetration Pricing

Penetration pricing involves setting a low initial price to quickly attract customers and gain market share. This strategy is particularly useful for new products or brands entering a competitive market. The key benefits of this approach include:

- **Market Entry**: Lower prices can attract customers to try a new product, making it easier to penetrate a saturated market.

- **Economies of Scale**: Increased sales volume can lead to cost savings in production and distribution.
- **Customer Loyalty**: Once customers are acquired, they may remain loyal even if prices are gradually increased later.

However, penetration pricing may result in initial losses and requires careful planning to avoid long-term negative impacts on profitability.

9.3.2 Skimming Pricing

Skimming pricing sets a high initial price to maximize revenue from early adopters willing to pay a premium. This strategy is suitable for innovative or unique products with little competition. Key points to consider include:

- **Early Revenue**: High prices can generate substantial early profits and recover the research and development costs.
- **Perceived Quality**: Premium pricing can enhance the perceived quality and exclusivity of a product.
- **Gradual Price Reduction**: Prices can be lowered over time to attract more price-sensitive customers as competition increases.

Skimming pricing works best when there is a strong brand presence and the product offers unique features that justify the higher price.

9.3.3 Psychological Pricing

Psychological pricing leverages human psychology to make prices appear more attractive. Common techniques include:

- **Charm Pricing**: Setting prices just below a round number, such as $9.99 instead of $10.00, in order to create the perception of a better deal.
- **Prestige Pricing**: Using round numbers, like $50.00, to convey luxury and high quality.
- **Price Anchoring**: Displaying a higher original price next to a discounted price to highlight the savings.

Each of these techniques can significantly influence consumer purchasing decisions and enhance sales without changing the actual cost structure.

9.3.4 Bundle Pricing

Bundle pricing involves offering multiple products together at a discounted price. This strategy can increase sales volume and provide additional value to customers. Key advantages include:

- **Increased Perceived Value**: Customers perceive that they are getting more for their money, thereby encouraging larger purchases.

- **Inventory Management**: Bundling can help to clear excess stock of less popular items.
- **Cross-Selling Opportunities**: Promoting complementary products together can increase overall sales.

Effective bundle pricing requires careful selection of products that complement each other and provide genuine value to the customer.

9.3.5 Discount Pricing

Discount pricing offers products at a reduced price for a limited time in order to stimulate demand. This strategy includes:

- **Seasonal Discounts**: Offering discounts during specific seasons or holidays to boost sales.
- **Clearance Sales**: Reducing prices to clear out old inventory.
- **Volume Discounts**: Providing lower prices for bulk purchases to encourage higher sales volume.

While discount pricing can attract price-sensitive customers and increase short-term sales, it must be used strategically to avoid eroding brand value and profitability.

9.3.6 Premium Pricing

Premium pricing sets higher prices to reflect the superior quality, exclusivity, or brand reputation of a product. This strategy works well for luxury and specialty items. Considerations include:

- **Brand Image**: High prices can reinforce the perception of quality and exclusivity.
- **Target Market**: Premium pricing targets customers who are willing to pay more for perceived value and status.
- **Consistency**: Maintaining consistent quality and brand messaging is crucial to justify premium prices.

Premium pricing can lead to higher margins and a loyal customer base, but it requires a strong brand and clear differentiation from competitors.

9.4 FACTORS INFLUENCING PRICING DECISIONS

Pricing decisions in the food industry are influenced by innumerable factors that can affect both the cost structure and the market dynamics. Understanding these factors is crucial for developing a pricing strategy that not only covers costs but also maximizes profitability and aligns with market conditions. The key factors that influence pricing decisions include cost of production, market demand, competition, brand positioning, and regulatory environment

9.4.1 Cost of Production

The cost of production is the foundation of any pricing strategy. It includes all expenses incurred in manufacturing and delivering a product to the market. Key components are:

- **Direct Costs**: Costs directly tied to production, such as raw materials, ingredients, and direct labour.
- **Indirect Costs**: Overheads and administrative expenses, such as utilities, rent, and salaries of non-production staff.
- **Fixed and Variable Costs**: Fixed costs remain constant regardless of production volume, while variable costs fluctuate with changes in production levels.

Understanding and managing production costs is essential in ensuring that the price set covers all expenses and achieves the desired profit margin.

9.4.2 Market Demand

Market demand significantly impacts pricing decisions. The relationship between price and demand is often represented by the demand curve, which shows how changes in price affect the quantity demanded. Key considerations include:

- **Price Elasticity**: It measures how sensitive consumers are to price changes. Products with high elasticity show significant changes in demand with price fluctuations, while inelastic products show minimal changes.
- **Consumer Preferences**: Understanding consumer preferences and trends can help to predict demand and set prices accordingly.
- **Seasonality**: Seasonal variations can influence demand, necessitating adjustments in pricing during peak and off-peak periods.

Analyzing market demand helps businesses to set prices that optimize sales volume and revenue.

9.4.3 Competition

The competitive landscape plays a crucial role in pricing decisions. Companies must consider competitor prices and strategies to remain competitive. Key aspects include:

- **Competitor Pricing**: Analyzing competitor prices helps to determine whether to price above, below, or at par with them.
- **Market Positioning**: Deciding on a pricing strategy that reflects the brand's market position, whether as a premium, a mid-range, or a budget option.
- **Market Saturation**: In highly competitive markets, aggressive pricing may be necessary to gain market share, while niche markets may allow for higher prices.

Understanding the competitive environment enables businesses to position their products effectively and respond to competitive pressures.

9.4.4 Brand Positioning

Brand positioning influences how customers perceive a product and its value, which in turn affects their willingness to pay. Key factors include:

- **Brand Image**: A strong, positive brand image can justify higher prices as customers associate it with quality and reliability.
- **Value Proposition**: Clearly communicating the unique benefits and value of the product helps to justify its price.
- **Customer Loyalty**: Established brands with loyal customer bases can maintain higher prices without significantly impacting demand.

Effective brand positioning aligns pricing with the perceived value, enhancing profitability, and customer satisfaction.

9.4.5 Regulatory Environment

The regulatory environment can impact pricing decisions through various policies and regulations. Key considerations include:

- **Price Controls**: Government-imposed price controls, such as price ceilings, can restrict pricing flexibility.
- **Taxes and Tariffs**: Taxes on raw materials, production, and sales, as well as tariffs on imported goods, can increase costs and affect pricing.
- **Food Safety Regulations**: Compliance with food safety and labelling regulations can add to production costs, influencing pricing decisions.

Staying informed about regulatory changes and ensuring compliance helps to avoid legal issues and additional costs.

9.5 PRICING MODELS AND TECHNIQUES

In the dynamic landscape of the food industry, implementing effective pricing models and techniques is crucial for optimizing revenue and maintaining competitiveness. Businesses can employ various pricing models and techniques to adapt to market conditions, consumer behaviour, and technological advancements.

9.5.1 Dynamic Pricing

Dynamic pricing involves adjusting prices in real time based on market demand, competition, and other external factors. This model is particularly useful in environments with fluctuating demand, such as online retail. Key features include:

- **Real-Time Adjustments**: Prices are updated frequently to reflect current market conditions.

- **Data-Driven Decisions**: Advanced algorithms and data analytics are used to predict demand and set optimal prices.
- **Flexibility**: Allows businesses to capitalize on high-demand periods and clear excess inventory during low-demand periods.

Dynamic pricing maximizes revenue by aligning prices with market conditions, but it requires sophisticated technology and continuous monitoring.

9.5.2 Freemium Pricing

Freemium pricing offers a basic version of a product for free while charging for premium features. This model is common in digital products but can be adapted for the food industry, such as offering free samples or trial versions. Key benefits include:

- **Customer Acquisition**: Free offerings attract a large customer base, increasing brand awareness and market penetration.
- **Upselling Opportunities**: Encourages customers to upgrade to paid versions for additional features or higher quality.
- **Customer Loyalty**: Free users can become loyal customers and brand advocates.

Freemium pricing works best when there is a clear value proposition for the premium version and a strategy to convert free users to paying customers.

9.5.3 Subscription Pricing

Subscription pricing involves charging customers a recurring fee to access a product or service. This model is gaining popularity in the food industry through meal-kit services, subscription boxes, and delivery plans. Key advantages include:

- **Predictable Revenue**: Recurring payments provide a steady and predictable revenue stream.
- **Customer Retention**: Subscriptions foster long-term customer relationships and loyalty.
- **Value-Added Services**: Additional benefits, such as exclusive content or personalized offerings, enhance the value of the subscription.

Subscription pricing requires delivering consistent value to retain subscribers and prevent churn.

9.5.4 Geographic Pricing

Geographic pricing sets different prices for the same product in different locations based on factors such as cost of living, regional demand, and local competition. Key considerations include:

- **Regional Costs**: Accounting for variations in distribution, taxes, and tariffs across regions.
- **Market Conditions**: Adjusting prices based on local demand and competitive landscape.
- **Consumer Perceptions**: Ensuring price differences are perceived as fair by customers.

Geographic pricing allows businesses to optimize prices for local markets while maximizing overall profitability.

9.6 CONSUMER BEHAVIOUR AND PRICING

Understanding consumer behaviour is fundamental to setting effective prices in the food industry. Consumers' perceptions of value, their sensitivity to price changes, and their purchasing habits play a critical role in terms of how they respond to different pricing strategies. Psychological and behavioural aspects influence consumer decisions, helping businesses tailor their pricing strategies to better meet customer needs and drive sales.

9.6.1 Price Sensitivity and Elasticity

Price sensitivity, also known as price elasticity of demand, measures how consumers respond to price changes. Products can be categorized based on their elasticity:

- **Elastic Products**: Demand changes significantly with price variations. These products are often non-essential or have many substitutes.
- **Inelastic Products**: Demand remains relatively stable despite price changes. These products are usually essential or have few substitutes.

Understanding the elasticity of your products helps to determine how much you can adjust prices without significantly affecting demand. Strategies for managing price sensitivity include:

- **Promotions and Discounts**: Temporary price reductions can stimulate demand without permanently lowering prices.
- **Tiered Pricing**: Offering different versions of a product at various price points can cater to different segments of price-sensitive consumers.

9.6.2 Perceived Value and Quality

Consumers often associate price with quality and value. Higher prices can suggest better quality, while lower prices might imply lower quality. Key factors influencing perceived value include:

- **Brand Reputation**: Strong, well-known brands can command higher prices due to established trust and perceived quality.
- **Product Differentiation**: Unique features, superior ingredients, or exclusive benefits can justify higher prices.
- **Marketing and Communication**: Effective advertising and clear messaging about the product's benefits enhance perceived value.

Aligning pricing with perceived value helps to attract customers willing to pay a premium for quality and ensures that prices reflect the product's true value.

9.6.3 Impact of Brand Loyalty

Brand loyalty significantly influences pricing power. Loyal customers are less sensitive to price changes and more likely to continue purchasing despite higher prices. Building brand loyalty involves:

- **Consistency**: Delivering consistent quality and customer experience reinforces trust and loyalty.
- **Customer Engagement**: Engaging with customers through loyalty programmes, personalized offers, and excellent customer service strengthens relationships.
- **Emotional Connection**: Creating an emotional connection through storytelling, brand values, and community involvement fosters deeper loyalty.

Leveraging brand loyalty allows businesses to maintain higher prices and reduce the impact of competitive pricing pressures.

9.6.4 Psychological Pricing Techniques

Psychological pricing techniques leverage cognitive biases to make prices more attractive. Common methods include:

- **Charm Pricing**: Under this pricing technique, price is fixed just below the whole number in order to get some psychological effect from the consumer perspective. For example, setting prices just below a round number, such as $9.99 instead of $10.00, creates the perception of a better deal. This strategy is highly effective for daily consumable products.
- **Price Anchoring**: Displaying a higher original price next to a discounted price highlights the savings and makes the lower price seem more attractive. For example, if a jacket is marked as $300 but discounted to $220, customers perceive greater value because the discount makes the new price seem like a significant saving. Although the discounted price indicates the true value of the product.
- **Decoy Pricing**: Introducing a higher-priced option to make the mid-range option appear more reasonable and attractive. For instance, if a small popcorn is priced at $3, a medium at $6.50, and a large at $7, most customers might initially choose the small option as it seems more affordable.

These techniques can significantly influence consumer perception and purchasing behaviour without altering the actual value proposition.

9.7 PRICING FOR DIFFERENT MARKET SEGMENTS

In the food industry, a 'one-size-fits-all' pricing strategy rarely works. Different market segments have unique needs, preferences, and willingness to pay, necessitating tailored pricing approaches. We must explore how to effectively segment the market and develop pricing strategies that cater to the specific characteristics of each segment, including retail, wholesale, and online markets.

9.7.1 Retail Pricing

Retail pricing targets end-consumers for purchasing products for personal use. Key considerations include:

- **Consumer Demographics**: Understanding the age, income, lifestyle, and preferences of your target consumers to help in setting appropriate prices.
- **Store Location**: Prices may vary based on the geographic location of the retail store, considering the cost of living and local competition.
- **Product Placement**: Premium shelf space or promotional displays can justify higher prices due to increased visibility and perceived value.

Strategies for Retail Pricing

- **Everyday Low Pricing (EDLP)**: Offering consistently low prices to attract price-sensitive shoppers and build long-term loyalty.
- **Promotional Pricing**: Using discounts, coupons, and limited-time offers to drive traffic and boost sales during specific periods.
- **Tiered Pricing**: Providing multiple product versions at different price points to cater to diverse consumer segments.

9.7.2 Wholesale Pricing

Wholesale pricing is aimed at businesses purchasing in bulk for resale or large-scale use. Key considerations include:

- **Volume Discounts**: Offering lower prices for larger order quantities to incentivize bulk purchases.
- **Long-Term Contracts**: Establishing fixed pricing agreements for extended periods to secure stable revenue and build strong relationships with buyers.
- **Cost Efficiency**: Streamlining production and distribution processes to reduce costs and pass savings on to wholesale customers.

Strategies for Wholesale Pricing

- **Cost-Plus Pricing**: Adding a standard mark-up to the cost of production to ensure profitability while remaining competitive.
- **Negotiated Pricing**: Customizing prices based on the buyer's order volume, purchase history, and long-term partnership potential.
- **Incentive Programmes**: Providing additional discounts or benefits for early payment, long-term commitments, or meeting specific purchase targets.

9.7.3 Online Pricing

Online pricing caters to consumers shopping through digital platforms. Key considerations include:

- **Price Transparency**: Consumers can easily compare prices online, so competitive pricing is crucial.
- **Dynamic Pricing**: Leveraging real-time data to adjust prices based on demand, inventory levels, and competitor pricing.
- **Personalization**: Using customer data to offer personalized prices, discounts, and promotions based on browsing history and purchase behaviour.

Strategies for Online Pricing

- **Flash Sales**: Offering significant discounts for a short duration to create urgency and boost sales.
- **Freemium Models**: Providing basic products for free with the option to upgrade to premium versions for a fee.
- **Subscription Plans**: Offering recurring delivery services or membership programs with exclusive benefits and discounts for subscribers.

9.8 CHALLENGES AND SOLUTIONS IN FOOD PRICING

Effective pricing in the food industry involves navigating several challenges, each with its own set of complexities. This section explores key challenges such as market fluctuations, cost management, and ethical considerations, providing practical solutions to address these issues.

9.8.1 Dealing with Market Fluctuations

9.8.1.1 Challenge

Market fluctuations, including changes in consumer demand, economic conditions, and competitive dynamics, can significantly impact pricing strategies. These fluctuations can create uncertainty and make it difficult to set and maintain optimal prices.

9.8.1.2 Solution

To manage market fluctuations, businesses should adopt a flexible pricing strategy that allows for adjustments based on real-time data and market conditions. Techniques such as dynamic pricing and frequent market analysis can help businesses respond to changes swiftly. Additionally, employing forecasting tools and maintaining a buffer stock can mitigate the impact of demand spikes or supply shortages. Regularly reviewing and adjusting prices based on market trends and consumer behaviour ensures that pricing remains competitive and aligned with market conditions.

9.8.2 Managing Costs and Margins

9.8.2.1 Challenge

Rising costs of raw materials, labour, and production can squeeze profit margins and complicate pricing decisions. Fluctuations in costs can make it challenging to maintain consistent pricing while managing profitability.

9.8.2.2 Solution

Effective cost management is crucial for maintaining healthy profit margins. Businesses should implement strategies such as cost control measures, supplier negotiations, and efficiency improvements to manage expenses. Techniques like cost-plus pricing, where a standard mark-up is added to production costs, can help ensure that prices cover costs and achieve desired margins. Additionally, analyzing cost structures and exploring alternative sourcing options can help mitigate the impact of cost increases on pricing. Regularly reviewing cost data and adjusting prices as needed can help maintain profitability while keeping prices competitive.

9.8.3 Addressing Ethical Considerations

9.8.3.1 Challenge

Ethical considerations, including fairness, transparency, and sustainability, are increasingly important in pricing decisions. Businesses must balance profitability with ethical practices, ensuring that pricing strategies reflect social responsibility and consumer expectations.

9.8.3.2 Solution

To address ethical considerations, businesses should adopt transparent pricing practices that clearly communicate the basis of pricing to consumers. Implementing fair trade practices and sourcing sustainably can enhance the ethical standing of a company's products and pricing. Additionally, businesses can incorporate social responsibility into their pricing strategies by supporting community initiatives or offering fair pricing for low-income consumers. Engaging with stakeholders and addressing consumer concerns about pricing practices helps build trust and aligns business practices with ethical standards.

9.9 FUTURE TRENDS IN PRICING

As the food industry continues to evolve, so too do the strategies and technologies that shape pricing decisions. Keeping abreast of future trends is essential for businesses looking to remain competitive and responsive to market changes. We will discuss emerging trends in pricing, driven by technological advancements, changing consumer behaviours, and shifts in the global market.

9.9.1 Technological Advancements

Technology is playing an increasingly critical role in pricing strategies. Key advancements include:

- **Artificial Intelligence (AI) and Machine Learning**: AI and machine learning algorithms analyze vast amounts of data to predict demand, optimize pricing, and adjust prices in real time based on market conditions. This enables dynamic pricing strategies that can respond to fluctuations in supply and demand with unprecedented precision.
- **Big Data Analytics**: The use of big data allows businesses to gather and analyze detailed information about consumer behaviour, preferences, and purchasing patterns. This data-driven approach enables more accurate pricing decisions and personalized pricing strategies.
- **Blockchain Technology**: Blockchain can enhance transparency and traceability in the supply chain, leading to more informed pricing strategies based on real-time data about product origin, quality, and authenticity.

9.9.2 Changing Consumer Behaviours

Consumer behaviours are continuously evolving, influencing how businesses set their prices. Key trends include:

- **Sustainability and Ethical Consumption**: Increasingly, consumers are willing to pay a premium for products that are sustainably sourced and ethically produced. Pricing strategies that reflect these values can attract environmentally conscious consumers.
- **Personalization**: Consumers expect personalized experiences, including customized pricing. Businesses can leverage customer data to offer tailored discounts, promotions, and pricing structures that cater to individual preferences and purchasing history.
- **Experience Over Product**: Consumers are prioritizing experiences over products. Pricing strategies that bundle products with unique experiences or value-added services can enhance perceived value and justify higher prices.

9.9.3 Shifts in the Global Market

Global market trends also impact pricing strategies in the food industry. Key shifts include:

- **E-commerce Growth**: The rapid growth of e-commerce is transforming traditional pricing models. Online platforms enable businesses to implement dynamic pricing, personalized offers, and global reach, but also require careful management of price transparency and competition.

- **Global Supply Chain Dynamics**: Changes in global supply chains, such as disruptions due to geopolitical events or pandemics, can affect production costs and pricing strategies. Businesses need to be agile in adjusting prices to reflect these changes.
- **Emerging Markets**: Expanding into emerging markets presents opportunities and challenges. Pricing strategies must be adapted to local economic conditions, consumer purchasing power, and competitive landscapes.

9.9.4 Regulatory and Economic Factors

Regulatory and economic factors continue to shape pricing strategies in the food industry. Key considerations include:

- **Regulatory Compliance**: Staying compliant with local and international regulations, such as price controls, taxation, and trade policies, is essential for setting appropriate prices.
- **Economic Fluctuations**: Inflation, currency exchange rates, and economic cycles influence production costs and consumer purchasing power. Pricing strategies must be flexible to adjust to these economic conditions.

9.10 CHAPTER SUMMARY

In conclusion, pricing strategies play a crucial role in the food industry, influencing profitability, consumer behaviour, and competitive positioning. By understanding and applying various pricing models, businesses can effectively align their pricing with market demands and organizational goals. Key strategies such as value-based pricing, dynamic pricing, and psychological pricing offer different advantages depending on the market context and consumer preferences. Additionally, staying adaptable and leveraging technological advancements can help businesses optimize their pricing decisions and respond to market changes.

At last, it can be said that a well-crafted pricing strategy not only drives sales and profitability but also strengthens brand positioning and customer loyalty. By integrating the insights and techniques discussed, businesses can navigate the complexities of pricing and achieve sustained success in the competitive food industry.

CASE STUDY

Whole Foods Market – Premium Pricing Strategy for Organic Products

Whole Foods Market is well known for emphasizing premium natural and organic items. By providing sustainable and ethically sourced items at a premium price, the firm has developed a devoted consumer base. Maintaining market status despite having premium pricing has been achieved through emphasis on product quality, environmental sustainability, and ethical sourcing on the part of Whole Foods Market. This case study probes how Whole Foods uses premium pricing and the factors that have been instrumental in ensuring its success.

Background

Founded in 1980, Whole Foods Market has evolved to become a leading player in the world's natural and organic food markets. Its acquisition by Amazon in 2017 provides it with better access to robust logistical and technological resources. Yet, even after acquiring this new organization, Whole Foods manages to sustain its premium pricing strategy which will entice customers willing to pay more to enjoy organic and sustainably sourced products. Its price strategy is directly aligned with its brand image, which is of high-quality, ethical, and healthy offerings for consumers who value such attributes in their food choices.

Premium Pricing Strategy at Whole Foods

Premium pricing entails establishing elevated costs relative to rivals, based on the premise that consumers would invest more for perceived enhanced quality, exclusivity, or ethical advantages. Whole Foods Market implements this approach by emphasizing the significance of organic items, exceptional product quality, and a dedication to sustainable operations.

***Organic and Natural Product Focus*:** The core of Whole Foods Market's value proposition is its selection of organic, non-GMO, and additive- and chemical-free goods. Given that customers believe organic food to be safer and healthier, Whole Foods may charge more for its goods thanks to this positioning.

***Perceived Value and Brand Loyalty*:** Whole Foods has effectively developed a brand that appeals to customers who are concerned about their health and the environment. Even though its prices are higher than those of rivals like Walmart or Trader Joe's, the firm is able to retain consumers due to their strong brand loyalty. Customers of Whole Foods Market are prepared to spend extra because they believe the items are better and more in line with their ideals.

***Ethical Sourcing and Sustainability*:** When compared to its rivals, Whole Foods Market stands out for its dedication to sustainable operations, fair trade, and ethical sourcing. Farmers and suppliers who adhere to strict ethical and environmental standards provide many of its goods. Buyers feel good about shelling out more cash since they know it goes towards supporting environmentally-friendly farming methods.

Factors Contributing to the Success of Premium Pricing

Targeting a Specific Demographic

Whole Foods has found a group of customers who care about health, the environment, and doing the right thing in food processing when making food. Most of the time, these customers are ready to pay more for high-quality items that stand for what they believe in. Whole Foods has been able to keep its high prices without losing a great deal of market share by focusing on this small group of customers.

Brand Positioning

Whole Foods has established itself as the store to visit when looking for sustainable, ethically produced, organic goods. The corporation has been able to justify premium pricing and generate an impression of greater quality because of its brand image. Their brand positioning has been further strengthened by their dedication to ethical standards in everything from sourcing to packaging.

In-Store Experience

Whole Foods shops are better than most other stores: they are well-designed, have knowledgeable staff, and have a lot of unique organic goods. This better shopping experience gives the customer more value, which is another reason for the higher prices.

Challenges of Premium Pricing

Competition

Whole Foods confronts growing competition as additional supermarket chains, including Walmart and Kroger, grow their organic product lines. These rivals provide cheaper organic goods, which may reduce Whole Foods' market share. It is harder to stick to a premium pricing strategy when rivals offer comparable goods for less money.

Market Perception of High Prices

Since its high pricing have often drawn criticism, Whole Foods has been dubbed "Whole Paycheck." The organization has difficulties in expanding its consumer base due to this unfavourable impression. Although the firm has lowered some of its prices following its acquisition by Amazon, its premium pricing approach remains essential to its identity.

Economic Downturns

Whole Foods may find it challenging to maintain its premium pricing policy during economic downturns, when customers may put money above quality. Therefore, in these financially unpredictable times, the firm must find a way to keep its competitive edge while continuing to provide high-quality products and services.

Key Lessons for the Food Industry

***Differentiation is Key*:** One of the reasons Whole Foods can maintain pricing power is because it has differentiated itself from competitors. With organic, sustainable, and ethically sourced offerings, the company has created a unique value proposition for its target market.

***Brand Loyalty Can Offset High Prices*:** Whole Foods proves that brand loyalty can carry a company through higher price points. Customers who believe there is value in being high quality, sustainable, or ethical will switch less for price even if it represents an advantage.

***Continuous Innovation*:** Since the food industry experiences continuous innovation in terms of products and pricing strategies, companies must develop themselves to face this change. Whole Foods Market responded to the competition by lowering the price cut policy with Amazon and increasing the private labelled products to offer more affordable organic products.

Whole Foods Market's premium pricing approach has been crucial in sustaining its competitive edge in the organic food sector. Whole Foods has established a devoted clientele prepared to pay a premium by portraying itself as a purveyor of high-quality, ethically produced items. The corporation must consistently innovate and confront problems presented by competition and market volatility to maintain its pricing strategy over the long run.

Discussion Questions

1. How does Whole Foods Market justify its premium pricing strategy in a highly competitive grocery market? What elements of its business model make this strategy sustainable?
2. What are the major challenges Whole Foods faces with its premium pricing strategy, especially as competitors like Walmart and Kroger introduce similar organic products at lower prices?
3. As Whole Foods expands internationally, what adjustments might it need to make to its pricing strategy in different geographic markets? Would a premium pricing model work equally well in all global markets?
4. Whole Foods Market emphasizes ethical sourcing and sustainability as key components of its value proposition. Do you think these elements alone are enough to justify premium pricing?
5. How might future trends in consumer behaviour, such as increased demand for convenience or digital shopping experiences, impact Whole Foods' pricing strategy? Should they adjust their model to remain competitive?

CHAPTER 10

Food Labelling Regulations and Compliance

LEARNING OBJECTIVES

After completing this chapter, you should be able to:

- Understanding regulatory frameworks.
- Recognize and describe the essential components of food labels.
- Understand the procedures that food producers must follow to ensure compliance with labelling regulations.
- Analyze current consumer trends related to food labelling and assess their impact on labelling practices.
- Investigate the role of emerging technologies in food labelling, such as smart labels, blockchain, and artificial intelligence (AI).
- Predict future trends and potential changes in food labelling regulations.
- Evaluate the legal, financial, and reputational consequences of non-compliance with food labelling regulations.

10.1 INTRODUCTION

Food labelling is one of the most critical aspects of food manufacturing because it is the first point of contact that the producer encounters with the consumer. Among the information provided on the food packaging is the identity, the nutritional value, the ingredients, and safety. Food labelling, therefore, is more than simply the role played in complying with the compliance requirements; it becomes a key tool for the protection of the consumer, public health, and the making of choices.

Food labelling regulations reflect values, expectations, and consumer health priorities; thus, they vary widely from one country to the next. In short, regulations were aimed at providing consumers with information, in particular, nutritional content,

DOI: 10.1201/9781003473305-10

the existence of allergens, and manufacturers' claims, so that the buyers make informed decisions regarding what they buy. Healthy consumerism increases scrutiny over the labels that are found on foods as the call for transparency and accuracy in marketing practice makes the practice more intense.

Meanwhile, the food landscape has been transformed through more novel labelling solutions which would achieve more transparency and traceability through technological innovation. Smart labelling technologies, such as QR codes and blockchain, provide consumers with ample information about the products they consume, thereby reinforcing the demand for accountability in food production.

This chapter goes into detail about regulation and compliance in food labelling. It gives a comprehensive overview of the prime constituents of food labels, the regulatory authorities involved, the legal standards that guide labelling activities, and the effects of compliance on food producers. The importance of understanding the relationship between regulation, technology, and consumer behaviour in facilitating stakeholders to navigate the challenges facing the process of food labelling while upholding public health and safety can be understood fully.

10.2 KEY COMPONENTS OF FOOD LABELS

Food labels guide both the consumers and the producers in making suitable purchases for dining consumption and ensuring one's health. Not knowing the key components of food labels, therefore, is necessary for both consumers and producers. Every bit of information on a food label has a purpose – that of communicating essential information, guaranteeing transparency and upholding regulatory standards (Figure 10.1).

10.2.1 Product Name and Description

The product name is the brand shown on the label, and this indicates the exact nature of the food commodity.

Both the product name and description must be straightforward and not misleading, as this helps consumers quickly understand what they are purchasing. In some areas, terms like "natural" or "fresh" are regulated to prevent misleading claims.

10.2.2 Ingredient List

The ingredient list breaks down the constituents of what is contained in the product, ordered descending by weight.

This transparency enables consumers to identify specific ingredients, including allergens and additives. It also assists individuals with dietary restrictions, such as vegetarians or those with food allergies, in making informed choices.

10.2.3 Nutritional Information

A nutrition information panel is a defined portion of the food label, showing the nutritional content per serving. The item may encompass macronutrients like

- Cereals containing gluten; that is, wheat, rye, barley, oats, spelt or their hybridized strains and products of these;
- Crustacea and products of these;
- Eggs and egg products;
- Fish and fish products;
- Peanuts, soybeans and products of these;
- Milk and milk products (lactose included);
- Tree nuts and nut products; and
- Sulphite in concentrations of 10 mg/kg or more.

Figure 10.1 Major categories of foods that should be listed on the label.

Source: CAC, 2010.

calories, fats, carbohydrates, and proteins; micronutrients in the form of vitamins and minerals; and other pertinent dietary constituents, such as fibre, sugars, etc. (see Chapter 11). Proper nutrition labelling empowers customers to manage diets effectively and select healthier options.

10.2.4 Allergen Statements

Mandatory food allergen labelling plays a crucial role in ensuring consumer safety. Most regulations would require the clear identification of common allergens such as nuts, dairy, soy, gluten, and shellfish inside the ingredient list or as part of a separate allergen statement. This is essential in ensuring that those with food allergies or intolerance avoid potentially life-threatening reactions.

10.3 REGULATORY AUTHORITIES

Regulatory authorities are key in the establishment, implementation, and regulation of food labelling legislation to ensure fairness, safety, and transparency in food production. In this regard, there is a need to understand the role and jurisdiction of these bodies by any food manufacturer or stakeholders about food labelling regulations.

10.3.1 Global and Regional Regulatory Bodies

Global organizations operate the provisions and policies at a worldwide level concerning the labelling of foods. The organizations are in place to protect health conditions among consumers and fair trade commodities, harmonizing labelling across different countries.

10.3.1.1 Codex Alimentarius Commission (CAC)

The Food and Agriculture Organization (FAO), in a joint effort with the World Health Organization (WHO), created the Codex Alimentarius Commission (CAC).

This organization sets international standards for food safety and quality. The Codex provides internationally recognized guidelines on food labelling to ensure that consumers receive accurate and adequate information about the products purchased. Even though Codex standards are not legally binding, many countries use or refer to them while formulating national regulations. The commission fosters the harmonization of food labelling laws that facilitate international trade while protecting public health.

10.3.1.2 European Food Safety Authority (EFSA)

The European Food Safety Authority (EFSA) focuses on food safety regulations within the European Union (EU). Further, it has the role of reviewing and evaluating risks in the food chain in addition to offering scientific advice to EU policymakers. EFSA recommendation plays a significant role in labelling regulations within the scope of the European Union, especially with regard to nutrition labelling, allergens, and health claims. Under EFSA, Regulation (EU) No. 1169/2011 clearly outlines food labelling regulations among all member states within the European Union to standardize accurate and informative labelling.

10.3.1.3 World Health Organization (WHO)

Although the World Health Organization (WHO) always prioritizes global health, it would signify food standards with the assistance of other organizations, such as the FAO. The WHO's contributions are often in the area of healthier diets through influencing labelling practices concerning nutrition and health claims, which are set to reduce the global burden of non-communicable diseases.

10.3.2 National Regulatory Authorities

Country-specific national regulatory authorities implement and enforce food labelling laws at the country level by ensuring food manufacturers comply with certain standards that protect consumers. Ingredient disclosures, nutritional labelling, and health claims are examples of regulations developed and enforced by these authorities.

10.3.2.1 Food and Drug Administration (FDA) – United States

In the U.S., the Food and Drug Administration (FDA) is the primary body responsible for food labelling regulations. The FDA offers truthfulness, lack of deception, and adequacy of food-label information to provide consumers with guidance to make responsible choices. The FDA strictly follows regulations established under the Federal Food, Drug, and Cosmetic Act and the Nutrition Labeling and Education Act provisions regarding the comprehensive labelling of nutrition statements as well as lists of ingredients. Additionally, the FDA monitors claims of "organic," "natural," and health-related statements to prevent practices by some vendors that falsely imply healthier products.

10.3.2.2 Food Standards Australia New Zealand (FSANZ)

The focus on Food Standards Australia New Zealand (FSANZ) regulation of food labelling in Australia and New Zealand is on consumer protection and public health. According to the regulation from FSANZ, the Food Standards Code provides that the appearance of a food label has to include standard aspects: a nutrition information panel, an ingredient list, and allergen declarations. The regulation also influences what claims can be made and that these claims are substantiated and do not mislead.

10.3.2.3 Food Safety and Standards Authority of India (FSSAI)

The FSSAI (Food Safety and Standards Authority of India) oversees food labelling in India, emphasizing the importance of clear and accurate information for consumers. Its guidelines mandate that labels feature comprehensive ingredient lists, nutritional information, and allergen warnings. The authority also enforces regulations concerning health claims and organic labelling. Under the Food Safety and Standards Act (FSSA) of India, producers are required to comply with stringent regulations designed to prevent misleading information and safeguard public health.

10.3.3 Local Regulatory Authorities

While this is overseen by national authorities, local regulatory bodies also play an essential role in ensuring food labelling compliance. The specific mandates or the need for more targeted regulations often fall to regional health concerns or localized production practices.

10.3.3.1 State and Local Health Departments – United States

In the United States, state and local health departments are typically joint enforcers of food labelling regulations. Some of these departments may inspect, accept consumer complaints, and take enforcement action against food producers that fall within their jurisdictions to comply with federal and state requirements related to food labelling. Most importantly, local authorities play a vital role in monitoring small-scale food producers and retailers since local health requirements need to exist.

10.3.3.2 Provincial Health Agencies – Canada

Provincial agencies in Canada also complement the work of the national Canadian Food Inspection Agency, which enforces the standards of food labelling laws at the provincial level. They can enforce region-specific rules, if any, to cater to regional food preferences and dietary needs.

10.3.4 The Role of Regulatory Authorities in Ensuring Compliance

Regulatory authorities have several key responsibilities ensuring compliance with food labelling laws. These include the following:

Developing Standards: The science behind food label development standards is linked to the health concerns of the public and consumers. The regulatory agencies put those on a food label, which should appear, how such information should be presented, and for which claims need evidence.

Monitoring Compliance: Relevant authorities or agencies examine, audit, and review food products to ensure that such foodstuffs comply with the labelling requirements. The ingredient list, nutrition facts, or allergen declaration will be verified for accuracy.

Enforcing Regulations: Corrective actions can be imposed upon the companies concerned, such as a call for warning, fines, product recall, or lawsuits against the concerned companies that do not comply with labelling standards.

Educating Stakeholders: The regulatory bodies also educate the stakeholders in labelling. These regulatory bodies would educate food producers and manufacturers on labelling requirements while ensuring that food consumers could easily understand the labels for the right choice-making.

10.3.5 Expiration Dates and Batch Numbers

Expiration dates signify the shelf life of products and their safety, directing consumers on when the item should be consumed to maximize its quality. Batch numbers enable traceability of products in case of recalls, which enables manufacturers and regulatory authorities to track all food safety issues efficiently.

10.3.6 Claims

Food labels often put forward claims that strongly affect the behaviour of consumers. These claims vary from health claims such as "low fat" or "high fibre" to sustainability claims like "organic" and "fair trade." Some claims are subjected to regulations, whereas others need to be supported with scientific evidence. Knowing more about these regulations helps both consumers and producers.

10.3.7 Country of Origin and Traceability

Labelling requirements for the country of origin affect consumer trust and purchase decisions. Information on origin, for example, can create a perception of quality, safety, and environmental impact. Traceability, or the ability to trace what ends up on consumers' plates to its source, is becoming a growing requisite of food safety and sustainability.

10.4 FOOD LABELLING LAWS AND STANDARDS

Food labelling laws and standards are considered critical for accurate, clear, and reliable information to be offered to consumers as they purchase products.

Regulations vary from country to country and also regionally as this reflects cultural values and practices, public health issues, and economic concerns. The laws and standards for food labelling are discussed below, showing their implications for the producers and the consumers.

10.4.1 Differences in Food Labelling Laws Across Countries

The legal framework related to food labelling differs from one country to another. Although most countries have enacted their exclusive legislation, some have followed international requirements provided by organizations like Codex Alimentarius.

United States: Food labelling in the United States is mainly set by the FDA, which enforces strict, harmonized rules for nutritional labelling, ingredient declaration, and allergen declaration. The foundation of nutritional labelling standards in the United States is set by the Nutrition Labeling and Education Act, which requires that disclosure of specified nutrient contents and standardized serving sizes.

European Union: The European Union (EU) has a comprehensive set of regulations regarding food labelling, mainly under Regulation (EU) No 1169/2011 concerning the provision of food information to consumers. With regard to food labelling, this regulation conveys information on nutrition declarations, allergen labelling, and specific rules on claims and indications of origin.

India: In India, food labelling falls under the Food Safety and Standards Act (FSSA). The country's Food Safety and Standards Authority of India issues regulations that enforce clear and accurate labelling. Its requirements include ingredient listings, nutritional information, and allergen declarations.

10.4.2 Specific Laws Related to Health Claims, Food Additives, and Allergens

Food labelling laws typically deal with product information on different aspects, mainly about the product's health claims, food additives, and allergens.

Health Claims: Legislation on health claims dictates what can be said about the health benefits of a food product. In the United States, any claims should be scientifically supported and approved by the FDA. In the European Union, health claims are limited under the Health Claims Regulation and would be based on scientific substantiation.

Food Additives: Regulations on food labelling usually require a declaration of any food additives in an ingredient list and their purpose. In the United States, the FDA is in charge of the country's food additives through the Federal Food, Drug, and Cosmetic Act (FFDCA). The European Union has regulatory processes regarding food additives, including their approval and labelling.

Allergen Labelling: Allergen declaration is of paramount importance given the high incidence of food allergies. Labelling regulations require a clear declaration of allergens. In the United States, the Food Allergy Labeling and Consumer Protection Act (FALCPA) requires significant allergens to be declared clearly on the label. The EU requires that allergens be declared in the list of ingredients and separate information about allergens.

10.4.3 Labelling Requirements for Genetically Modified Foods

As public awareness and concern regarding genetically modified (GM) foods have grown, many countries have implemented specific labelling requirements for these products.

United States: The FDA does not require the labelling of GM foods, except under specific circumstances, when GM foods are not substantially equivalent to their non-GM counterparts against nutritional content or safety. The new National Bioengineered Food Disclosure Standard requires that manufacturers of processed foods disclose the presence of bioengineered ingredients.

European Union: The European Union has a more significant regulation that requires foods containing GM ingredients to be labelled. Whenever such a foodstuff contains more than 0.9% of a GM substance, this must be visibly identified in the product's package so that customers know what they are buying.

10.4.4 Organic Labelling Standards

Organic labelling standards ensure that products marketed as "organic" meet specific criteria set by regulatory authorities.

United States: Organic certification is administered by the USDA's National Organic Program. Production practices and use of synthetic fertilizers and pesticides must be strictly complied with in order to acquire a stamp labelled organic. A product shall be labelled "100% organic" only if it contains all organic ingredients, while a product shall be labelled organic if it contains at least 95% organic ingredients.

European Union: The EU rules on labelling organic food are equally strict. The EU Organic Logo guarantees that products have been produced by EU standards for organic production. These standards include bans on the application of synthetic chemicals as well as animal welfare requirements.

10.4.5 Consumer Protection and Legal Implications

In essence, food labelling laws and regulations focus on consumer protection against misleading or incomplete information on labels. A failure to observe such regulations may cause the most stringent legal measures, heavy fines, the withdrawal of products in all markets, and also reputational damage. Transgressing these standards may lead to litigation against consumers or regulatory agencies as a result of making false or misleading claims from the labels. Thus, proper labelling practices are taken into consideration, which ensures the consumer's confidence and later legal defence routes. The bodies regulating them inspect, audit, and carry out monitoring systems during the enforcement of labelling law. By doing so, penalties can be charged for violations meant to deter misleading labelling practices.

10.5 COMPLIANCE WITH FOOD LABELLING REGULATIONS

For food manufacturers and producers, the guarantee of consumer demand, market accessibility, and brand is to be ensured for labelling compliance. This section elaborates on how compliance may be obtained, the importance of accurate labelling, the effects of non-compliance, and best practices in managing labelling requirements.

10.5.1 Procedures for Ensuring Compliance with Labelling Laws

Food producers must implement several key procedures to ensure compliance with food labelling regulations:

Understanding Applicable Regulations: Producers need to be aware of the laws and regulations concerning labelling on the products they are producing, which is unique for the regions where they operate. It involves national laws, regional requirements, and other related international standards applicable to their business.

Conducting Ingredient and Nutritional Analysis: The heart of compliance is ingredient listings and nutrition facts. Ingredients are supposed to be analyzed so that manufacturers can ensure that all ingredients end up on the labels correctly and accurately, with proper nutrition facts; third-party labs can be very useful when checking the accuracy of those details.

Training and Education: Employees who participate in product design, marketing, and quality control must also be trained and educated constantly. Employees need to be informed of labelling requirements and the best practices to avoid errors, and possibly non-compliance cases.

Regular Audits and Reviews: Periodic audits on the procedures followed during labelling help identify areas that need improvement, and thus the rules in place are maintained strictly. Internal reviews may also evaluate the correctness of the labels applied, to ensure no inconsistency in legal requirements.

10.5.2 Importance of Accurate Labelling

Accurate labelling is not only a regulatory requirement but also a critical factor in consumer trust and brand loyalty. Here are some reasons why accurate labelling is vital:

Consumer Safety: The proper labelling of products allows consumers to know the allergens, nutritional content, and ingredients for better choices in what they buy. Misleading or wrong labelling can cause serious health risks to sensitive consumers with food allergies or specific requirements.

Building Brand Reputation: More opportunities to gain the trust of customers are made possible if the brands have a focus on transparency and accuracy in labelling. Quality and honesty create a reputation and, soon thereafter, more loyalty from customers and a good perception of the brand.

Legal Protection: Given the fact that an organization that observes labelling laws is immunized from a suit over false labelling, firms ranked high in compliance are most likely to protect themselves against false advertisement or consumer deception.

10.5.3 Risk Management and Legal Implications of Incorrect Labelling

Non-compliance with food labelling regulations can result in severe consequences for manufacturers. Understanding these risks is essential for effective risk management:

Financial Penalties: Regulatory bodies can also levy enormous amounts of money in the form of fines on offending organizations against labelling regulations. Fines can be calculated according to the nature and level of seriousness of the infraction; besides, recall or remedial action can incur further costs.

Product Recalls: The regulatory bodies may also recall the products if the product fails to meet the pre-set regulations for the protection of public health. This will significantly lead to a loss of sales as consumers lose confidence in the brand.

Litigation: Companies can become susceptible to litigation actions from consumers or competitors at a later time when those inaccuracies have been proven to have caused harm based on false advertisements. It could then bring out severe financial liability and reputational damages.

10.5.4 Best Practices for Managing Labelling Compliance

To navigate the complexities of food labelling regulations effectively, producers should adopt best practices for managing compliance:

Implement a Labelling Management System: Since labelling management software ensures that information is up-to-date and according to requirements, the creation, updating, and checking of labels become easier and less time-consuming.

Engage Legal and Regulatory Experts: Reach out to legal and regulatory experts to navigate the many legalities involved in food labelling laws and the potential pitfalls.

Stay Informed of Regulatory Changes: Public health guidelines on food labelling regulation might change to coincide with changes in consumer demand. Regulatory changes would inform companies of the changing practice trends with which they need to comply.

10.5.5 Technological Innovations in Food Labelling

Technological advancements forming the new landscape of food labelling include improvements in transparency, traceability, and consumer engagement. This section examines innovative technologies that are shaping the future of food labelling, their benefits, and their implications for producers and consumers.

10.5.5.1 Smart Labels

Smart labels rely on digital technologies that allow users to know much more about their food. The tags also carry Quick Response (QR) codes, Near-Field

Communication (NFC) tags, or Radio-Frequency Identification (RFID) chips, which are scanned or tapped by a device on a smartphone or any other device of this kind. Full information about a product is provided through smart labels: nutrition facts, origin, manufacturing processes, and even recipes. Higher transparency enables consumers with more information to make wise choices about products. It raises the rate of customer loyalty. Smart labels improve the traceability of products from the supply chain. Therefore, it allows consumers to know where their food comes from and traces its path from the farm to the table, thereby enhancing confidence in the safety and quality of food.

10.5.5.2 Blockchain Technology

Blockchain is changing food labelling in the sense that it is not only secure but clear and unchangeable chronology for any transaction within this supply chain. Indeed, a transaction or piece of data goes into a block that gets aligned with a previous block; hence it becomes impossible to change. Blockchain makes business in food labelling even more transparent for the reason that consumers can trace the whole history of this product through the supply chain. It even checks assertions over aspects such as organic certification and fair trade, as well as sustainability claims over the products. A blockchain means that, in the event of food safety incidents, products can quickly be traced and unsafe ones identified and notified, thereby effectively reducing health risks and the costs incurred in recalls.

10.5.5.3 Augmented Reality (AR) and Virtual Reality (VR)

AR technologies (AR filters, AR-enabled packaging, AR Menus) and VR technologies (virtual restaurant tours, sensory analysis, and VR stimulation in new product development) have emerged as new and innovative tools for enhancing consumer contact with food labels. Using AR applications, customers can have an interactive experience when scanning a food label. For example, clients can access videos and recipes as well as educational content about the product. It will make the shopping experience more engaging and informative. Educational purposes can also be utilized for VR technology, whereby one can experience the origins of their food and understand sustainable practices within a virtual environment. Such an experience may lead a consumer to build a deeper connection with his food choices.

10.5.5.4 Artificial Intelligence (AI) in Labelling Compliance

The labelling process has also undergone this development in integrating artificial intelligence (AI) and machine learning into their processes to ensure compliance and optimize labelling practices. AI can analyze product labels in real-time, ensuring adherence to regulations and standards. This automated verification minimizes human error and streamlines the labelling workflow. AI can predict trends among consumers, thus understanding what the consumer wants next and helping strategize manufacturers with their labelling approach. Companies can thus optimize their labels to meet changing market demands by analyzing data from consumer interactions.

10.5.5.5 Environmental Considerations

As consumers place more and more emphasis on sustainability, technological progress in food labelling further emphasizes the reduction of environmental impacts. New materials and printing technologies allow the production of ecologically friendly labels recyclable or biodegradable. Companies can base their innovation on these developments to communicate their sustainability efforts on the label, speaking well to the environmentalist consumer. Some companies now put carbon footprint information on the label, allowing consumers to have much better knowledge of how their food purchase might impact the environment. This again aligns with the upsurge in consumer demand for sustainable and responsible food choices.

10.6 CONSUMER TRENDS AND LABELLING

Consumer trends now force their influence onto food labelling, such that it is no longer simply about transparency of labelling but about health consciousness and ethics. The more aware and involved consumers are, the more their expectations evolve over the information being transmitted through food labelling. This section will discuss current consumer trends in food labelling, what such trends mean for producers, and the problems in relation to this.

10.6.1 Growing Demand for Transparency

Transparency in food labelling is becoming increasingly crucial for consumers who want to understand the origins, ingredients, and production methods of the products they buy. There is a growing interest in knowing where food comes from and how it is made. Labels that explain where ingredients are sourced, especially if they emphasize local and sustainable practices, attract consumers who value quality and care for the environment. With greater awareness of health, consumers demand not only transparent and comprehensive nutrition content on the label of food but also formats that allow them to deliver a snap judgement of its health value. This accounts for the popularity of simplified systems of nutrition labelling, such as the traffic light system.

10.6.2 Health-Conscious Labelling Trends

Increasing health consciousness in consumers seeking more knowledge from the label regarding health-related aspects of the product creates opportunities for labels that highlight desirable attributes, such as "high in fibre," "low in sugar," or "rich in vitamins," but poses dangers to products emphasizing higher-than-average levels of sugars, fats, or other added artificial ingredients that are subject to possible negative backlash and lower sales. Functional foods, or products supplemented with higher levels of nutrients or other health benefits, represent a fast-growing trend that intensifies the demand for labelling to describe those benefits. Labels must therefore accurately portray the health claims and functional ingredients to suit the expectations of the consumer.

10.6.3 Ethical and Sustainable Labelling

Consumers grow more concerned about the ethical dimensions of their food choices, and there is demand for labelling those practices that they deem sustainable. Organic and fair trade products are in high demand as consumers demand that their food be produced ethically and sustainably. Certified labels related to organic farming, the humane treatment of animals, and fair trade also reflect what consumers are looking for in a brand that they believe takes social and environmental responsibility seriously. Labels referring to the environmental impact of a product-related carbon or water footprint are also seen to be gaining momentum. Many consumers will be willing to make more environmentally-friendly choices if brands are transparent.

10.6.4 Challenges in Balancing Regulations and Consumer Demands

Food labelling regulations are not always something that can be aligned with the demands of consumers for transparency and detail. For example, regulations may restrict claims producers can make or the format in which nutritional information needs to be presented, hindering their ability to meet such expectations of the consumer while remaining compliant. The literate consumer will stay away from the false promise or the "greenwashing." It means brands have to be attractive but truthful and based on data to avoid consumer distrust, which, ultimately, leads to litigation.

10.6.5 The Role of Education and Awareness

Consumer education plays an important role in helping individuals make informed choices based on food labels. Improving labelling literacy – helping consumers read and understand labels—is indispensable. The educational campaigns will enable the consumer to make better decisions for healthier and more responsible lifestyles through correct information. For instance, some forms of interactive labelling involve mobile phone applications that will scan and decode the information on the label, enhancing the consumers' knowledge and even helping them bond with the product. Such tools will bridge the gap of information between labelling and consumers' understanding.

10.7 CHAPTER SUMMARY

Consumers would be engaged in discussions on the labelling requirements, making the standards more responsive and relevant. Food labelling combines multifaceted aspects of the food industry itself, meaning compliance with regulations, the expectations of consumers, and the advances in technology. Through this chapter, we have been able to discern the critical components of food labelling, the roles that the regulatory authorities play, and an evolving landscape informed by consumer trends and innovations. A food label is an important tool which can inform a customer about the products they buy to guarantee transparency and safety. They allow customers to make enlightened decisions informed by information on the nutritional aspects of a product, ingredients sourced, and similar value-oriented aspects. Food labelling has to be controlled, but the regulations are heterogeneous, and country-dependent. Manufacturers ought not to be prosecuted and shall bring their goods into the market knowing the relevant laws and standards in those regions. Food labelling undergoes a metamorphosis with smart labels, blockchain, and AI technologies. The innovations make compliance monitoring efficient, increase transparency and traceability and therefore assist producers and consumers alike. More emphasis on sustainability, tailor-made nutrition, and harmony around the globe is going to change the fate of labelling styles. Stakes need to observe and be flexible enough to change with this revolution.

CASE STUDY

The Labelling Crisis of Kellogg's Kashi Products

Kellogg's Kashi brand was launched into the natural organic food space, and the brand was operated with offerings of a variety of cereals, granola bars, and snack items. The brand presented itself as a non-artificial brand, using slogans like "All Natural," thereby promoting GMO-free and organic food products. This attracted consumers who requested healthier foods, more transparency in labelling, and an aversion to GMOs and synthetic ingredients.

However, in 2011, the brand's reputation was tested when an independent study showed that some of the ingredients in Kashi products were genetically modified or synthetic – directly opposing the company's "All Natural" claims. That labelling crisis resulted from confusion and quite obscure regulation in food labelling laws and the ambiguity of terms such as "natural."

Compliance Issue

Kellogg's was thrown into question when Kashi products labelled "All Natural" were shown to contain many synthetic ingredients and GMOs. The controversy raised questions over the ambiguous use of the term "natural" because this term did not have any legal definition at the time under U.S. law. The company stated that it had met the FDA's criteria; however, consumers felt they had been misled by the labelling.

False advertisements regarding the brand led to the bringing of a class-action lawsuit against Kashi in 2011. The accusation was that consumers are misled into purchasing natural foods untampered with artificial additives or genetic alteration. Thus, the deception that undermined all the trust Kashi had gained among its consumers ended.

Regulatory and Legal Action

Although the term "natural" had not been legally defined, the lawsuit contended that using it to describe products with synthetic and genetically modified ingredients was misleading. Kellogg's settled the lawsuit in 2014 for $5 million, promising to withdraw the "All Natural" label from any products that included synthetic ingredients or GMOs. The reimbursement of affected consumers was part of the settlement and triggered changes in how Kashi labelled its products.

Although the FDA had no regulations with regard to using the term "natural" for such products, this case drew attention to the need for clearer regulations on food labelling in cases where consumers make purchasing decisions based on such claims.

Consequences and Repercussions

Financial Losses: The case was settled in court, and Kellogg had to pay $5 million in attorneys' fees. In addition to this, the company incurred costs related to the removal and rebranding of mislabelled products, legal fees, and marketing efforts to rebuild consumer trust.

Damage to Brand Reputation: Customers were deceived into believing through these claims on the labels. The most damaged, however, was the target consumer group for Kashi: the health-conscious consumers who still opt for natural ingredients. Loss of trust followed; sales nosedived.

Increased Scrutiny on Labelling Practices: The events also brought to light the way the general problem of false labelling in food manufacturers is a concern if truth and honesty claims in the food packaging have to be made serious. A similar pressure fell on other brands within the natural foods sphere, forcing the change across the industry regarding labelling.

Corrective Actions

In response to the crisis, Kellogg's took several corrective measures to address consumer concerns and prevent future issues:

Revised Labelling Practices: The company would no longer use the term "All Natural" in areas where the products had genetically modified or synthetic ingredients. The company ensured that labelling on health claims was clear and this met consumer expectations.

Expansion of Non-GMO Product Lines: The company enlarged its product lines into non-GMO products and clearly labelled items, which would meet the strict non-GMO criteria to regain trust with consumers.

Consumer Education: Kashi has implemented a consumer education programme to make the ingredient sourcing process and production more transparent to raise consumer awareness. The company engaged the brand in educating the customers on what constitutes organic, non-GMO, and natural products so that they can make an informed decision.

The labelling problem in connection with Kellogg's Kashi brand calls attention to transparency and accuracy in food labelling. Some technically legitimate but misleading claims can also harm a brand, financially and reputationally. Once again, the case throws into prominence the requirement for more transparent regulatory guidelines on such terms as "natural," which have confused consumers and producers alike. After the corrective action and refocusing on transparency, Kashi was in a position to rebuild trust, but it remains a cautionary tale for food producers.

Discussion Questions

1. How did the "All Natural" label on Kashi products affect consumer trust, and why is transparency in food labelling so important in maintaining a brand's reputation?
2. The term "natural" has no formal definition under U.S. law. Should the FDA and other regulatory bodies establish clearer guidelines for terms like "natural"? How might this improve consumer understanding and trust?
3. In this case, Kellogg's Kashi faced both financial and reputational losses due to mislabelling. What are the long-term implications of misleading food labels for companies in the food industry?
4. What steps can companies take to rebuild consumer trust after a labelling controversy? Evaluate the effectiveness of Kashi's response to the labelling crisis and suggest any additional measures they could have implemented.
5. Given the rise of health-conscious consumers and the demand for transparency, what changes should be made in food labelling regulations to prevent similar issues from occurring in the future?

BIBLIOGRAPHY

CAC. (2010). General Standard for the labelling of prepackaged foods. In *FAO*. Retrieved October 5, 2024, from https://www.fao.org/fao-who-codexalimentarius/sh-proxy/es/??lnk=1&url=https%253A%252F%252Fworkspace.fao.org%252Fsites/?%252Fcodex%252FStandards%252FCXS%2B1-1985%252FCXS_001e.pdf

CHAPTER **11**

Nutritional Labelling Requirements and Health Claims

LEARNING OBJECTIVES

By the end of this chapter, you will be able to:

- Understand the regulatory frameworks governing nutritional labelling and health claims.
- Evaluate the impact of health claims on consumer behaviour and food marketing.
- Analyze the challenges and risks associated with nutritional labelling.
- Evaluate how technologies like blockchain, artificial intelligence (AI), and personalized nutrition are transforming the future of food labelling.
- Investigate the role of sustainability and environmental impact labelling in influencing consumer choices.
- Discuss the implications of clean labelling and the rise of consumer demand for natural, minimally processed products.

11.1 INTRODUCTION

Today, consumers have become increasingly conscious of the types of food and their general effects on health. Nutritional labelling assists consumers in making decisions based on highlighting their required nutrition. It acts as an intermediary between foodstuff manufacturers and the public, clarifying crucial points of nutritional information such as calorie content, macronutrient composition, or vitamin and mineral levels.

A label extends beyond information relating to the nutritional value of food; it is also a marketing tool whereby firms exploit the trend towards the consumption of nutritious and functional foods. However, this balancing act, involving factual and regulatory-compliant information, has placed food manufacturers in a dynamic environment of unpredictable regulation and public concern.

DOI: 10.1201/9781003473305-11

Nutritional labelling standards generally differ from country to country. This means that food marketers simultaneously encounter both challenges and opportunities. The Food and Drug Administration (FDA) in the United States, for example, requires some items to appear on the nutritional label; in the European Union, these matters are regulated by the European Food Safety Authority (EFSA). In other regions, there are further variations. Compliance with this range of standards make it necessary for corporations to create consumer trust and avoid legal consequences.

In claims relating to food health, manufacturers can include the benefits existing in their product. Such claims are important because, whether related to nutrient content claims or health benefit claims, they help form a crucial part of consumer decisions. That being said, such claims are strictly governed to avoid misleading consumers or exaggerating the benefits of their products.

With the food industry fast becoming dynamic, nutritional labelling today represents one of the critical approaches towards tackling diet-related diseases, obesity, and heart disease. Proper and understandable nutrition information contributes to a more informed choice, healthier lifestyles, and trust with consumers.

This chapter follows a discussion of the complexity of nutritional labelling requirements and health claims, regulation of the practice, influence on consumer behaviour, challenges for the industry, and future trends.

11.2 KEY NUTRITIONAL LABELLING COMPONENTS

The most prominent features of any nutritional label are the listings of macronutrients, which include proteins, fats, and carbohydrates. Macronutrients are all the important energy sources in food, considered necessary for maintaining all bodily functions. Nutritional labels generally present the amount of each macronutrient per serving size so consumers can monitor their intake and apply a dietary change if necessary.

11.2.1 Macronutrients: Proteins, Fats, and Carbohydrates

The most prominent feature of any nutritional label is the listing of macronutrients—proteins, fats, and carbohydrates (Figure 11.1). These are the primary energy sources in food and are critical for maintaining bodily functions. Nutritional labels typically display the quantity of each macronutrient per serving size, allowing consumers to track their intake and make dietary adjustments as needed.

Proteins: These are often linked with muscle health and recovery. Many people are seeking a diet that contains enough protein. Athletes, along with those with certain health goals, pay extra attention to this part of the label.

Fats: Fats are divided into subcategories, including saturated fats, trans fats, and unsaturated fats (both polyunsaturated and monounsaturated). It would be beneficial to know about these ingredients, as excessive intake of unhealthy fats results in heart disease and other health problems. Labels emphasize “low-fat” or “trans-fat-free” items for the health-conscious consumer.

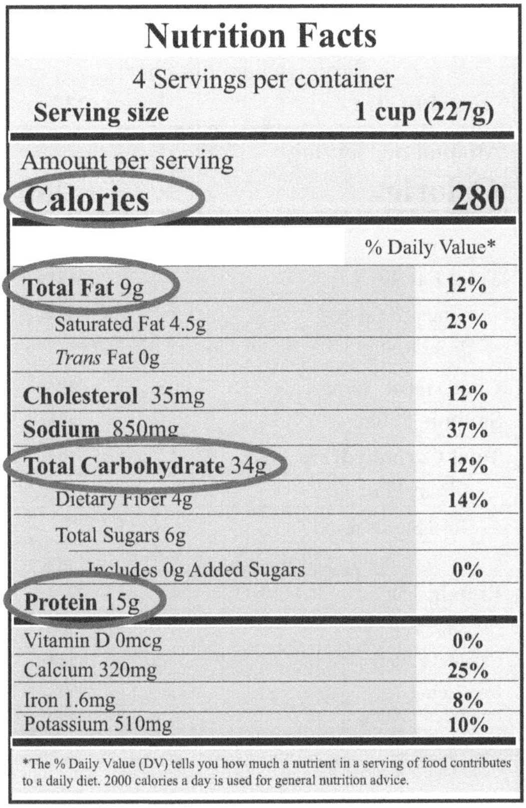

Figure 11.1 Labels of macronutrients.

Source: The Authors.

Carbohydrates: The total listed on a label usually comprises total carbohydrates, broken down into fibres and sugars. For most existing customers, the less sugar the better; thus, the quantity of sugar in the product is always a marketing play for customers who struggle with obesity and even diabetes. Consumers will also look for an appropriate level of fibre because this enables healthier digestion and reduces the possibility of developing some diseases.

11.2.2 Micronutrients: Vitamins and Minerals

Micronutrients are equally important, but are required in smaller amounts. The nutritional contents label on food packaging usually focuses attention on the number and amount of important vitamins and minerals, such as Vitamin A, Vitamin C, calcium, and iron (Figure 11.2). These micronutrients play very essential functions in health and wellness, helping to facilitate healthy body functioning, maintain immune function, and promote bone health, among other functions. To assist consumers in deciding,

Nutrition Facts	
4 Servings per container	
Serving size	**1 cup (227g)**
Amount per serving	
Calories	**280**
	% Daily Value*
Total Fat 9g	**12%**
Saturated Fat 4.5g	**23%**
Trans Fat 0g	
Cholesterol 35mg	**12%**
Sodium 850mg	**37%**
Total Carbohydrate 34g	**12%**
Dietary Fiber 4g	**14%**
Total Sugars 6g	
Includes 0g Added Sugars	**0%**
Protein 15g	
Vitamin D 0mcg	**0%**
Calcium 320mg	**25%**
Iron 1.6mg	**8%**
Potassium 510mg	**10%**

*The % Daily Value (DV) tells you how much a nutrient in a serving of food contributes to a daily diet. 2000 calories a day is used for general nutrition advice.

Figure 11.2 Labels of micronutrients.

Source: The Authors.

within their customary, day-to-day needs, how a food product fits, labels generally include the percentage of daily recommended intake for most vitamins and minerals. This helps consumers make a quick assessment as to whether or not a product is high or low in specific nutrients.

11.2.3 Additional Information: Calories and Serving Size

One of the most visible parts of the nutritional label is the calorie content, usually displayed at the top of the label. The figure for calories measures the energy that food contains, and knowing this is important to anyone who tracks their weight or caloric intake. Food packaging also identifies serving size; in many cases, this can be misleading because the serving size might not be representative of the actual quantity regularly consumed at one time. Knowing both calories and serving size is crucial to making proper comparisons between food products. Serving per container indicates how many servings are in a package. Thus, it becomes clear to which amount of the product the nutrition facts apply.

11.2.4 Allergen Information

Allergen labelling is a critical safety component on food packaging, particularly for individuals with food allergies or sensitivities. In most countries, it is a common practice to clearly declare the presence of common allergenic ingredients on food labels. The adoption of allergen labelling contributes to the protection of consumers from undesirable reactions as well as improving the trust in a brand's commitment to safety.

11.2.5 Ingredient List

The ingredient list is another vital part of nutritional labelling. Ingredients are listed in descending order of the amount they represent in the food product, meaning communicating to consumers what is the most prevalent ingredient (see Figure 11.3). This is particularly useful for those who may wish to avoid certain ingredients, artificial additives, preservatives, and high-fructose corn syrup but also must meet additional dietary needs, such as vegetarian or gluten-free.

11.2.6 Nutrient Reference Values (NRVs) and Daily Value Percentages

Nutritional labels often provide Nutrient Reference Values or equivalent units that tell a consumer how much of any given nutrient they should take on a daily

Figure 11.3 Ingredients and allergen content.

Source: The Authors.

Nutrition Facts	
4 Servings per container	
Serving size	**1 cup (227g)**
Amount per serving	
Calories	**280**
	% Daily Value*
Total Fat 9g	**12%**
Saturated Fat 4.5g	**23%**
Trans Fat 0g	
Cholesterol 35mg	**12%**
Sodium 850mg	**37%**
Total Carbohydrate 34g	**12%**
Dietary Fiber 4g	**14%**
Total Sugars 6g	
Includes 0g Added Sugars	**0%**
Protein 15g	
Vitamin D 0mcg	**0%**
Calcium 320mg	**25%**
Iron 1.6mg	**8%**
Potassium 510mg	**10%**

*The % Daily Value (DV) tells you how much a nutrient in a serving of food contributes to a daily diet. 2000 calories a day is used for general nutrition advice.

Figure 11.4 Labels of additional information.

Source: The Authors.

basis. It would display the percentage daily value (%DV), suggesting how much of the daily nutritional recommended amount is found in one serving of the product (see Figure 11.4). This uniform reference helps consumers quickly compare and contrast the nutritional content of the food in the larger context of their diet.

11.2.7 Front-of-Pack Labelling

An emerging trend in the food sector is front-of-pack labelling, which still provides clear, simple nutrient information at a glance. Systems such as Traffic Light Labelling or Health Star Rating summarize nutritional information in a format which is easier to apply (Figure 11.5). With such systems, consumers can easily compare products. One major reason is that in order to enable consumers to choose easier healthier options without having to decrypt the entire label, it might primarily focus on calories, fat, sugar, or salt compared to anything else.

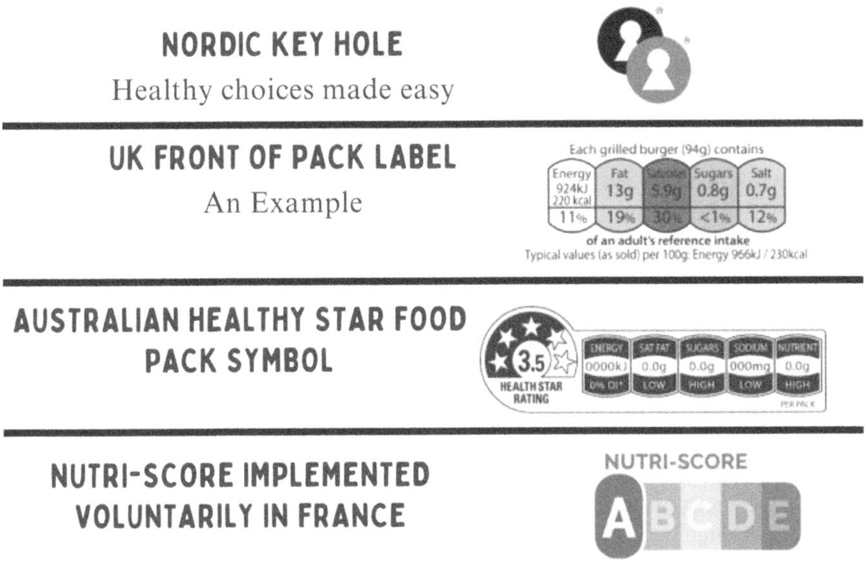

Figure 11.5 Examples of front-of-pack labels from different countries (Norway, the UK, Australia, and France).

Source: The Authors.

11.3 NUTRITIONAL LABELLING REGULATIONS AND COMPLIANCE

Nutritional labelling varies significantly from country to country, but there has been an international push towards uniformity of information presented to consumers. Regulations protect consumers against misleading claims and allow them to make informed dietary choices. Governments and regulatory bodies strictly enforce guidelines on the way food companies must present nutritional information, but these vary from region or country to country. This is an introduction to the basic legislation governing nutritional labelling, the difficulties of compliance, and the penalties that will be incurred for non-compliance.

11.3.1 Regulatory Bodies and Their Roles

Global organizations establish and enforce laws concerning nutrition labelling. While such details vary, the point of this movement seems to be similar worldwide, ensuring that consumers can receive reliable and up-to-date information on the food products they are willing to purchase. Below are some of the prominent regulatory organizations and their mandates:

Food and Drug Administration (FDA) – United States: The FDA is responsible for setting the standards for food labelling in the U.S. It mandates listing specific nutritional components contained in the packaged commodity, such as calorie count, serving size, and amounts of macronutrients. These also include health claims that need to be supported by science.

European Food Safety Authority (EFSA) – European Union: Most food products sold in pre-packaging must show a detailed composition of nutritional information under EU regulations, under the strict standards and guidelines of Regulation (EC) No 1924/2006 on nutrition and health claims. EFSA is involved as the scientific advisor to policy and supervises food labelling standards in the EU.

Codex Alimentarius Commission – International: The Codex Alimentarius, a collection of internationally recognized food standards, guidelines, and codes of practice, serves as a reference for many countries in setting their food labelling laws. This standard follows and promotes fair trade practice as it assists in ensuring consumer health by setting guidelines on safe and transparent food labelling.

Food Standards Australia New Zealand (FSANZ) – Australia and New Zealand: In Australia and New Zealand, FSANZ is the enforcement agency that develops food standards on labelling. Moreover, FSANZ regulates the nutritional panels, ingredient lists, and allergen declarations. Lastly, FSANZ supervises the implementation of voluntary front-of-pack labelling schemes, such as Health Star Rating.

11.3.2 Core Regulatory Requirements

Each regulatory body may have slight differences in labelling requirements, but there are fundamental components that are typically mandated across most jurisdictions:

Nutrition Facts Panel: This would typically appear on the back or side of a food package and contain calories, the percentage of macronutrients, and number of micronutrients. It will also include serving size, number of servings per container, and total calories per serving. Formats often become standard for the font size to use, size, and arrangement to view clearly.

Health Claims Regulations: Any health-related claim on food labelling should be substantiated by science or established regulatory approvals. For instance, the FDA reviews all claims advising a consumer that a product may reduce the risk of disease or help to prevent it in the United States, whereas in Europe, similar claims are reviewed by EFSA. Health claims fall into different categories:

- Nutrient Content Claims (e.g., "Low Fat" or "High in Fibre")
- Structure/Function Claims (e.g., "Supports Immune System Health")
- Disease Risk Reduction Claims (e.g., "May Reduce the Risk of Heart Disease")

Ingredient List: Ingredients must be listed in descending order by weight. This provides transparency and ensures that consumers are fully aware of what they are consuming. Visibility of allergens should be made clear so that allergic reactions are not triggered.

Serving Size and Portion Control: Accurate serving size information is critical, as it affects the interpretation of nutritional data. Fraudulent or misleading serving sizes mislead consumers, whereby they are likely to consume more calories or nutrients than intended. Agencies set regulations on how to calculate serving sizes to avoid exploitation of such information by food manufacturers.

11.3.3 Challenges in Global Harmonization

One of the most significant challenges faced by multinational food companies is regional diversity in requirements for nutritional labelling. Though there have been efforts towards an international approach to standardization, for example, through Codex Alimentarius, regional approaches to labelling exist that differ from one another. For example, manufacturers must indicate added sugars on a product in the U.S., whereas this is not exactly required in the European Union.

Labelling Formats: Some countries demand more detail about specific nutrients, while others are much less taxing on what format their labelling should be in. This means companies have to use different labels for different markets, meaning that a lot of complexity and expense added to compliance.

Health Claims: Countries have different thresholds for the scientific substantiation of health claims. What may be deemed an approved health claim in one country might be classified as misleading or scientifically unsound by another. That is a challenge for companies that aspire to sell health across borders.

11.3.4 Penalties for Non-Compliance

Failure to comply with legislation on nutrition labelling will hence attract toughened measures within the food manufacturers. For example, the fines will range from product recalls depending on the jurisdiction while, in extreme cases, the company will face a lawsuit. Common non-compliance issues include:

Misleading Health Claims: If a product's health claim is found to be false or exaggerated, a regulator can impose fines, direct the company to remove its product from stores, or require modifications in labelling. In 2010, the European Court of Justice barred a yogurt brand from marketing by disapproving the health claims of the said yogurt.

Incorrect Nutritional Information: Data for the nutritional constituents, like false calorie content or level of macronutrients, will attract some form of penalty or product recall. Inaccurate labelling further jeopardizes a company's reputation and hampers consumer confidence.

Failure to Declare Allergens: Failure to declare or properly label allergens in products will expose critical health risks for consumers facing severe allergic conditions. Undeclared allergens in such products are usually treated as serious offences in most countries, hence recalling these products or lawsuits.

11.3.5 Enforcement Mechanisms

While focused on random inspections, audits, and consumer complaint investigations regarding compliance with the nutritional labelling requirements, in other countries, the regulatory agencies may inflict severe sanctions according to the level of offence. The other ways of controlling the situation include the recalling of products, public warning notices, or injunctions in order not to prevent consumers from being deceived or intimidated by hazardous food products.

Figure 11.6 Different types of health claim.

Source: The Authors.

11.4 HEALTH CLAIMS ON FOOD LABELS

Health claims on the food label have been a potent marketing tool for food manufacturers (Figure 11.6). Their health claims enable companies to be able to speak about the potential benefits of their products, especially in targeting health-conscious consumers. Nonetheless, the use of health claims is strongly regulated because, while false information can be perpetuated if used, credible science backs any claim about the health benefits of a particular food. In the next section we discuss the various kinds of health claims, the regulatory environment in which they exist, and behavioural and marketing implications for consumers.

11.4.1 Definition and Types of Health Claims

Health claims on food labels refer to statements that link a food product or a nutrient it contains to certain health benefits, such as improved bodily function or reduced risk of disease. These claims fall into several distinct categories, each with its regulatory requirements:

11.4.1.1 Nutrient Content Claims

- These claims describe the level of a particular nutrient in a food product. They often use terms like "low," "high," "reduced," or "free" to quantify the nutrient content.

For example, "Low Fat," "High Fibre," or "Sugar-Free" are common nutrient content claims.

- Regulatory Note: These claims must meet specific regulatory thresholds. For example, in the U.S., the FDA defines "Low fat" as products containing 3 grams or less of fat per serving.

11.4.1.2 Structure/Function Claims

These claims concern the effect that a nutrient or dietary ingredient has on a body's normal structure or function. They have nothing to do with preventing disease but can say, for instance, "Calcium builds strong bones" or "Fibre helps regulate digestion."

Like other nutrient content or disease claims, for structure/function claims, pre-market approval from regulatory agencies is not necessary, but they cannot be false or misleading. In the U.S., they must include a disclaimer that the claim has not been evaluated by the FDA.

11.4.1.3 Disease Risk Reduction Claims

These claims should convey that a food or a nutrient might be associated with the reduced risk of a particular disease or health-related condition. For example, "may reduce the risk of heart disease" or "lowers cholesterol levels."

These claims are very regulated and need high scientific evidence to pass. For example, the FDA and EFSA approve the claims that have been brought out based on clinical research and allow only what is supported by significant scientific consensus.

11.4.2 Regulatory Framework for Health Claims

Appropriate regulation governs the health claims of the regions so that no consumer of relevant products is misled by any false or overstated claims. Regional health claims regulations are followed as below:

United States (FDA): All health claims in the United States are pre-approved by the FDA, which decides whether there is scientific support for the claim. Even disease risk reduction claims require "majority scientific consensus" to be in place before they can be used on labels. Structure/function claims can be used without ever being pre-approved, but they should carry disclaimers.

European Union (EFSA): This authority has been licensed to permit health claims through Regulation (EC) No. 1924/2006 in the EU. Once these claims have successfully passed the EFSA scientific evaluation test, it is allowed for use. It is put on the public register. As such, all member states will guarantee openness and compliance.

Canada (CFIA): The Canadian Food Inspection Agency regulates health claims in Canada. Like both the U.S. and the EU, it requires scientific substantiation for disease risk reduction claims but is more lenient on structure/function claims provided they are not misleading.

Australia/New Zealand (FSANZ): FSANZ administers the Food Standards Code. This government agency is in charge of health claims monitoring. The regulation of health claims is very strict, and, in Australia/New Zealand, both "general-level" claims such as the content of nutrients, and "high-level" claims regarding the reduction of disease fall under the regulation of FSANZ.

11.4.3 Scientific Substantiation of Health Claims

Health claims, particularly disease risk reduction claims, must be supported by substantial scientific evidence. For this reason, the regulatory authority considers clinical trials, epidemiological studies, and meta-analyses of such evidence submitted together with the claim application. In most regions, independent scientific panels will scrutinize the data submitted.

Key Requirements for Approval:

- Consistency in research findings across multiple studies.
- The quality and design of the studies (e.g., randomized controlled trials).
- Biological plausibility of the claim (i.e., the mechanism by which the nutrient affects health).

Even structure/function claims, which do not need pre-approval, must be grounded in scientific research, though the standards for substantiation are often lower than for disease risk claims.

11.4.4 Examples of Misleading Health Claims and Regulatory Action

While health claims indeed work as excellent marketing tools, sometimes these come with quite serious legal challenges if mishandled. Regulatory agencies frequently enforce action against companies making false or misleading health claims. Below are some prominent examples:

Kellogg's "Frosted Mini-Wheats" Case (2009): The U.S. Federal Trade Commission litigated Kellogg for false advertising relating to health claims that their Frosted Mini-Wheats cereal improves the attentiveness of children by 20%. Kellogg settled and changed its campaign on the product.

Danone's "Activia" Yogurt Case (2010): Danone argued that its Activia yogurt reduced the indigestion caused by probiotics. In April 2010, the FTC decided that the evidence was too weak to validate the claim and ordered Danone to pay $21 million in fines and to withdraw particular claims from packaging.

These cases underscore the importance of complying with regulatory standards and ensuring health claims are not overstated or misleading.

11.5 IMPACT OF NUTRITIONAL LABELS ON CONSUMER BEHAVIOUR

Usually, nutritional labels and health claims have played a significant role in deciding which product to purchase by the consumer. With an increase in health awareness among customers, they tend to rely on this information alone to gauge the nutritional value of food items and whether the food item meets their dietary requirements. This chapter highlights how these nutritional labels and health claims decide consumer behaviour, the psychological mechanism, and, finally, the overall food marketing strategy.

11.5.1 The Role of Nutritional Labels in Consumer Decision-Making

The most critical information through which consumers can make a better choice about food will be from nutritional labelling. A study has found that consumers who use such labels compare products based on calories, sugar, fat, and nutrient content. However, to what extent consumers will benefit from nutritional information might change with the combination of health awareness, education, and dietary preferences.

Nutrient-Specific Focus: Consumers tend to be concerned about nutrient-specific information. Thus, if the consumer seeks pounds, they will be more concerned with calorie content, the consumer concerned about heart health will be more concerned with fat or sodium content, and the health-conscious consumer will care about added sugars, fibre, and protein content.

Use of Serving Size Information: As consumers read calorie and nutrient content, research concludes that they are unaware of the serving sizes. The perception of size compared to the actual size makes the customer miscalculate how much they are taking. This transparency of the serving size provides an important foundation in the legislation.

Simplified Labelling Preferences: Many consumers who do not have much time or knowledge of nutrition would like to see labelling in a simplified format, like a front-of-pack label. These might include, for example, the Traffic Light System or Health Star Rating. These labelling systems give an immediate and easy overview of the nutritional quality of the product, thereby making informed choices without requiring analysis from a full nutrition facts panel.

11.5.2 The Role of Health Claims in Shaping Consumer Perception

Health claims, such as "high in fibre" or "supports immune health," have an enormous impact on consumer perception and purchase decisions. For example, these sorts of claims can evoke very positive emotions concerning the food product, even when consumers have a pretty weak idea of the brand and overall nutritional profile of its products. But health claims can do much more than nudge consumer perceptions of particular foods – they often nudge perceptions about what is healthy to eat.

Health Halo Effect: The most significant behavioural phenomenon related to health claims has undoubtedly been the "health halo" effect. Food bearing a positive health claim has led consumers to make the whole product healthy, even though it may contain other unhealthy ingredients – one example is high sugar or fat content. For example, a breakfast cereal that claims to be "high in fiber" may lead someone to believe it is a healthy choice, yet the large amount of sugar it may contain can make all the difference between proper food choice and overconsumption.

Increased Perceived Value: Products bearing health claims will be perceived to be of a higher value because of this implied connection. Research findings indicate a greater willingness to pay premium levels for any product supporting health claims that help ward off disease, promote weight loss, or contribute towards better performance will result in more frequent business positioning within the health-conscious market.

Trust and Scepticism: While health-related claims are a great tool in building consumer trust for the brand, they also potentially expose it to scepticism. Consumers in developed markets, especially those with better nutrition knowledge, would be suspicious of such statements or consider them marketing gimmicks. If there is transparency and supportive scientific findings then a lot of the trust can be protected. Conversely, if such health-related claims are hyperbolic or not supported by facts, there may be some damage to brand reputation.

11.5.3 The Effect of Front-of-Pack Labelling Systems

Front-of-pack labelling systems have emerged as a solution to help consumers quickly assess the nutritional value of food products. Several countries and regions have implemented different front-pack-labelling systems. Some of them are mentioned below.

Traffic Light System (UK): In the UK, the Traffic Light Labelling system is graded by colour-coded classifications of three critical nutrients: fat, sugar, and salt. High should be red, moderate yellow, and low green. A few studies have shown that this system has led to healthier consumers making decisions because the nutritional information is given in easy-to-understand forms. The more red, the more likely that consumers will avoid the product and the more green, the more likely shoppers are to see this product as a better choice.

Health Star Rating (Australia/New Zealand): The rating system, which indicates a simplified score for the nutritional value of food, classifies it within a range of 0.5 to 5 stars, therefore, making it possible for consumers to have a simple indication of how healthy the product would be. This impacts consumer behaviour by encouraging them to purchase lower-fat and lower-sugar products.

Nutri-Score (EU): The Nutri-Score system in Europe uses a colour-coded scale ranging from A (healthiest) to E (least healthy), summarizing the nutritional quality of a product. Consumers will not only make healthier choices, but they can also read an easy-to-understand guide from Nutri-Score. Consumers perceive products rated A and B as being healthier and will, in general, prefer those over less healthy options.

11.5.4 The Role of Nutritional Labels in Shaping Health-Conscious Markets

With the increasing public awareness of health and wellness, nutrition labelling and health claims would, within the context of 'health-conscious markets,' have been developed more significantly. It is because of these primary concerns about health, environmental sustainability, and ethics that people increasingly call for clearly accurate and transparent labelling of products.

Consumer Segmentation: The fact that an ever-growing proportion of consumers develop specific dietary needs, such as being vegan, gluten-free, or organic, for which they are prepared to pay a premium if only they can be matched with a product that reflects their values; nutritional labels are critical to helping consumers navigate the marketplace so that they can make purchasing decisions that help them attain their health objectives.

Marketing Opportunities: Nutritional labelling and health claims are a rich source of differentiation opportunities for food marketers, tapping into that growth space driven by the health trend. Those brands that can communicate the health basis behind the product in ways that stay within regulatory boundaries will be well-positioned to take a share in this growth health-conscious market.

11.6 CHALLENGES IN NUTRITIONAL LABELLING

While nutritional labelling has proven to be a useful tool for consumers as well as for food manufacturers, it is created at the cost of several challenges that need to be overcome for it to exhibit clarity, transparency, and accuracy. Further pressure is therefore going to mount on food companies to provide both current legislation and market demand with increasingly detailed reliable information. We will be discussing in this chapter some of the key issues with nutritional labelling and consumer interpretation and complexity in regulation or technology.

11.6.1 Consumer Misunderstandings and Interpretation Issues

Although there is an abundance of information that reveals the nutritional content of foodstuffs, consumers remain confused as to how to interpret the data presented on packaging. There are several reasons for this: low nutritional literacy, confusion over serving sizes, and a tendency to focus on single nutrients rather than an overall nutritional profile of the product.

Nutritional Literacy: Most consumers are inappropriately educated to interpret even the simplest terms on a nutrition label. Terms like "daily value," "percent of daily intake," and "balance of macronutrients" confuse the uninformed consumer. It either leads to the misinterpretation of nutritional information or it is not used at all when a decision is made in a purchase.

Serving Size Confusion: The largest area of consumer misunderstanding is the serving size. Nutrition fact panels are formatted to show how many nutrients are present per serving, but most consumers do not realize how many servings are present in a package. For instance, if the consumer believes they are getting more calories or more nutrients from a food product because the package appears bigger than they thought the product would be, they will consume more than they should.

Focus on Single Nutrients: The customer will concentrate on a single nutrient, or more usually two (such as sugar or fat) while ignoring the overall nutritional value of the product. In a low-fat product, this is interpreted as being healthy while its content also contains high sugar or sodium levels. This tremendous focus on single nutrients has led to unhealthy food choices among customers, principally because it puts their perception on the wrong track.

11.6.2 Controversies around Health Claims

Health claims are very useful in marketing, but can sometimes be too broad or misleading; therefore, the suspicion of the consumer and the pressure from the regulatory bodies are bound to arise. That a product may advertise as "heart-healthy" or "immune-boosting" must be scientifically supported, although even approved claims are subject to more than one interpretation.

Overemphasis on Health Claims: Overemphasizing health claims would be interpreted in terms of shouting at the top of one's lungs about healthy product consumption while downplaying other unwanted features, such as high levels of sugar or sodium. This is likely to create an aura of false hope that creates either overconsumption or unrealistic expectations on the part of consumers. For example, some cereals labelled as high in fibre contain highly abundant sugars. Nevertheless, the fact that there is a positive statement about these outweighs the negative statements about them.

Vagueness in Claims: Health claims are improper; hence, consumers can hardly make any sense of what benefits the product. Phrases like "boosts energy" or "helps support digestive health" can be used to communicate a wide range of messages, and, in the absence of proper context or substantiation, these claims may cause consumers to be confused or unconvinced. Elsewhere, statutory requirements are enacted in the sense that such claims are couched in good scientific proof, but that does not stop the words themselves from obfuscating the line between advertisement and fact.

11.6.3 Balancing Transparency with Marketing Objectives

Another challenge is achieving the right mix between marketing aims and transparency. While labelling is meant to allow customers, clearly and honestly, to understand the food they are buying, it is also a means by which food manufacturers market. The outcome is to over-rely on detail description to convey high detail and to over-rely on simplification for high appeal.

Overload of Information: To meet regulatory requirements and satisfy consumer demand for transparency, food manufacturers often include a great deal of information on their packaging. As consumers buy less and food manufacturers create more, the fine print on ingredient labels and elaborate nutritional breakdowns floods consumers with information. Meeting the demand for more information with user-friendly design remains an open-ended challenge for food companies.

Simplification vs. Accuracy: Manufacturers tend to oversimplify the nutritional information required for display on the front of the package simply so that the labels appear more attractive and easier to read. It sometimes misses important information. Front-of-pack labelling schemes that employ symbols and ratings may create a general, overarching feeling about the food product's healthfulness, but do not explain its entire nutritional profile. This lets consumers make easier choices but also can lead to oversimplification or, in some cases, a false perception.

11.6.4 Regulatory Compliance in a Global Marketplace

For multinational food companies, ensuring compliance with nutritional labelling regulations across different countries and regions is a complex task. Each market has its own set of labelling laws, and, while there are efforts to harmonize international standards, significant differences remain.

Varying Regional Requirements: Countries have their specifications regarding nutritional labelling regulations, including information to be declared, serving size definitions, and the approval in terms of health claims. For example, the US has mandated the manufactures to list added sugars, while EU does not have such mandate. This will mean that multinationals will have to adapt their packaging and marketing strategies to unique markets, and this will increase the cost and complexity.

Harmonization Challenges: Globally harmonized food standards are in place, but complete alignment cannot be achieved by international bodies such as the Codex Alimentarius Commission as their priority will vary from one region to another, along with cultural habits and public health issues. Therefore, such a gigantic market creates a challenge for international brands as they have to maintain the same label according to market requirements but not brand inconsistency.

11.6.5 Consumer Demands for Greater Transparency

Today, consumers are much more aware of how the food they consume would impact health, the environment, and society. This same reason is also a factor behind increased demand for transparency in labelling about ingredients, sourcing practices, and whether additives or allergens are contained within them.

Clean Label Movement: Transparency in ingredients has emerged as the new leading trend away from artificial additives or overly processed foods: this has seen to the developments of the so-called "clean label" movement. Consumers want to know what's going on in their food, and therefore they are demanding clear, simple, and honest labelling. That has, however, presented food manufacturers with the opportunity to reformulate according to "clean-label" standards without being compelled to compromise on taste, quality, and price.

Demand for Sustainability Information: Consumers ask for information about the environmental impact related to their food choices. This includes carbon footprint, the use of water, and sourcing practices – for example, organic, and fair trade. Even here, where some begin to label sustainability-related information, there is no regular practice set in place that standardizes this data, thereby making informed decisions difficult for consumers.

11.7 CHAPTER SUMMARY

Nutritional labelling and health claims are getting more importance day by day in the food industry as consumers seek more transparency, health benefits, and sustainability from the products. Businesses have to follow the winds of change in terms of technology, compliance concerning regulatory standards, and such calls for transparency and accuracy of information. This will eventually manifest in the direction of personalized nutrition. Consumers will then be provided with diet advice customized to their health profile. Digital tools will play a very significant role in delivering more complete and personalized information regarding nutrition through QR codes and AI-powered labels, enabling consumers to make healthier choices that align with the objectives of their health.

Along with clearer labelling on the front of the packaging, these systems will multiply and help consumers judge faster about the nutritional quality of products. Visual cues like the Traffic Light Labelling system and the Nutri-Score will help consumers make quicker and healthier decisions. The rapidly growing "clean-label" movement will drive a need for greater transparency, and consumers will focus on simple, natural ingredients and origin transparency from the food they are consuming. Sustainability labelling, carbon footprint, and water usage can also all be a strong differentiator for the eco-conscious who want to make an environment-friendly choice.

However, these opportunities also create challenges. Food manufacturers work to make the labels as simple as possible, but this does not abolish the need for correctness and alignment in the information placed on them with changing regulations. Health claims would come under much stricter demands from regulatory agencies for them to provide stronger scientific evidence. Keeping ahead of these changed regulations will hold the key to business in upholding consumer confidence and avoiding legal pitfalls. Moreover, firms have to align their marketing objectives with the dictates of transparency so that health claims and labelling practices are honest and clear to the consumers themselves.

More control for consumers, but at what cost? The necessity to educate consumers on what exactly the average level of food knowledge on the label implies will arise. Only when tailor-made nutrition becomes the way of the future should personalized dietary advice, made available to match the exact health needs of each individual, make food use more rewarding and also healthier. The future labels for nutrition and health claims are in a dynamic shift towards transparency, personalization, and sustainability within the food industry. Companies embracing these trends will do well to thrive within these competitive markets and fit into the consumers' necessities for informed, health-conscious, and environmentally-conscious purchasing choices.

CASE STUDY

The "Smart Choices" Programme – A Lesson in Misleading Nutritional Labelling

In 2009, several major food companies in the United States, including General Mills, Kraft Foods, and Kellogg's, launched the Smart Choices Program, a voluntary front-of-pack labelling initiative. The programme aimed to help consumers make healthier food choices by identifying products that met specific nutritional criteria with a green checkmark labelled as a "Smart Choice." The programme was developed in collaboration with nutrition and health experts, industry leaders, and academic researchers. The stated goal was to create a standardized system that simplified nutritional information and supported consumer education.

The Controversy: Despite its intended purpose, the Smart Choices Program faced immediate criticism from both public health advocates and regulatory agencies. Many products bearing the "Smart Choices" label were far from being considered healthy by general standards. For example: Froot Loops, a sugary cereal with 12 grams of sugar per serving, was labelled a "Smart Choice." Other processed snacks, high-calorie frozen meals, and sugary beverages also received the designation.

Critics argued that the nutritional criteria for the label were overly lenient and allowed highly processed, sugar-laden products to appear as health-conscious options.

The backlash grew when consumer advocacy groups, including the Center for Science in the Public Interest (CSPI), accused the programme of being a marketing tool rather than a genuine effort to improve public health.

Regulatory Response: The controversy drew the attention of the U.S. Food and Drug Administration (FDA) and the Department of Agriculture (USDA). The FDA issued a warning, stating that it would closely monitor front-of-pack claims to ensure compliance with federal food labelling regulations. The agency expressed concerns that the Smart Choices label could mislead consumers about the nutritional value of the products. Under mounting pressure, the programme was voluntarily suspended by its organizers in late 2009.

Key Lessons Learned

Transparency and Rigorous Standards are Essential: The failure of the Smart Choices Program highlighted the need for robust and transparent nutritional criteria for health claims. Criteria perceived as favouring corporate interests over public health undermine trust.

Consumer Trust is Fragile: Once the programme's credibility was questioned, consumers and health advocates lost trust in the label. This eroded its effectiveness and damaged the reputation of participating brands.

Regulatory Oversight is Critical: The FDA's involvement underscored the importance of regulatory oversight in ensuring that health claims and labelling practices do not mislead consumers. Programmes such as Smart Choices can only succeed when aligned with existing regulations.

Simplified Labels Must Balance Clarity and Accuracy: While consumers appreciate simplified labelling, oversimplification – especially when it misrepresents nutritional quality – can have the opposite effect.

The Smart Choices Program serves as a cautionary tale for companies and industry groups developing nutritional labelling systems. It underscores the importance of science-backed criteria, transparency, and alignment with consumer expectations and regulatory standards. Front-of-pack labelling has the potential to positively influence consumer behaviour, but only when it genuinely reflects the healthiness of a product.

Discussion Questions

1. Do you think the companies behind the Smart Choices Program acted unethically by labelling products like sugary cereals as "smart choices"? Why or why not?
2. How does misleading nutritional labelling affect consumer trust in food brands and the overall food industry? What steps can companies take to rebuild this trust?
3. What role should government agencies like the FDA play in regulating voluntary labelling programmes? Should stricter regulations always be enforced, or is there value in industry-led initiatives?
4. How could the nutritional criteria for the Smart Choices Program have been improved to avoid backlash? What specific changes would you propose?
5. Simplified front-of-pack labelling is designed to make it easier for consumers to make healthier choices. What are the potential risks and benefits of such systems? How can companies ensure these labels are both clear and accurate?

CHAPTER 12

Sustainable Practices in Food Marketing

LEARNING OBJECTIVES

After reading the chapter, you should be able to:

- Define sustainability within the context of food marketing and its importance for businesses and consumers.
- Identify the key elements of sustainability.
- Evaluate the role of sustainable sourcing and ethical practices.
- Discuss the impact of eco-friendly packaging on waste reduction.
- Discuss the strategies food companies use to reduce their carbon footprint.
- Evaluate the challenges companies face in balancing profitability with carbon reduction goals and regulatory requirements.
- Assess the importance of waste reduction initiatives in food marketing.
- Explain the importance of transparency in building consumer trust.
- Understand how sustainability influences consumer behaviour and purchasing decisions.
- Recognize the financial, regulatory, and logistical challenges companies face when adopting sustainable practices.

12.1 INTRODUCTION

Sustainability has become a critical factor in the food industry, not only influencing the way products are produced, packaged, and distributed but also create a long term impact in the minds of the customers. Today, the marketplace is shaped more by environmental and social considerations by the consumer who considers the long-term impact of his purchasing decisions. Consumers have become accustomed to demanding only brands that align with their values concerning sustainability and ethical practices. This can be seen as either a challenge or an opportunity for food companies, in which profitability will weigh against insatiable demand for eco-friendly and socially responsible practices, is the new scenario.

Sustainable food marketing is more than 'greenwashing' or 'window dressing' that makes empty claims about environmental awareness. Rather, it is a holistic

DOI: 10.1201/9781003473305-12

approach towards reducing the side effects on the environment and society, while simultaneously ensuring health sustainability and economic viability. Sustainability in food marketing can indeed be multidimensional, simultaneously encompassing perhaps a reduction in carbon footprint through ethical sourcing, minimization of waste, and eco-friendly packaging solutions.

The concept of the triple bottom line – people, planet, and profit – has become a guiding principle for the visionary companies. Businesses have come to realize that it not only makes their brand image better to deal with environmental and social issues but also becomes vital for the long-term sustainability of a business. Consumers want to know where, how, and what practices are used in the production, processing, or transportation of food. They require the knowledge of these practices to help bring about positive change.

For these reasons, the food industry uniquely positions its companies to be at the forefront of a sustainability movement: through the reduction of food waste and improvement in resource efficiency, through ethical labour practices, and through the promotion of regenerative agriculture, among others. Sustainable focus-based companies will appeal to all sorts of consumers, but also help improve operational efficiency, reduce costs, and stimulate innovation at the same time.

Throughout this chapter, we will look at the different sustainable dimensions of food marketing. In doing so, we will identify how companies are adopting changes, implementing more ecological packaging solutions, and effectively reducing waste. We will also analyze how brands leverage sustainability to reach their consumers, talk about their challenges in this context, and analyze the future trends that are expected to move the game forward in this dynamic environment.

12.2 SUSTAINABLE SOURCING AND ETHICAL INGREDIENTS

One of the most distinctive characteristics of food marketing is sustainable sourcing, which directly affects a company's footprint in terms of both the environment and society. To put it concisely, it refers to the responsible sourcing of ingredients and raw materials that help preserve ecosystems, contribute to local communities' well-being, and promote good labour practices. The sustainability consumer might find this information sufficient to ensure that their food has been sourced responsibly. Proper stewardship of the environment and adherence to specific ethical standards can help differentiate a brand from its competition.

12.2.1 Define Sustainable Sourcing

Sustainable sourcing in the food industry is particularly more about minimizing environmental and social damage related to agriculture, fishing, and other food production processes. These include protecting biodiversity, reducing deforestation, conserving water resources, and ensuring that farmers and workers involved in the supply chain receive fair wages and can provide good working conditions. Sustainable sourcing must also include practices focused on soil health and the decrease in both chemical fertilizer and pesticide use while maintaining long-term viability for food producers and ecosystems.

For example, sourcing coffee beans from farm-established shade-grown agriculture preserves biodiversity and forest cover; similarly, sourcing seafood from fisheries that use sustainable fishing prevents the over-exploitation of marine ecosystems.

12.2.2 Certifications and Standards for Ethical Sourcing

To assure consumers of their sustainable sourcing practices, many food brands pursue third-party certifications. These certifications provide an independent verification that the products meet specific environmental and social criteria. Among the most widely recognized certifications in the food industry are the following:

Fair Trade Certification: This certification ensures the producers of developing countries to receive a fair price for their products and are supported in improving their livelihoods. It also promotes ethical labour practices, gender equality, and community development.

Rainforest Alliance Certification: This certification focuses on protecting biodiversity, ensuring sustainable land use, and improving the livelihoods of farmers and workers. Products with the Rainforest Alliance seal come from farms that prioritize environmental conservation and social responsibility.

USDA Organic Certification: Organic farming practices eliminate the use of synthetic fertilizers, pesticides, and genetically modified organisms (GMOs). USDA Organic certification assures consumers that the food is grown in a way that supports soil health, biodiversity, and ecological balance.

Marine Stewardship Council (MSC) Certification: This certification applies to seafood products and ensures that fish are sourced from well-managed, sustainable fisheries. MSC-certified products help maintain healthy ocean ecosystems and prevent overfishing.

These certifications are increasingly important for consumers who want to make informed choices and support brands that reflect their values. Brands that display these certifications on their packaging benefit from enhanced trust and credibility in the marketplace.

12.2.3 Challenges in Achieving Sustainable Sourcing

While many brands are fully committed to sustainable sourcing, it is difficult to meet 100% sustainability. The supply chain for some ingredients is often complex and combines several intermediaries, regions, and regulatory environments. For example, high commodity demands such as palm oil, coffee, and cocoa have caused destruction and habitat loss through global demands on these products, despite efforts to buy such goods sustainably.

In addition, responsible and sustainable sourcing is often more expensive; this may, for example, increase prices for smaller or emerging brands. Some certifications and third-party audits are costly and many companies must balance the quest for sustainability goals against the obligation to deliver profitability. Climate change is also heightening the risk to the security and quality of key ingredients for sourcing, making urgency and the work of attaining sustainability more challenging.

12.2.4 The Impact on Brand Perception

Sustainable sourcing benefits the environment and communities, as well as having a positive impact on the perception of the brand. Today, consumers know where his food comes from and also the practices behind his favourite brand. There has been some evidence that consumers do spend more on products ethically sourced and which have less of a strain on the environment. Brands that will speak to the world about their source practices and express commitment to sustainability gain a loyal and trusting heart among green-oriented consumers.

However, there is always the risk of over-promising and making unsubstantiated claims about sustainability efforts. In any case of greenwashing, an inevitable reputational blow can arise from there being inconsistencies between the words of a brand and its actual practices being exposed by consumers or advocacy groups.

12.2.5 The Future of Sustainable Sourcing in Food Marketing

Sustainable sourcing will emerge as an essential element in food marketing as consumers continue to demand more ethical products. The strongest loyalty and market differentiation will go to the most proactive brands in adopting practices of sustainable and transparent sourcing. Another aspect is the new technology that is constantly being introduced, especially the influence of blockchain on traceability and transparency, in order to improve and provide verifiable proof in the case of sourcing claims made by a brand.

While the future looks bright for regenerative agriculture, carbon-neutral farming, and zero-waste supply chains, companies are on their way to furthering these actions in pursuing reduced environmental footprints. The strategic embedding of sustainable sourcing will be an important factor for food brands as they seek to raise their flag as leaders in this more sustainable future.

12.3 ECO-FRIENDLY PACKAGING SOLUTIONS

Packaging is an essential criterion for sustainable food items, reflecting not only the environment but also consumer perceptions. In this aspect, because concerns over the environment develop progressively with time, this forces the food industries to reduce the environmental footprint of packaging. Presently, consumers require food or food packaging with attributes that are biodegradable, compostable, or recyclable to reduce waste and pollution. According to the Statista report, most consumers believe that compostable packaging is the most sustainable packaging type to achieve sustainability (Figure 12.1). This section also elaborates on innovation in sustainable packaging, the problem of companies finding ways of executing such changes, and how packaging affects consumer behaviour and brand perception.

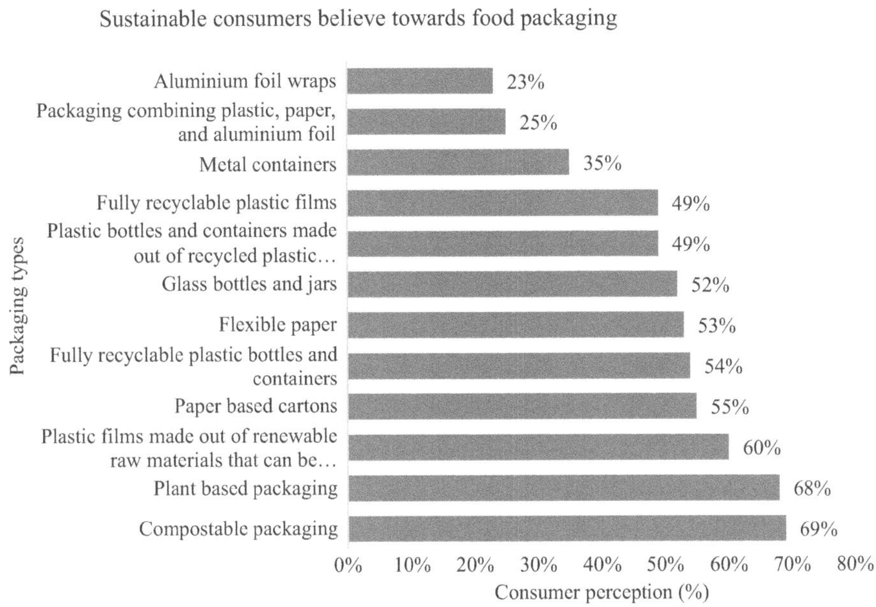

Figure 12.1 Sustainable consumers' beliefs regarding different food packaging, 2023.

Source: Statista, 2024b.

12.3.1 The Environmental Impact of Traditional Packaging

Plastic is a large of the mainstream traditional packaging materials because of the material's strength and low price, and, above all, because of its effectiveness in avoiding from contamination and spoilage. Plastic packaging, although it has many varied applications, especially in the case of single-use plastics, has a terrible environmental impact as it takes hundreds of years to decompose. The majority of it goes into landfills or ends up in the oceans, thereby promoting pollution and posing a danger to marine life. The perception of single-use plastic concern also varied based on the generation, with baby boomers being more concerned about pollution than other generations (Figure 12.2).

Every year, the food industry creates millions of tons of plastic waste through its use of packaging. This has led to a concerted response from governments and consumers alike, calling for the development and introduction of more sustainable alternatives. Single-use plastics are becoming increasingly regulated, and, therefore, many brands are revisiting their packaging strategies in the face of the strict eco-friendly standards laid out before them.

12.3.2 Innovations in Sustainable Packaging

To address the environmental challenges associated with packaging, many food companies are looking for innovative solutions that can reduce, reclaim, and

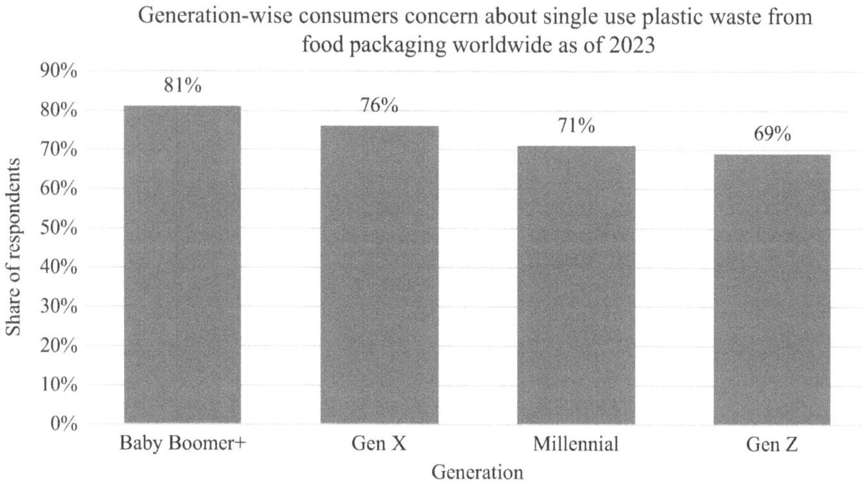

Figure 12.2 Global concern about plastic food waste from food packaging, 2023.
Source: Statista, 2024f.

significantly minimize the use of non-renewable resources. Among the most promising eco-friendly packaging solutions are:

Biodegradable Packaging: In this instance, the packaging material is taken from natural resources; these include corn starch and sugarcane combined with cellulose. The product degrades itself in the environment with no harmful effects. It is used widely in the food processing industry, packaging fresh fruits, snacks ans beverages. The biodegradable packaging market value is increasing continuously, and is expected to reach around US$809 billion by 2032 (Figure 12.3).

Compostable Packaging: Compostable packaging is designed in such a manner that it decomposes into non-toxic components that can enhance the quality of the soil during industrial or household composting. Compostable packaging differs from biodegradable packaging because its decomposition involves specific conditions (Figure 12.4).

Recyclable Packaging: Materials such as certain types of plastics, glass, and metals can be recycled. This opens an opportunity to create new products in the form of packaging from recycled materials. Recycling saves waste, but it requires a good recycling infrastructure and the incorporation of consumers (Figure 12.5).

Edible Packaging: One of the latest trends in sustainable packaging allows consumers to eat the packaging along with the product. Edible packaging is made of natural ingredients such as seaweed, rice, or wheat and eliminates waste. The global market size of this packaging was US$1.4 billion in 2022, and it is expected to reach US$5.26 billion by 2032 at a 14.2% Compounded Annual Growth Rate (CAGR) (Figure 12.6).

Minimalist Packaging: Another area in which companies are not only using sustainable materials but also reducing the number of packaging materials themselves, minimalist design on their products, using less material and minimizing the use of unnecessary layers or parts. A report published by SNS Insider (2024) highlights that the market value of this packaging was US$13.60 billion in 2023 and that it is expected to reach US$23.37 by 2031 with a CAGR of approximately 7%. This highlights that the market is growing rapidly and may be an opportunity for food marketers to use this packaging technique in the future.

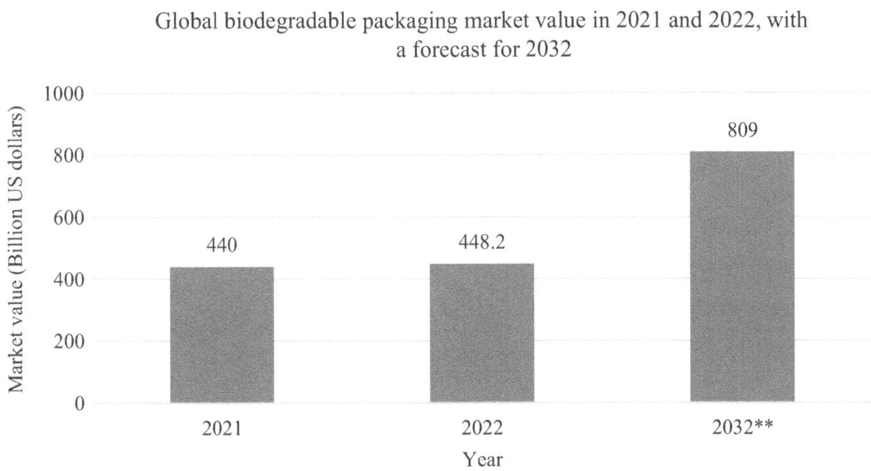

Figure 12.3 Global biodegradable packaging market (2021–2032).

Source: Statista, 2024a.

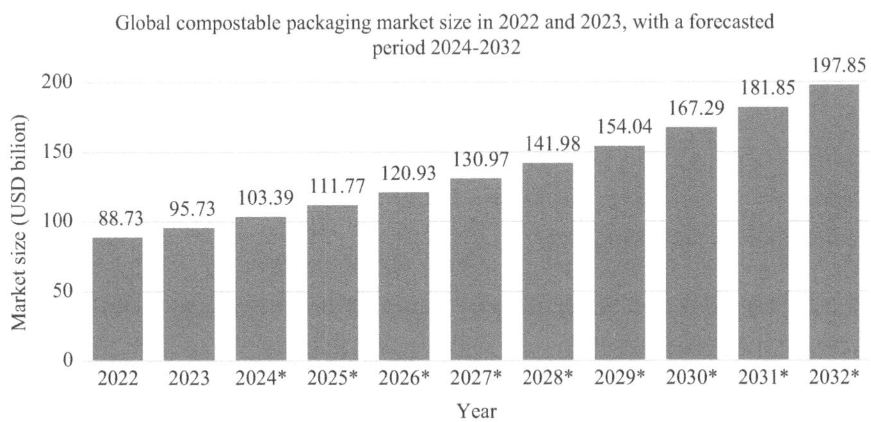

Figure 12.4 Global compostable packaging market size (2022–2032).

Source: Towards Packaging, 2024a.

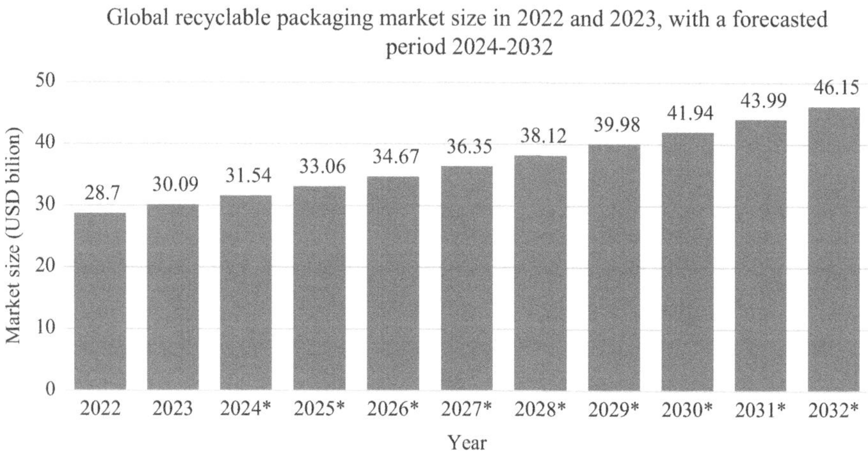

Figure 12.5 Global recyclable packaging market (2022–2032).

Source: Towards Packaging, 2024b.

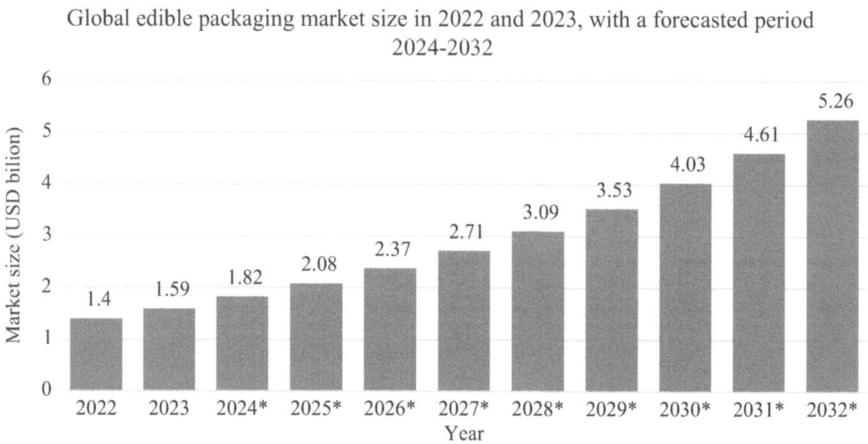

Figure 12.6 Global edible packaging market (2022–2032).

Source: Towards Packaging, 2024c.

12.3.3 The Cost and Scalability of Sustainable Packaging Solutions

One of the greatest hurdles for such companies in adopting eco-friendly packaging is the cost involved. The sustainable packaging material can be either biodegradable plastics or materials that are compostable. Such packaging materials could be more expensive to produce than conventional plastics. This places a financial burden on smaller companies, making it a very difficult task for brands working under very tight margins.

Another dimension is scalability for sustainable packaging solutions. Large corporations will have more opportunities to invest in new packaging technologies and large-scale productions than smaller companies. Additionally, many compostable and biodegradable packages require special routes for disposal – such as industrial composting facilities or other special sites that are not available within certain regions and therefore limit their practical effectiveness.

12.3.4 Consumer Behaviour and Packaging Preferences

From the above discussion it is clear that packaging is a very important factor in determining consumer opinion and consequently customer buying behaviour. Environmental impact can be a very significant reason for consumers when deciding which products or brands to buy. The latest studies have revealed that customers are willing to pay more for eco-friendly packed products, especially among the younger generation who desire to be environmentally friendly (Figures 12.7 and 12.8).

Consumer education and activation will also help the success of sustainable packaging. For instance, whereas recyclable packaging does mark a step forward in reducing waste, recyclability requires active recycling on the part of the consumer, which is not assured in most cases. Similarly, compostable packaging may only achieve its advertised success if consumers both have the right infrastructure set up to compost and also know what to do with the packaging.

It is through marketing and consumer education that consumers become engaged with sustainable packaging. Brands communicating the environmental benefits of their packaging and indicating how best to dispose of it may develop stronger relationships with the environmentally-savvy consumer. Packaging that adds functionality with eco-friendliness – whether resealable, portion-controlled, or simply innovative – will only further add to loyalty.

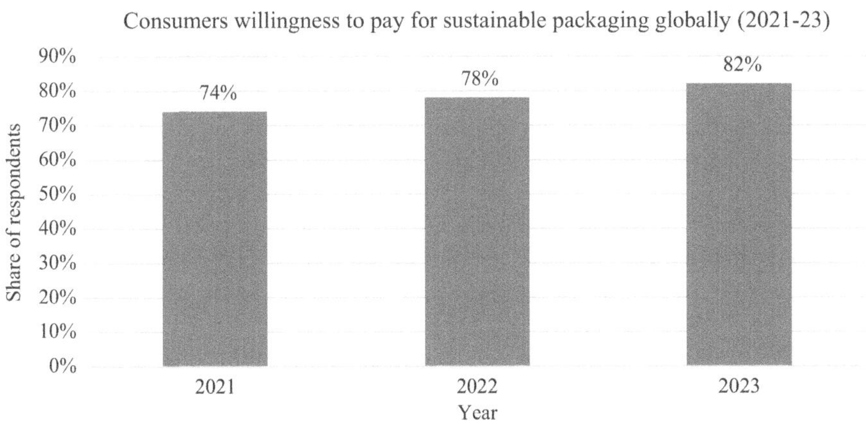

Figure 12.7 Consumers' willingness to pay for sustainable packaging.

Source: Statista, 2024c.

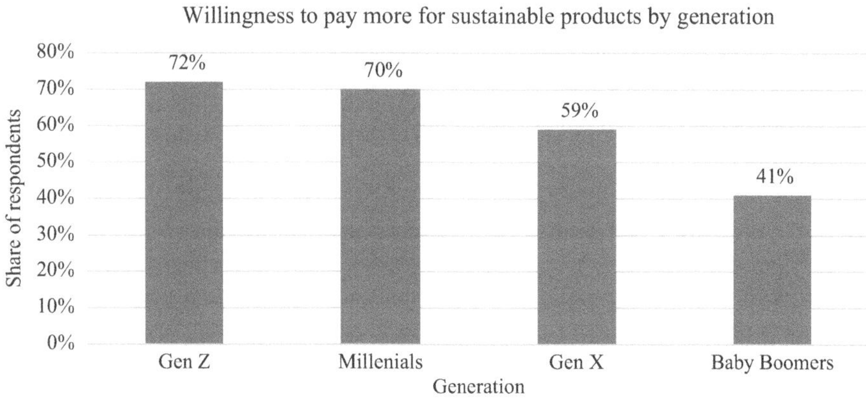

Figure 12.8 Share of consumers willing to pay more for sustainable products in the Asia-Pacific region, 2023.

Source: Statista, 2024d.

12.3.5 The Future of Eco-Friendly Packaging

Mounting environmental concerns will increase the demand for sustainable packaging. Governments around the world are tightening up their regulations regarding plastic waste, and consumers' responses have included an increased preference for environmentally responsible brands in terms of purchasing decisions. Those companies who respond with appropriate investments in eco-friendly packaging solutions will be better placed to navigate these fast-growing markets and will end up fortifying their brands' more sustainable images.

Technological advancements will be a prime catalyst in the future of green packaging. Innovations in new generation materials science, such as fully biodegradable plastics and advanced recycling technologies, will propel the next big shift in sustainable packaging solutions. Companies will increasingly focus on ways to design packaging that can be reused, repurposed, or returned to the supply chain given the growing momentum towards the development of the circular economy.

12.4 ENERGY EFFICIENCY AND REDUCING CARBON FOOTPRINT

A reduced carbon footprint for food production is now at the top of the agenda in a drive towards sustainability. The food industry has links with agriculture to manufacturing, packaging, and distribution. All these areas contribute to emissions of greenhouse gases. Companies are beginning to become sensitive to their carbon footprint as climate change becomes a widely known concern. They're making some efforts by installing energy-efficient practices, adopting renewable energy sources, and optimizing their supply chains. This section of the chapter highlights the strategies employed by food companies to achieve more energy efficiency while reducing emissions in order to secure a more sustainable future.

12.4.1 Carbon Emissions in the Food Production and Supply Chain

In other words, the carbon footprint of food products would encompass the aggregate emissions of greenhouse gases (GHG) at different steps along the lifecycle of a food product raw material extraction through farming and processing, packaging, transportation, and, finally, disposal. Agriculture is a significant source of global GHG emissions, being responsible for about 25%, primarily because of changes in land use, livestock, and fertilizer use. Manufacturing, packaging, and distribution also make significant contributions to the overall environmental footprint of a food product.

Food companies aiming to lower their carbon footprints must address multiple stages of the supply chain. By optimizing energy use, reducing waste, and transitioning to more sustainable agricultural practices, companies can significantly reduce their environmental impact while appealing to eco-conscious consumers.

12.4.2 Improving Energy Efficiency in Manufacturing

One of the most effective means to reduce carbon emissions in food companies is the enhancement of energy efficiency in the manufacturing process. This will reduce the carbon footprint as these companies reduce their consumption and use less fossil fuels due to the adoption of efficient energy technology.

Equipment and Facilities Upgrading: This can be achieved by replacing energy-intensive machinery with newer machines, and develop heating, ventilation, and air-conditioning systems that save much of the energy used in food processing and packaging. Automated energy management systems can easily enable companies to monitor and optimize energy use to determine where improvements are needed.

Improving Manufacturing Operations: Putting into place efficient ways of using energy, such as heat recovery systems, can reduce the energy wasted by a business. This can be done, for example, through utilizing waste heat generated in one section of the manufacturing operation to power other parts of the operation.

Reducing Water Usage: Water, an important input in food manufacturing, requires large amounts of energy for pumping, heating, and cleaning. All three of these processes use smaller amounts of water as well as less energy to treat and manage it more because of better technologies, fewer instances of water reuse throughout the process, and greater efficiency in processing.

12.4.3 Renewable Energy Adoption

Another possible strategy for carbon reduction in food production would be through the adoption of renewable sources of energy. With companies pioneering renewable energy through the use of solar, wind, and bioenergy, this may power most operations at a more sustainable level.

Solar and Wind Energy: The most commonly utilized renewable energy sources, both solar and wind energy are excellent applications for food processors who rely on them to conduct business. Installing solar panels on the rooftops of manufacturing facilities or wind turbines around production facilities generates clean, renewable power that displaces fossil fuels.

Bioenergy: Other companies continue to invest in bioenergy, which refers to energy produced from organic products, such as waste from crops and food streams, and animal by-products. These bioenergy systems convert the organic materials into electricity or fuels, thereby providing renewable sources of energy while lowering the costs of waste disposal.

Power Purchase Agreements (PPAs): In other cases, where companies cannot produce renewable energy at the location, they enter into power purchase agreements (PPAs) to source clean energies from renewable sources. PPAs enable companies to fund schemes of renewable energy and also reduce their reliance on nonrenewable sources of electricity.

12.4.4 Carbon-Neutral and Net-Zero Targets

Increasingly, food companies will commit themselves to ambitious, carbon-neutral net-zero goals, in which the amount of carbon emitted by operations is balanced on an equivalent basis with the amount of carbon removed from the atmosphere. In this quest for carbon neutrality, they will reduce energy, invest in renewables, and offset the rest through carbon credits.

Carbon Offsetting: If it proves impossible to completely eliminate the emission, companies can acquire offsets of carbon that finance projects that reduce or capture GHGs, such as projects in reforestation, renewable energy, or methane capture. It helps companies offset their residual emissions and hence contribute to global carbon reduction efforts.

Net-Zero Commitments: 'Net zero' means that the whole value chain of a company from production to distribution and beyond results in no net increase in greenhouse gases. Effective net zero requires considerable changes in sourcing ingredients by companies, in the way they produce and distribute goods, and in how they interact with suppliers. Leading brands have committed to timelines of reaching net-zero emissions by 2030 or 2050.

12.4.5 Optimizing Distribution and Transportation

Companies also lessen carbon emissions through the lessening of distribution and transportation activities of their food products. The transportation of food is considered to be a large emitter of GHG gases when the product travels over long distances and is transported via heavy-energy modes, including trucks, ships, and airplanes.

Length of Supply Chain: Another easy strategy for reducing transportation emissions is sourcing locally as it shortens the supply chains. Companies source ingredients from regional farmers and producers, thereby reducing their carbon footprint while helping local economies and reducing the amount of long-distance transportation.

Environment-Friendly Transport: Companies are increasingly investing in much more environmentally-friendly transport options such as electric vehicles and hybrid delivery fleets. Electric vehicles for deliveries over short distances would cut emissions significantly. This can have a significant impact on urban areas since diesel trucks generate a considerable amount of pollution.

Efficient Logistics and Route Optimization: This also reduces emissions due to improved logistics systems. Route optimization technologies rely on data analytics that deliver fuel-optimum routes for delivery, resulting in shorter distances to be travelled and lower levels of fuel consumption.

12.4.6 Consumer Response to Carbon Labelling

As a result of growing awareness and care among consumers pertaining to the environmental footprint of their food choices, more brands than ever are adopting carbon labelling that provides greater transparency to the carbon footprint of a product. Carbon labels indicate the amount of CO_2 emitted during the production and delivery process of a product, which makes for better consumer choices based on the environmental footprint.

Carbon labelling would affect and shape the consumers' actions, especially from environmentally-conscious buyers who aim to reduce their carbon footprint. However, the effectiveness of carbon labelling depends on clear communication and consumer education, as many people are still unfamiliar with how to interpret carbon data on food products.

12.4.7 The Role of Government and Industry Collaboration

Government regulations and industry partnerships make significant contributions to the creation of easier access for food companies in order to decrease their carbon footprint. Policies that encourage firms to adopt renewable energy, taxation on carbon production, and sustainability reporting frameworks all promote investment in cleaner technologies and considerably more energy-efficient practices.

Industry alliances and collaborations, such as Science-Based Targets initiative (SBTi), are facilitating companies in pursuing ambitious emission reductions in line with the goals outlined in the Paris Agreement. Through food companies, each can share best practices and drive large-scale changes towards reducing the carbon impact on their industry.

12.5 WASTE REDUCTION IN THE FOOD INDUSTRY

In recent times, the focus of food businesses seeking to adopt greener practices has been on waste reduction. The industry of food from production to consumption contributes hugely to waste, which causes several pollution issues, such as GHGs, the inefficient use of resources, and overfilled landfills. Much food waste could improve sustainability while opening up environmental and economic benefits through the introduction of circular economy principles. This section discusses the strides that companies take to prevent food waste by upcycling, optimizing their supply chain, and reducing packaging waste.

12.5.1 The Global Issue of Food Waste

Globally, nearly one-third of all the food produced is lost or wasted, amounting to a figure of approximately 1.3 billion tons of food every year. This waste occurs various stages of the food supply chain, from farm to fork. In developing countries, food loss occurs mainly at the earlier stages of production and distribution because of inadequate infrastructure and an absence of facilities for storage. By contrast, in developed countries, most post-harvest losses occur at retail and consumer levels as a result of overproduction, poor handling, and consumer behaviour.

Food waste has immense environmental implications. Food waste sent to landfills leads to the later release of methane; methane is a very potent greenhouse gas, and is suggested to be a significant contributor to climate change. Moreover, water, energy, and labour inputs into growing the food are all wasted. Therefore, food waste reduction becomes an integral part of any strategy which is concerned with minimizing environmental impacts.

12.5.2 Strategies for Reducing Waste in the Food Industry

Food companies are employing various strategies from production right up to retail, and even beyond the supply chain, to eradicate wastage. From this, companies not only cut down on their environmental footprint but also find greater efficiency and profitability in their processes.

Upcycling-Reproducing: The new methods of minimizing food waste are called upcycling-reproducing by-products or surplus ingredients to create new and valuable products. Spent grains from breweries can be processed into high-fibre flour; vegetable peels can be turned into snacks or animal feed. Upcycling not only minimizes the companies' waste output but also generates new revenue from material previously discarded as waste.

Surplus Food Donation: Many food companies donate surplus food products to food banks and charities. These programmes make sure that much edible food does not go into landfills. Through partnerships with organizations like Feeding America and Food Rescue, it has re-routed millions of pounds of surplus food towards consumers.

Efficient Supply Chain and Inventory Management: Improved control over inventory and the optimization of supply chains are essential measures for minimizing the waste of food at both the production and retail levels. Data analytics can more accurately predict demand, reduce overproduction, and ensure that perishable goods are delivered and sold before they go to waste. This will further enhance businesses' ability to track its inventory in real time, including identifying inefficiencies in the supply chain and then acting accordingly to minimize the likelihood of product spoilage.

Innovation in Packaging: Prolonging shelf lives minimizes food waste. Innovation in packaging technology is one of the key features through which food wastage is avoided. Vacuum packs, Modified Atmosphere Packing, and re-closure packs succeed in keeping the food for a longer duration due to the fact that only a few resultant products can go bad at a time. Smart packaging technologies, which are being developed today, include using sensors to send an alert when the food starts to turn.

Consumer Education and Behaviour Change: Most food waste in developed countries is attributed to improper storage, over-purchasing, or (quite often) incorrect interpretation of expiry dates. This is one of the major issues in which consumer education and behaviour come into play. Food companies are now creating educational campaigns on food waste, including tips on the effective use and storage of foods. Better labelling of expiration dates, for example, "Best before" against "Use by" dates, would also help in decreasing confusion and ensuring food is not unnecessarily discarded early.

12.5.3 The Role of Packaging in Minimizing Waste

Packaging is probably one of the most double-edged players in the food industry: it may foster and encourage waste, while also possibly contribute to saving it. Packaging waste generates problems, but packaging innovation can save food waste by offer better protection of food products.

Portion-Controlled Packaging: Portion-controlled packaging is one way in which consumer-level food waste could be reduced. These offer smaller, pre-measured amounts of food. This is a very important development for snacks, ready-to-eat meals, and beverages. By having the company package with proportionate volumes to the consumers' requirements, such an approach loses the possibility of over-production.

Resealable Packaging: The wrappings could be sealed after use; this will avoid waste since it maintains the freshness of the product over time. This resealable bag, container, or wrap also helps consumers store food safely for longer periods, thereby reducing the possibility of it being discarded.

Edible Coatings and Biodegradable Packaging: Edible coatings applied to fresh produce enhance shelf life through the formation of a protective barrier created to delay ripening. Biodegradable and compostable packaging options promote a more environmentally friendly alternative in terms of the burden created by packaging waste, with the packaging naturally breaking down when it is disposed of.

12.5.4 Government Initiatives and Regulations

Governments and regulatory agencies worldwide are paying more attention to addressing the issue of wasting food by implementing policies and initiatives that encourage businesses and consumers to take action to reduce such waste. Some countries require supermarkets and restaurants to donate excess food that would otherwise be discarded. For example, in France, a law passed in 2016 bans supermarkets from dumping unsold food, forcing them to donate it instead.

Many other countries have now legislated to set targets for the reduction of food waste and policies that are targeted at cutting waste at different points in the supply chain. The European Union's Farm to Fork Strategy, for example, sets targets on food loss and waste reduction that are aligned with broader goals of sustainability. Similarly, the US Department of Agriculture and Environmental Protection Agency has pledged to reduce food waste by 50% by 2030.

This regulation does play a role in generalizing how the expiration of the term or best before dates are labelled; it should reduce food waste related to consumer misunderstanding of terms such as "Best before" and "Use by." Moves that encourage consistency and clarity make it easier for the consumer to know what is good to eat, and when.

12.6 TRANSPARENCY AND TRACEABILITY IN THE SUPPLY CHAIN

Today, the elements of transparency and traceability are key components of sustainability that both drive consumer demand and regulatory pressure in the food industry. Consumers are demanding greater knowledge about the origin of the food they're purchasing, how it is produced, and whether that food meets not only high ethical standards but also environmental requirements. This forms the basis of trust with the consumer; more importantly, however, it becomes a critical enabler for ensuring compliance with sustainability goals, increasing efficiency, and decreasing risks of fraud or contamination in food companies. This section covers the significance of transparency and traceability, technologies that enable such practice, and the challenges with which companies grapple while implementing these practices.

12.6.1 The Demand for Transparency

Consumer expectations have really changed within the last few years. Major consumers now want to know where their food is coming from, and such trends are attributed mostly to younger, environmentally-conscious consumers who believe in a sustainable and ethical approach to their buying processes. Consumers want to know not if their products are chemical-free or sourced ethically, but rather what is the extent of their footprint on the environment. They also seek assurance that companies adhere to fair labour practices, animal welfare standards, and environmental stewardship.

Transparency in the supply chain is about providing this visibility – making clear where ingredients are sourced, how they are processed, and how the food reaches consumers. In fact, transparency in food can establish a much more positive brand reputation and loyalty, especially at a time when the unconvinced consumer is sceptical of vague or unsubstantiated sustainability claims.

12.6.2 The Importance of Traceability

Traceability refers to the tracking movement of food products from production, processing, distribution, and through the latter part of the retail. A good traceability system will also allow the firms to trace the product journey so that they can ascertain authenticity regarding origin, safety, sustainability, and ethical sourcing. Traceability becomes important for several reasons:

Food Safety: Traceability will allow companies and regulators to trace back easily in the case of contamination or food-borne illness, quickly identifying and separating the affected products, and therefore minimizing health risks and reducing the extent of recalls. This practically decreases possible damage and helps maintain consumers' confidence in the brand.

Regulatory Compliance: Many governments and international organizations have issued directives that require food businesses to put traceability systems in place to ensure the safety and quality of the end product. The EU General Food Law recommends traceability systems by food businesses "one step forward and one step back" tracing their products within the supply chain.

Sustainability Verification: Today, consumers are giving more importance to sustainability, traceability will enable companies to provide verifiable proof of their claims regarding environmental and ethical considerations. Companies that have some way of demonstrating that their products are sustainably sourced, organic, or fair trade through traceability systems have a competitive advantage in the marketplace.

12.6.3 Technologies Enabling Transparency and Traceability

Technological advancements are making it easier for food companies to achieve greater transparency and traceability. From blockchain and Internet of Things (IoT) devices to data analytics and cloud computing, these technologies are transforming the food supply chain.

Blockchain Technology: Among the most promising tools available to support enhancement in traceability for the food industry, one such promising tool seems to be blockchain. This is basically a decentralized digital ledger used in the tracking of transactions in a very secure, and therefore immutable, way. From farming operations, manufacturers' sites, and retailers' outlets all the way down to consumers' tables, every single step in food traceability can be recorded and made available to the entire supply chain of stakeholders. Every transaction, such as the harvesting of crops or food processing, is recorded in the blockchain, so nobody may get involved in tampering with the journey of the product.

Blockchain offers several benefits for traceability:

- All participants in the supply chain can access the same information, ensuring that data is consistent and transparent.
- Because blockchain is decentralized and tamper-proof, it is nearly impossible to alter or falsify data, reducing the risk of fraud.
- Blockchain streamlines the traceability process by eliminating the need for intermediaries and manual data entry, reducing delays and errors.

Internet of Things (IoT) and Sensors: The Internet of Things (IoT) means only the network of devices that collect and exchange data. In this context, however, in the food industry, products can be monitored in real time by sensors embedded in these IoT devices, which record temperatures, and humidity, among other conditions in storage and transportation; this information is then transmitted to stakeholders to ensure the proper handling of food.

For example, the temperature of perishables in transit can be monitored through IoT sensors so that they do not exceed the safety level. In an instance in which the system deems the temperature variation unsafe, suppliers or distributors are notified so that they can correct the situation even before food spills.

QR Codes and Consumer Engagement: Today, QR codes, scanned via the smartphone, are used to empower consumers to get detailed information related to the products that they buy. The scanning of a QR code printed on the packaging of the product enables consumers to get detailed information about the origin of the product, environmental impact, certifications, and other farmer or producer details about the product. This direct contact with the consumers helps in promoting greater transparency while promoting their ability to speak a narrative regarding the sustainability of the products.

Data Analytics and Cloud Computing: Data analytics and cloud-based systems improve traceability through the storage and management of large amounts of data the firms generate about the supply chain in that it is easy to trace products, monitor compliance and build improvements on elements. Sophisticated data analytical tools would then be used to analyze the data to figure out patterns, predict demand, and enhance supply chain efficiency.

12.6.4 Challenges in Implementing Transparency and Traceability

Although the advantages of transparency and traceability are easy to understand, their deployment across a global supply chain is difficult. The following challenges are among those faced by companies:

Cost and Complexity: A strong traceability system is relatively cost-intensive, as it demands investment in technology, infrastructure, and training. Smaller firms or companies operating in less-advantaged regions with respect to technological infrastructure may find the cost of blockchain, IoT, or advanced data analytics to be quite prohibitive. Moreover, traceability integration across a varied supply chain consisting of different suppliers, distributors, and retailers further complicates the process.

Data Standardization: The other challenge of traceability is data standardization at every point along the chain of supply. Most suppliers and manufacturers use different systems and formats, which creates interoperability difficulties. Since standards cannot be easily communicated when they are not shared, information cannot easily flow along the chain of supply.

Privacy and Security Concerns: This multitude of needs for transparency is, however, balanced by the concomitant privacy concerns, such as allowing sensitive third-party data or sensitive information into the open domain, which increases security risks by letting out proprietary business information or making the supply chain vulnerable to cyber-attacks.

Supplier Engagement: What is required is the involvement of suppliers that make all stakeholders along the supply chain – from small-scale farmers to large distributor – completely transparent. Suppliers will be hesitant to disclose certain types of information because they think that this would compromise their own advantage, data privacy, and the cost of technology introduction. Thus, any traceability system requires building trust and rewarding supplier participation so that the initiative can be achieved successfully.

12.7 SUSTAINABILITY MARKETING AND CONSUMER BEHAVIOUR

Sustainability is currently the buzzword of the food industry. It has become one of the driving forces behind consumer purchases. The eco-friendly consumer looks for such products and brands they can associate their sustainability values with. Thus, more than ever, companies are focusing on sustainability marketing – that is, using it as a tool in order to build trust and generate loyalty for their brand as well as differentiate themselves in a really competitive marketplace. This section examines how sustainability affects consumer behaviour, how brands market their sustainability efforts, and the also the risks associated with misleading claims, or "greenwashing."

12.7.1 Consumer Attitudes towards Sustainability

Consumer attitudes towards sustainability have changed dramatically over the past decade. Increasingly in reports and studies, in market research, there is evidence that a greater proportion of consumers, primarily millennials and Gen Z, now base their buying decisions on considerations connected with environmental and ethical impact. Among the factors determining many decisions between two foods are the environmental impacts of the production, packaging, and transport of the respective goods.

Several factors drive consumer interest in sustainable products:

Environmental Issue: Due to an increasing awareness of environmental concerns such as climate change, deforestation, water shortages, and plastic pollution, consumers look for better ecologically-friendly products. Now consumers are interested in owning products coming from low carbon-emitting companies and brands which adopt the responsible use of sustainable packaging and source material.

Ethical and Social Responsibility: Many consumers are now driven by the social and ethical issues related to the food supply chain: fair labour practices, the welfare of animals, and the treatment of smallholder farmers. This includes more attractive products certified with labels such as fair trade, organic or cruelty-free representing an ethical-oriented consumer with 'value' in sourcing.

Health and Wellness: There is another very close association between health and wellness related to sustainability on both environmental and social grounds. The consumers also placed the sustainable products higher in terms of quality and nutritional value. Foods made with a lesser number of chemicals, pesticides, and artificial additives appear healthier in the eyes of consumers.

Transparency and Trust: Today's consumer is looking for total transparency with regard to a brand's efforts towards sustainability. There is a higher probability that they will support companies that provide clear, verifiable information about the origin, production, and packaging processes of their products. Brands that are transparent about their struggles and achievements in terms of overcoming their challenges with regard to sustainably achieving their goals are viewed as more trustworthy and honest.

12.7.2 Sustainability as a Marketing Tool

As consumer interest in sustainability increases, more food brands are using their environmental and social initiatives as key marketing strategies. Sustainability marketing involves highlighting a company's efforts to reduce its environmental footprint, support ethical practices, and contribute to social causes. Good sustainability marketing not only improves a brand's image but also means that a deeper connection develops with consumers who share the same values.

Among the ways in which companies use sustainability in their marketing efforts are the following:

Labels and Certifications: Probably the most obvious way in which brands use to signal their sustainability efforts is through labels and certifications. Certifications such as Fair Trade, USDA Organic, Rainforest Alliance, and Carbon Neutral inform consumers know that, at the very minimum, their products meet standard environmental or ethical requirements. Labels also enable consumers to choose sustainable items off the shelf, which impacts how they ultimately make their purchasing decisions.

Eco-Friendly Packaging: One of the most visual criteria for a product is packaging; more brands are now packaging using eco-friendly packaging as one major selling point. Packaging that is recyclable, biodegradable, or renewable would significantly contribute to consumers' environmental sensitivities while reducing the brand's environmental footprint. Brands really take these features and display them boldly on packaging and all marketing materials (Figure 12.9).

Sustainability Stories and Campaigns: Storytelling is a mighty tool of sustainability marketing. Most brands use sustainability to craft interesting stories about their products. For example, a company might tell the story behind the smallholder farmers who grow a particular ingredient for its product or describe the regenerative farming they are using to restore health to the soil. Stories play a very important role in "humanizing" the brand and emotionally connecting with consumers.

Cause-Related Marketing: Some brands extend beyond self-promotion to complete their sustainability initiatives by undertaking cause-related marketing. This is manifested by being in partnership with non-governmental organizations (NGOs), donating portions of the profit to environmental or social causes, or launching campaigns to create awareness about sustainability issues. For instance, a brand can promote its efforts at reducing ocean plastics by advocating for beach clean-up operations or by manufacturing packaging using recycled ocean plastics.

Digital Marketing and Social Media: Digital marketing and social media have become equally important channels to share a brand's sustainability undertakings with its stakeholders. Publicizing key behind-the-scenes moments, including sustainability updates and real-time data on environmental impact, they share all this through Instagram, Facebook, YouTube, and many other platforms. Social media also offers an opportunity for brands to engage directly with consumers, respond to questions regarding sustainability, and address their concerns in a transparent, interactive way.

12.7.3 Consumer Response to Sustainability Claims

Consumers tend to want to buy from such sustainable brands, but their purchase behaviour can depend on many other variables too. The fact is that findings indicate consumers are quite willing to pay a premium for sustainable products, most often when they feel that the product delivers actual environmental or social benefits.

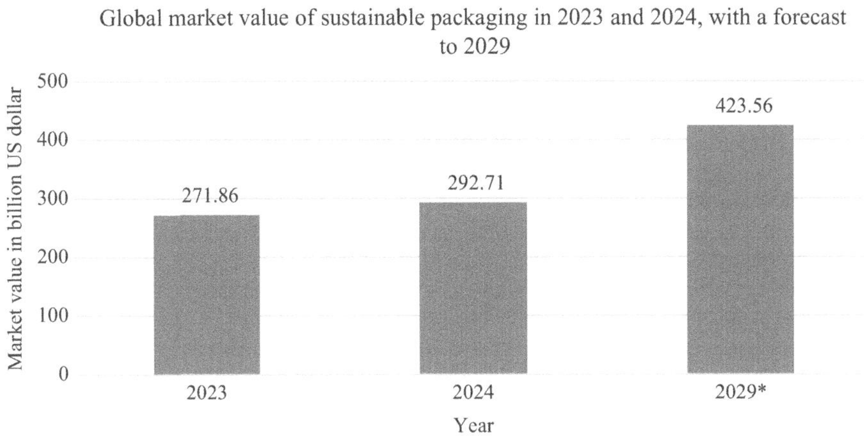

Figure 12.9 Global sustainable packaging market value (2023–2029).

Source: Statista, 2024e.

However, the effectiveness of sustainability marketing depends upon the following key elements:

Clarity: Sustainability-related communications should be clear and easy to understand. If a claim about sustainability is either unclear or too technical, the consumer will not know what that really means, reducing its impact on the purchase decision. Moreover, terms such as "eco-friendly" or "natural" can be misleading unless accompanied by detailed information or certifications.

Authenticity: Authenticity is a vital element in sustainability marketing. Consumers can easily detect an insincere or over-exaggerated claim, and, in such a case, the brands may lose their credibility. Authentic brands are those that communicate transparently about their sustainability journey, including challenges as well as progress over time. Consumers can easily believe in brands that authentically convey their sustainability practices.

Price vs. Value: Although consumers avow sustainability as a preference, price remains one of the most important variables. The greater the perception by the consumer that the premium for this extra value associated with sustainability is outweighed by what they are getting back, the higher the likelihood of selecting sustainable products. Only when the brand can effectively convey this value of sustainability will the consumer translate interest into purchasing power.

12.7.4 The Role of Certifications and Eco-Labels

As a result of all of this, certifications and eco-labels are crucial for building consumer confidence in developing knowledge that the claims made in the name of sustainability are real. Independent organizations verify that a product satisfies specific standards in any one of the environmental, social, or ethical areas. Seeing a certification on a product from a trusted source reassures a customer that the brand indeed owns its sustainability claims.

Among the most recognized certifications are:

Fair Trade: Ensures that producers in developing countries are paid fair wages and work under safe conditions, while also promoting sustainable farming practices.

USDA Organic: Certifies that products are grown and processed without synthetic fertilizers, pesticides, or GMOs and that farming practices promote biodiversity and ecological balance.

Rainforest Alliance: Focuses on protecting biodiversity and ensuring sustainable land use, while also supporting the livelihoods of farmers and workers.

Marine Stewardship Council (MSC): Certifies seafood products that come from sustainable fisheries, ensuring that fish stocks are not depleted and that marine ecosystems are protected.

These certifications not only help consumers make informed choices but also allow brands to differentiate their products in the marketplace.

12.8 BARRIERS TO IMPLEMENTING SUSTAINABLE PRACTICES

While the food industry is making strides towards more sustainability-in-life practices, many companies still face significant barriers to fully integrating the tenets of sustainability into their operations. Such barriers include the financial cost of transitioning to sustainable systems, managing complexity at the global level in supply chains, regulatory obstacles, and sceptical consumers. A proper understanding and negotiation of these barriers will be essential for companies seeking to adopt practices of sustainability while remaining profitable and competitive. This section explores the key challenges that food companies encounter and offers insights into how these obstacles can be overcome.

12.8.1 The Financial Costs of Adopting Sustainable Practices

One of the primary impediments to enhancing sustainability in the food industry is the initial cost of the necessary modifications. Making a shift towards sustainable sourcing, eco-friendly packaging, or renewable energy systems often involves significant capital investment, which is particularly taxing for small and medium-sized enterprises. For example, switching to organic farming or using sustainable palm oil will increase raw material costs; similarly, making a shift towards renewable energy or a change to manufacturing equipment involves long-term financial commitment.

For many companies, therefore, these costs must be weighed against short-term profitability. This can make it difficult for even more substantial organizations to sustain the initial outlay of sustainable investments; smaller companies with tighter margins will find it even harder to justify the cost, especially when consumer demand can be lumpy or sensitive to price. To overcome the financial burden, companies can explore partnerships, or subsidies and grants available for sustainable initiatives. An incentive could be in the form of tax credits to encourage investment in renewable energy or subsidies for sustainable farming. Companies can also introduce the changes gradually, beginning with minor and less expensive initiatives before moving towards broader sustainability goals.

12.8.2 Supply Chain Challenges in Sourcing Sustainable Materials

Yet another significant barrier to sustainability is the complexity of supply chains, especially those companies sourcing from around the world. Many food companies source ingredients and materials from all over the globe, with different regulatory environments, agricultural practices, and the availability of sustainable resources in each location. Consequently, full traceability and knowledge that all suppliers adhere to sustainability standards can prove extremely challenging, particularly if the small-scale farmers or suppliers are based in developing countries.

Sourcing responsibly often presents logistical difficulties. For example, companies committed to sustainable seafood are faced with scarcity due to overfishing and/or regulation, and organizations looking to acquire organic or fair-trade-certified products are often forced to work with new suppliers, thereby incurring higher costs. Furthermore, transportation and storage logistics can exacerbate the difficulties of sourcing perishable products. Companies can also collaborate with their suppliers, adopting a more cooperative approach, which will strengthen such relationships and ascertain that sustainability goals are met through the supply chain. Moreover, training and resources could be offered to suppliers in developing countries, by encouraging a long-term partnership. Sometimes companies have to manage supply lines by reducing the number of suppliers but more reliable and sustainable ones when they have to stay by the sustainability goals.

12.8.3 Navigating Regulatory Requirements

With food companies operating in multiple regions, the array of sustainability-related regulations and standards varies significantly from country to country and even market to market. While the European Union demands tough regulations on certifications like organic, packaging waste and carbon emissions, the regulations may be relatively lenient for other markets. Ensuring compliance with a different set of regulatory requirements in different markets can be cost-efficient and time-consuming.

Moreover, companies face pressure to comply with evolving regulations around food safety, labour practices, and environmental impact. These regulatory frameworks are continually being updated as governments introduce stricter requirements to address climate change and other sustainability challenges. Staying ahead of these regulatory changes requires constant monitoring, the adjustment of operational practices, and significant documentation to ensure compliance. Companies can build global sustainability strategies compatible with the most stringent regulatory requirements, ensuring compliance or exceeding levels across their entire sphere of markets. That would make it simpler to navigate all the different regulations and potentially position companies as leaders in sustainability. It engages the companies with associations and experts related to the industry to stay abreast of the latest ideas regarding regulatory directions, thereby enabling proactive compliance for future requirements.

12.8.4 Balancing Profitability with Sustainability Goals

Yet, while sustainability is growing in importance for consumers and corporations alike, it can be challenging to balance sustainable goals with profitability. Sustainable practices often involve costly initial investments, without immediate direct financial payback. For example, shifting to renewable sources of energy or energy-efficient machinery can save money in the long term, but the cost of entry may be straining financial reserves. Similarly, sourcing certified sustainable ingredients may cost more than traditional alternatives, keeping production expenses higher.

Another challenge is that while some consumers are willing to pay a premium for sustainable products, not all markets or demographics are equally committed to sustainability. In some cases, companies may struggle to recoup their sustainability investments if they are unable to pass these costs on to consumers through higher prices. To balance profitability with sustainability, companies can focus on initiatives that provide both environmental and financial benefits, such as energy efficiency measures that lower operating costs. Additionally, brands can communicate the value of their sustainability efforts to consumers, highlighting the long-term benefits of supporting ethical and environmentally responsible products. Some companies might even pursue tiered pricing structures that provide some premium sustainable products and others to make them more accessible to price-sensitive consumers.

12.8.5 Consumer Scepticism about Sustainability Claims

On the one hand, many customers express a desire for sustainability; on the other hand, companies' sustainability claims are visibly met with scepticism. Much of this might stem from the greenwashing anxieties, in cases when the company falsely or overstated the environmental practices. If sustainability claims are perceived as superficial or unfounded by consumers, then this can deteriorate brand reputation and even relationships.

These consumers may not even be sure how to audit sustainability claims when facing a multiplicity of eco-labels and certifications, many of which will likely be unknown or confusing. Consumers might find themselves in a situation where they are unsure whether the brand is really committed to sustainability or if it is merely greenwashing. To overcome consumer scepticism, a company has to provide clear and transparent reporting or third-party certification of its sustainability claims. In addition, straightforward sustainability goals challenges and even accomplishments can pave the way to trusting consumers. Brands are able to form more authentic relationships with their customers by acknowledging that they still continue to improve sustainably rather than claiming to be perfectly sustainable.

12.8.6 Overcoming Consumer Reluctance to Change

While many consumers agree with the theoretical concepts of sustainability, they do not always translate this agreement into action when making purchases. This is for a variety of reasons, including the belief that sustainable products are too costly and that individual actions will not actually make a difference or that customers like the familiar, less sustainable products.

The biggest challenge of the food industry with this pursuit for more sustainable products and consumption practices is changing consumer behaviour. As more sustainable options are becoming increasingly accessible, consumers now mostly aim for convenience, price, and brand loyalty. Companies can begin their endeavours in changing consumer reluctance to a more sustainable choice by first making those sustainable products accessible and affordable. Marketing strategies that empower the personal and social effects of the actions of sustainability also foster behaviour change. Incentives, discounts, reward programmes, or special offers on sustainable products are considered incentives that will encourage consumers to adopt more sustainable options.

12.9 CHAPTER SUMMARY

Sustainable food marketing practices are no longer optional but a need for long-term sustainability and success for companies in today's more environmentally-conscious world. As consumers, regulators, and stakeholders themselves continue to demand greater transparency, responsibility, and accountability from the food industry, it has given brands all the more reason to view sustainability as integral to every element of their business strategy. This ranges from sourcing sustainable ingredients and using bio-friendly packaging all the way through to reducing waste, improving energy efficiency, revealing transparency in supply chains, or, more generally, having enough steps towards reducing one's environmental impacts and promoting social responsibility.

While the path to sustainability is fraught with financial expense, complexity in supply chains, and pressure of regulations, there are clear advantages when companies invest in sustainability. Companies most capable of establishing trust with their consumers, enhancing brand loyalty, and meeting the shifting expectations of a market increasingly attuned to ethics and stewardship of the environment are those that commit to sustainability. Other than that, sustainability can improve the performance of businesses, increase operational efficiency, reduce costs over time, and access newly created market opportunities.

True sustainability involves much more than just splashy gestures or advertising jingles; actual change, transparency, credible third-party certifications, and continuous improvement are involved. The companies are supposed to do this while being authentic in their communications, showing what they could really change and not risk greenwashing. It is through honest communication with the consumer, choosing responsible suppliers, and making sustainability as integral to business strategy as profits, that food brands can genuinely impact achieving a more sustainable future.

The last word then is that sustainability in food marketing can be said to be about not meeting the demands of today's consumers but securing the health of our planet and society for future generations. Companies which do not prioritize sustainability will not only gain a competitive edge in the marketplace but also contribute to a more equitable, resilient, and environmentally sustainable food system.

CASE STUDY

Starbucks and the Push for Waste Reduction

The world's largest coffeehouse chain, Starbucks, has long been commended for its sustainability initiatives, especially focusing on waste reduction. With millions of beverages served across its global network on a daily basis, the company has tremendous problems in disposing of the waste generated from single-use cups, straws, lids, and other types of packaging. Over the years, Starbucks has committed to various environmental goals, including reducing the environmental impact of its packaging and promoting the use of reusable alternatives. However, balancing convenience with sustainability has proven difficult, especially when it comes to changing consumer behaviour.

The Plastic Straw Ban and Shift to Compostable Alternatives

Then in 2018, Starbucks made a very public commitment to eliminate plastic straws from all its stores in the world by 2020, in response to growing concerns over plastic pollution in oceans. The company set about introducing recyclable and compostable straw alternatives, while it developed a new recyclable strawless lid for cold beverages, which was again intended to reduce overall plastic waste. While these efforts were praised by environmental groups, the new lids actually used more plastic than the previous versions, drawing criticism for not being a true step forward in waste reduction.

In response to this, Starbucks started experimenting with various compostable and recyclable materials for the company's cups and lids. It also invested in research partnerships with leading environmental organizations to develop innovative, sustainable packaging materials that would break down naturally without harming the environment. This still isn't an easy feat: making sure that consumers dispose of such materials properly, especially in regions with meagre recycling and composting infrastructure.

Reusable Cups: Incentives and Challenges

One of Starbucks' earliest and most well-known sustainability efforts is its reusable cup programme. On that occasion, if customers bring in their own cups to get a beverage, Starbucks gives them a discount on it. This was an effort to reduce the amount of single-use cups left in landfills. In fact, if just 25% were to make use of this scheme, Starbucks estimated it could save billions of single-use cups every year. Consumer participation has been volatile, however. Although the green-conscious consumers were enthusiastic about the move, most consumers chose the easy way of using single-use cups. Inconvenience, forgetfulness, or health and hygiene issues, especially during the COVID-19 pandemic, have worked to limit the success of reusable cup uptake. Even when the basic incentive of money gets thrown into the pot, it would appear that changing consumer behaviour is the biggest challenge in taking reusables to scale.

Starbucks has developed other practices to promote reusable cups. In some markets, Starbucks implemented a borrow-a-cup programme, where the company offered an individual a reusable cup for a small, refundable deposit. The cups were then professionally cleaned and sanitized for the next use. Thus, while the pilot programme had much promise, it did raise logistical challenges in terms of handling the inventory and proper sanitation of the cups.

Partnership with Loop for Reusable Packaging

Starbucks has partnered with another circular economy platform Loop, which provides reusable packaging options. Under this joint collaboration, Starbucks customers were able to buy their drinks from certain stores using cups and containers that could be returned back to the store, washed, and reused. This is part of the much-broader commitment Starbucks has to introducing a circular economy model while minimizing waste and ensuring continually reused materials instead of being thrown away.

These notwithstanding, spreading such programmes to thousands of Starbucks stores globally is an extremely challenging task. Every market has different regulations, infrastructure, and consumer preferences, which complicates the rollout of a universal solution. Logistical difficulties in getting reusable containers collected, cleaned, and redistributed also add complexity to scaling the programme on a global level.

Food Waste Reduction Efforts

In addition to focusing on its packaging, Starbucks has also looked to minimize food waste in its stores. Its FoodShare programme ensures unsold food in its stores to be donated to local food banks and charities. Launched in 2016, the programme aims to divert unsold but still consumable food from landfills, addressing issues of both food waste and food insecurity. By 2020, the programme had donated millions of meals in the United States and Canada, and it was already prepared to roll into additional markets.

However, reducing food waste is certainly no easy task. Logistics pose major issues of how to get donations across different regional lines, keep perishables safe, and find those persons with whom they can consistently share their donations.

Nevertheless, Starbucks has been making tremendous strides. However, these efforts also contribute to the complexities of managing the scale of its operations, which can be a source of headaches for the company.

Balancing Convenience with Sustainability

One of Starbucks' biggest challenges is balancing convenience for its customers with its goals of sustainability. The company faces the dilemma of having customers who expect to have their coffee readily available in single-use packaging – a convenience for on-the-go customers – while at the same time promoting reusables and zero-waste stores. Despite the efforts of Starbucks to market its sustainability initiatives, it is difficult to change consumer behaviour on a large scale.

Additionally, even though Starbucks offers recyclable and compostable materials, most people do not know how to dispose of them properly, especially in areas where recycling infrastructure is limited. Recycling has different programmes in various cities and countries. Therefore, Starbucks still has difficulties in meeting its waste-reduction targets.

Starbucks' efforts towards zero waste focus on the ability and also the hurdles of taking sustainable practices to the food and drink industry. While Starbucks has achieved much through initiatives such as the elimination of the plastic straw, the promotion of reusable cups, and its collaboration with organizations like Loop, the most challenging problem it has faced is encouraging large-scale behaviour changes on the part of its customers. Besides this, logistical challenges associated with the globalization of these initiatives, including variations in recycling infrastructure and differences in market regulations, do not allow for easy universal success. As Starbucks continues to strive for innovative solutions and expands its sustainability efforts, it will inherently face the continued challenge of balancing consumer expectations for convenience with its commitment to doing the right thing for the environment. But the future of moving forward from this success depends on developing alternatives that are sustainable, and engaging the consumers meaningfully enough to bring about behavioural change and to ensure sustainability at the core of this brand identity.

Discussion Questions

1. What are the main challenges Starbucks faces in encouraging consumers to adopt reusable cups, despite offering financial incentives? How could the company improve the success of this initiative?
2. How effective do you think Starbucks' partnership with Loop is in reducing packaging waste? What are the potential barriers to scaling this programme globally, and how could Starbucks overcome them?
3. While Starbucks has introduced recyclable and compostable alternatives to plastic, the company has faced criticism regarding disposal challenges. How can Starbucks better educate consumers on proper waste disposal, especially in regions with limited recycling infrastructure?

4. In the case of Starbucks' food waste reduction efforts through the FoodShare programme, what logistical challenges do you think the company faces, and what strategies could be implemented to improve the programme's efficiency and effectiveness?
5. What role do you think consumer convenience plays in the success of Starbucks' sustainability initiatives, such as reducing single-use cups and straws? How can Starbucks balance customer expectations for convenience with its sustainability goals?

BIBLIOGRAPHY

SNS Insider. (2024). *Minimalist Packaging Market Size, Share & Growth Report 2031*. www.snsinsider.com. Retrieved October 6, 2024, from https://www.snsinsider.com/reports/minimalist-packaging-market-4227

Statista. (2024a, June 28). *Global biodegradable packaging market value 2021–2032*. https://www-statista-com.eu1.proxy.openathens.net/statistics/979152/global-biodegradable-packaging-market-value/

Statista. (2024b, June 24). *Consumers' perception sustainability of packaging types worldwide 2023*. https://www-statista-com.eu1.proxy.openathens.net/statistics/1468506/consumers-perception-of-sustainable-packaging/

Statista. (2024c, June 24). *Consumers' willingness to pay more for sustainable packaging worldwide 2021–2023*. https://www-statista-com.eu1.proxy.openathens.net/statistics/1467292/willingness-to-pay-for-sustainable-packaging-worldwide/

Statista. (2024d, September 18). *Willingness to pay more for sustainable products APAC 2023, by generation*. https://www-statista-com.eu1.proxy.openathens.net/statistics/1477446/apac-willingness-to-pay-more-for-sustainable-products-by-generation/

Statista. (2024e, July 19). *Global Market Value of Sustainable Packaging 2023–2029*. Retrieved October 6, 2024, from https://www.statista.com/statistics/1419940/global-sustainable-packaging-market-size/

Statista. (2024f, June 24). *Global concern about plastic waste from food packaging 2023, by generation*. https://www-statista-com.eu1.proxy.openathens.net/statistics/1472954/generational-concern-about-single-use-plastic-globally/

Towards Packaging. (2024a, June 3). *Compostable packaging market size, share, companies*. Retrieved October 6, 2024, from https://www.towardspackaging.com/insights/compostable-packaging-market-sizing

Towards Packaging. (2024b, June 3). *Recyclable packaging market Size*. www.towardspackaging.com. Retrieved October 6, 2024, from https://www.towardspackaging.com/insights/recyclable-packaging-market

Towards Packaging. (2024c, June 3). *Edible Packaging market Size*. www.towardspackaging.com. Retrieved October 6, 2024, from https://www.towardspackaging.com/insights/edible-packaging-market

CHAPTER 13

Technology in Food Production and Processing

LEARNING OBJECTIVES

By the end this chapter, you should be able to:

- Identify key technologies, such as automation, artificial intelligence (AI), nanotechnology, and biotechnology, and explain how they are applied to enhance food production and processing.
- Evaluate the advantages and challenges associated with automation.
- Understand how AI and data analytics improve decision-making, supply chain management, food safety, and predictive maintenance in food production.
- Understand the applications of biotechnology in food production:
- Understand how nanotechnology is applied to improve food safety, packaging, and shelf life through innovations like nanosensors and antimicrobial packaging.
- Identify and assess sustainable food processing methods.
- Discuss the role of sustainable practices in reducing food waste, conserving resources, and improving environmental impact.

13.1 INTRODUCTION

The food production and processing domain has greatly transformed due to technological changes that have affected every other aspect of modern life. Farm production and food processing, packaging, and delivery at the consumers' table have been streamlined by the developments into an efficient, safe, and responsive sector catering to the ever-growing populace through a global food industry. This pressure to produce more food sustainably and safely has possibly never been as high, given that the world population is supposed to reach nearly 10 billion by 2050. Through innovation under a scarcity of resources, food safety, nutritional enhancement, and environmental impact, the emphasis has fallen on technology at its core.

DOI: 10.1201/9781003473305-13

Initially, food was mainly handmade and dependent on the natural environment, labour-intensive without advanced machinery and processing. All this has changed with advances in agricultural techniques, machinery, and processing methods that significantly altered the landscape of food production while offering better yields, improved food safety, and reduced wastage. From hybridization to automation, and even artificial intelligence (AI), nanotechnology, and sustainable energy, countries are now modernizing the process of every step involved in the food supply chain.

This chapter reviews several technologies that could be developed for food production and processing. Technology has grown in importance for efficiency, reduced environmental impacts, and food safety through precision farming, smart agricultural tools, and the use of automation and robotics in food manufacturing. Furthermore, advances in biotechnology and artificial intelligence (AI) are not only optimizing production processes but also creating new possibilities for enhancing nutrition and food quality.

The next section narrates historical developments in the food industry, considering how they play those roles in taming modern day-to-day food challenges, changing the future of food production, and impacting emerging technologies will have on that future. In short, technological developments in the food industry show an ability to produce food using fewer resources, to ensure the population's nutritional needs are met while causing minimal degradation to the environment.

13.2 AUTOMATION AND ROBOTICS IN FOOD MANUFACTURING

Automation and robotics are transforming food manufacturing, playing an increasingly in contemporary food production processes, from development to packaging and, finally, distribution. Years ago, food was produced with extensive manual labour, which reduced scale, speed, and consistency. The developed automation technologies have overhauled the industry to achieve efficiency, improve product consistency and safety, and also reduce costs. Using automation, organizations enjoy several benefits, which are especially noticeable in repetitive operations like sorting, packing, even cooking, and controlling quality, through which companies can produce much greater quantities of food faster and with fewer errors.

13.2.1 The Rise of Automation in Food Production

Automation in food manufacturing involves the use of advanced machinery and control systems to carry out tasks that were traditionally performed by human workers. Automation ranges from automatic packaging lines to the use of robotic arms to handle fragile products. It allows standardization for production procedures, minimizes input errors, and cuts labour costs. Precision, speed, and hygiene are major concerns in high-volume productions; this type of technology is optimized.

For instance, packaged snack or soft drink manufacturers require automated filling systems, filling each unit up to the highest quality specifications for sealing and labelling. Automation makes these processes more efficient and waste-reducing products and avoids product defects and material usage.

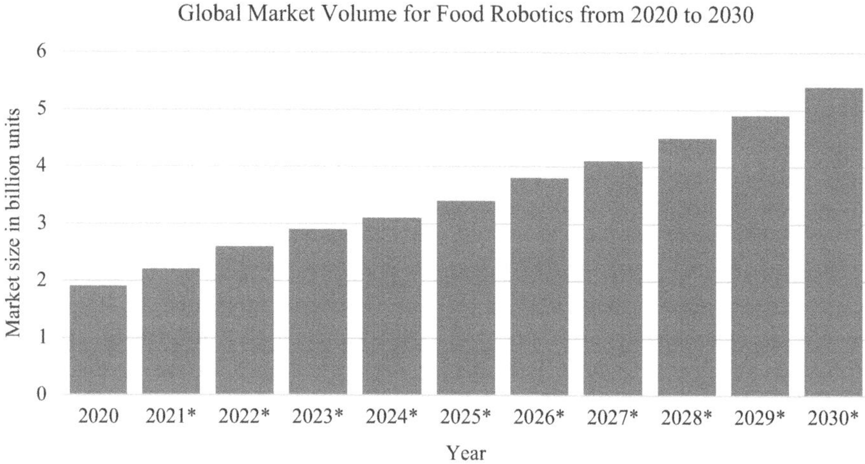

Figure 13.1 Global market unit volume for food robotics (2020–2030).

Source: Statista, 2022.

13.2.2 Robotics in Food Processing

Robotics is the key piece of equipment in food manufacturing automation because it means that most tasks can be performed faster and more accurately. The food robotics market was 1.9 billion units in 2020, but is expected to grow to 5.4 billion units in 2030 (Figure 13.1). This linear growth indicates that the demand for robotics increases on a daily basis and that businesses are taking advantage of this to gain competitive advantages in the market. The robotic systems chop vegetables and meat and sort and pack relatively fragile goods such as fruit, pastries or eggs. In addition, automated systems can imitate human movements, thereby functioning with complete accuracy while at the same time having minimal risk of contamination. For example, the packaging of foods uses robotic arms. These machines are equipped with sensors that may detect and pick a product of any shape or size without breaking or damaging it. For instance, in the baking industry, robots are presently programmed to decorate cakes with intricate designs or pack fragile products like cookies and muffins into trays without breaking them.

Apart from packaging, robotics is also applied to food preparation. Some fast-food chains and restaurants have started to employ robots to flip burgers, brew coffee, or assemble sandwiches. The machines can work at any shift, enabling quality consistency for the customer, with human resources freed to engage in more sophisticated work.

13.2.3 Benefits of Automation and Robotics in Food Manufacturing

The adoption of automation and robotics in food manufacturing brings numerous benefits to both businesses and consumers:

Increased Efficiency: Automation allows for faster production cycles and higher output, enabling companies to meet growing consumer demand. Robotic systems can work 24/7 without breaks, leading to continuous production and reduced downtime.

Improved Accuracy and Consistency: Human error plays a critical role in varying product quality; the automated system, by contrast, assures uniformity in food production. It is an important aspect in these industries that places a great deal of emphasis on precise measurements or specific cooking times.

Improved Food Safety: Robots minimize human touch with food, thus minimizing contamination risks. It can happen in the smallest possible area with critical hygiene controls that range from temperature-controlled rooms to sterile processing environments.

Cost Reduction: The high initial investment in automation technology translates into long-term savings. There also comes a reduction in the cost of labour, material waste, and product defects contributing to cost efficiency.

Worker Safety: Robotized hazards or extremely onerous operations increase levels of safety for the workers. Workers will never be exposed to the labour of repetition or dangerous activities such as working with heavy machinery or operating hot ovens, both of which can lead to accidents.

13.2.4 Challenges of Implementing Automation

Despite the many advantages, implementing automation and robotics in food manufacturing presents some challenges:

High Initial Costs: The cost of purchasing and installing automatic systems is very high. It is unaffordable to small and medium-sized enterprises.

Skill Gaps in the Workforce: As routine manpower jobs become automated, companies need technical specialists to repair those systems. Businesses can train existing employees or hire professionals to manage their automated systems.

Customization and Flexibility: The flexibility in customization is reduced because automation might work very well for high volumes but cannot match the need for hands-on work in smaller batches or products where there are more frequent changes.

Regulatory Compliance: The automation must be carried out in such a way as to comply with food safety regulations and quality standards. Companies should ensure that their automated systems are well-calibrated and maintained.

13.3 SMART FARMING AND PRECISION AGRICULTURE

Smart farming relies on current technologies to optimize crop and animal production. The application of smart farming practices has increased in recent years. A Statista report highlights that smart farming practices will significantly enhance different fields of agriculture (Figure 13.2). Instead of the uniform treatment of fields, smart farming offers tailor-made inputs by classifying interventions according to the needs of individual plants or sections. Companies can use real-time data to make sharper decisions regarding when to water, fertilize, or harvest crops with increased efficiency and sustainability.

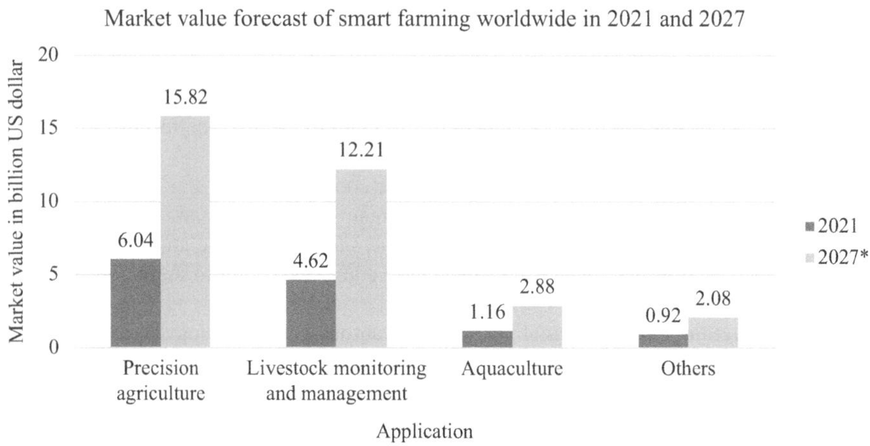

Figure 13.2 Smart farming market size worldwide, 2021 & 2027.

Source: Statista, 2023.

These techniques optimize resource use, particularly with regard to water, fertilizers, and pesticides. It also reduces ecological footprint, and increases crop yields. It will be even more relevant on a larger scale as the global challenges include climate change, water scarcity, feeding an ever-growing population, etc.

13.3.1 Key Technologies in Smart Farming

Several technologies are at the core of smart farming, each contributing to different aspects of farm management:

GPS-Guided Machinery: Tractors containing other machinery fitted with the crucial technology of GPS were among the first innovations in precision agriculture, allowing planting, fertilizing, and harvesting to take place more precisely. This technology ensures that farmers can work with high accuracy, reducing overlaps, gaps, and resource wastage. For example, GPS-guided tractors can follow predetermined routes with centimeter-level precision, ensuring that seeds are planted in optimal locations without redundancy.

Drones and Remote Sensing: Using camera-carrying and sensor-carrying drones, farmers can carry out the overhead monitoring of crop conditions and field status. From these aerial views and capturing data points on moisture levels in the soil, fields that are nutrient-deficient and stress on the plants are detected, through which the farmer can act in time to change farm activities. Similarly, by identifying areas or regions that face pest attacks, remote sensing can manage disease susceptibility or water scarcity.

Soil and Climate Sensors: These place sensors across fields to know what is happening in the field in real time to acquire data regarding moisture levels, temperature, and nutrient content. Based on this information, farmers can alter irrigation and fertilizer applications so that crops get adequate water and nutrients at the right time. The soil sensors help to enhance water use efficiency in water-scarce regions.

Variable Rate Technology (VRT): VRT allows variable applications of inputs like fertilizers, pesticides, and water depending upon varied requirements of different areas in a field. VRT systems can change resource allocations dynamically using sensor data assistance with soil samples and satellite imagery, reducing waste and providing healthier crops. The over-application of chemicals can be fatal for the environment and is also too expensive.

Internet of Things (IoT): The Internet of Things (IoT) can connect devices such as sensors and drones, including machines, to a central system through which one can collect data on farm management operations from any location. In addition, the IoT system can provide real-time information about crop conditions and weather to improve the equipment's efficiency, thereby helping farmers to respond to any change and optimize farm management. For instance, IoT-based irrigation may detect moisture levels in the soil and adjust the water flow so that excess water does not occur; crops are to be irrigated with the right amount of water.

Big Data and Analytics: Data is at the heart of smart farming. Crop performance, weather, and resource usage data enable farmers to make better decisions, thereby increasing yield production and lowering costs. Predictive analytics can even predict some future conditions, such as an infestation of pests or a shortage of water. This enables farmers to take precautionary measures beforehand. This data-driven approach enables precision in every aspect of farming, from planting to harvesting.

13.3.2 Benefits of Smart Farming

The adoption of smart farming technologies offers numerous benefits, for both farmers and the environment:

Increased Efficiency and Productivity: Data-informed decision-making can empower the farmer. Furthermore, better decisions led to improved end results, indicating more productivity using fewer resources. Precision agriculture stepped in with the correct interventions and the correct amount of water and nutrients at the most opportune moment.

Low Environmental Footprint: Smarter farming uses fewer fertilizers and pesticides, meaning that there is a smaller environmental footprints; it also consumes less water, which the farmers may reduce such risks as soil degradation, together with runoff, by improving long-term soil health and biodiversity using fewer applications when required.

Improved Resource Management: Precision agriculture fully exploits vital resources such as water and energy. In regions with insufficient water resources, smart irrigation can ensure the proper utilization of such elements, resulting in minimal water waste. Smart farming reduces pollution and the potential risk of water pollution since it reduces the application of chemicals.

Cost Saving: Although the initial investment is quite high in the case of smart farming technologies, the savings over time are relatively substantial. By employing advanced management practices, farmers can save inputs, meaning the application of fewer fertilizers, pesticides, and water ate the same time as increasing yields.

Better Decision-Making: Data on planting schedules and pest control real-time strategies are available to farmers. This approach, like the decisions themselves, will thus become much more data-driven and thus better-performed interventions that produce better crop yields or avert losses.

13.3.3 The Challenges of Implementing Smart Farming Technologies

Despite the many advantages of smart farming, there are several challenges to its widespread adoption:

High Initial Costs: The relatively high initial investment into precision agriculture technologies, equipment sensors, and software to top training gives the impression that smart farming is simply out of reach for small farm owners.

Technical Skills: Technical know-how will form the bedrock of the smart farming systems' operations. In this also, there is an expectation that farmers should be trained on use and data interpretation, and on how to assimilate these into daily activities. The implication is that there will be a steep learning curve, even in those places of low digital literacy.

Data Management: The more sensors, drones, and IoT devices employed by the farmer, the greater the volume of data. This requires strong software and technological infrastructures, which may be beyond the reach of small farms.

Connectivity and Infrastructure: The take-up of IoTs and other smart farming technologies will be hampered by the availability of internet connectivity, particularly the primitive kind found in most rural and remote areas. Consequently, if data and communication networks are not readily available, it will be impossible to attain the fullest advantages of precision agriculture.

13.4 FOOD SAFETY AND QUALITY CONTROL TECHNOLOGY

Combining food safety with quality standards can help to produce and process healthy food. Food production businesses employ sophisticated technologies to implement improved safety controls on foods and regulate quality according to sector standards as the expectations of consumers rise and legal frameworks tighten. Today, products are scrutinized from contamination detection to track their production in order to ensure proper assurance that food safety regulatory measures and quality standards are adhered to, from farm to table.

13.4.1 The Role of Technology in Food Safety

Food safety means the prevention of contamination and ensuring that food products are free from pathogens, chemicals, or other hazards that may cause illness. The most frequently used technologies in food safety focus on monitoring, detecting, and controlling the risk at each point of the manufacturing process. One of the basic components of modern food safety systems is the collection and analysis of data in real time to reduce the risk of contamination and enhance the response capability if there is a food safety issue. The major food safety technologies include the following:

Sensors and Real-time Monitoring: Such stages of food processing always need sensors that monitor real-time conditions of temperature, humidity, and even chemical content. To ensure that their products can be kept safely at bacterially safe temperatures, sensors monitor the temperatures in meat processing. Real-time monitoring systems allow food processors to identify real-time issues related to safety and take corrective action right on time.

The Detection of Contamination: Contamination detection technologies include spectroscopy, biosensors, and microbiological testing. It includes all the forms of contaminants like pathogens E. Coli or Salmonella, toxins, or harmful chemicals. Advanced detection systems can easily and quickly detect contaminants. This will save laboratory time and reduce the risk of outbreaks.

UV Light and Ozone Treatment: Excessive UV light and ozone treatments are applied to make surfaces, equipment, and even foods completely free from bacterial, viral, or fungal growth. These technologies provide an additional layer of safety without the need for chemical preservatives, making them ideal for minimally processed or organic foods.

13.4.2 AI and Machine Learning for Predictive Analytics in Food Safety

The efficiencies brought by artificial intelligence (AI) and machine learning change one's approach to food safety risks. Predictive analytics, which gives some indication of a pattern that may signal a food safety issue, allows its use in predictive measures to prevent hazards before they occur from history and environmental data and real-time monitoring data. For instance, AI systems can analyze data on temperature fluctuations, supply chain delays, and processing inconsistencies to predict where and when contamination might occur. As a result, the company can get timely interventions and reduce the occurrences of recalls or outbreaks of food poisoning. The accuracies in the detection of contaminants are enhanced with machine learning algorithms that can learn to pick out subtle patterns, which may not be observed by a human inspector. With this advancement in AI technology, the usage of food safety should increase significantly.

13.4.3 Blockchain for Food Traceability and Transparency

One of the aspects of this approach is blockchain technology, one of the newly emerging mighty tools designed to improve traceability and transparency in the food supply chain. A blockchain is defined as a digital ledger that decentralizes and records transactions securely and in a tamper-proof manner. By implementing blockchain technology into food production, it can garner complete details of every step along the farm-to-fork journey. The blockchain tracking of sourcing, production, processing, and distribution assures that food safety standards are followed in the entire supply chain. In case of a problem relating to food contamination or other food safety issues, blockchain allows both companies and the regulatory authority to trace back easily the source of the problem and whence it originated so that they can recall faster. Customers can obtain information about the origin, handling conditions, and certifications using QR codes or any other digital interface.

13.4.4 Quality Control through Automated Systems

In addition to food safety, another dimension that defines the success of a food company is product quality This focuses on how all food products could be standardized to have an attractive taste, texture, appearance, and nutritional content. Automated systems would ensure that product quality measurements are always consistent and objective and, hence, less prone to human errors.

Vision Systems: Camera and imaging technology are used in automated vision systems where food products are inspected for discoloration, irregular shapes, or packaging defects. This means that only the best quality reaches the consumers since those failing this benchmark reach them. It can be used in sorting fruits and vegetables and also in the production of bread and bakery and meat processing.

X-ray and Metal Detection: X-ray and metal detection systems are used to identify foreign objects, such as pieces of metal, plastic, or glass, that could compromise food safety or quality. These systems are often integrated into production lines to scan products before packaging, reducing the risk of contamination and ensuring product safety.

Nutritional and Chemical Analysis: Nutritional content may immediately be quantitatively measured Near-Infrared (NIR) spectroscopy. However, NIR cannot measure the moisture, fat, protein, and other major nutritional parameters. Thus, products will have labelled content and nutritional claims that will be correct by default.

13.4.5 Challenges in Implementing Food Safety Technology

While these technologies offer significant benefits, there are challenges to their widespread adoption:

Cost: Advanced food safety and quality control might be too expensive for most small and medium-scale enterprises. Purchase and maintenance of the systems could be very costly, and their upgrading may prevent a firm from making use of this technology.

Complexity: An entire food production system integrating multiple technologies, such as AI, sensors, and blockchain, is quite complex. This requires investments in training and technical support required for proper usage.

Data Management: There will be a generation of systems for real-time monitoring and sensor data and blockchain systems. Such data management requires a tremendous data infrastructure with analytical capabilities at a scale that is not readily reachable by smaller companies.

13.5 3D PRINTING IN FOOD PRODUCTION

Originally designed by the manufacturing world to revolutionize the production sector, 3D printing has evolved and continued in food production and innovation in the kitchen. Additive manufacturing is one of the terms used to describe creating 3D objects by depositing successive layers of material based on digital designs. The global market trend for 3D products and services indicates that 3D printing

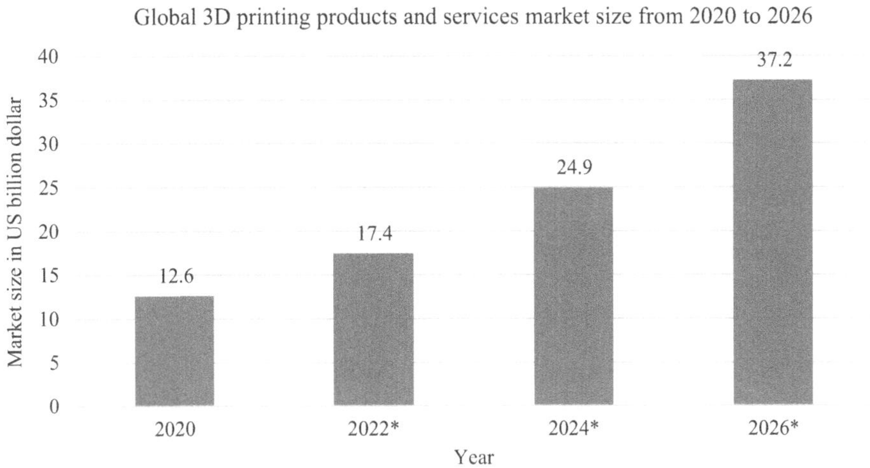

Figure 13.3 Worldwide 3D printing product and service market size (2020–2026).

Source: Statista, 2024.

is rapidly gaining in worldwide popularity (Figure 13.3). Thus, the application of 3D printing can be a significant tool for food marketing to reduce food waste and improve product efficiency.

13.5.1 Applications of 3D Printing in Food

The concept and application of 3D printing are created using "ink" made from edible agents in paste, puree, or powder which can build up intricate food structures layer by layer. This blueprint or digital model is translated into instructions which the printer can execute, extrude, or mould the food material to take on its desired form. 3D printing applications in food production touch all ends of the spectrum-high-end culinary experiences and address global food sustainability challenges.

Customized Culinary Creations: 3D printing allows the chef or the food producer to create elaborately designed models, which would be either impossible or too hard to present using the traditional cooking technique. This technology has been embraced by luxury restaurants to produce visually stunning dishes, chocolates, or pastries that enhance the dining experience. The ability to print complex shapes and textures offers chefs new creative possibilities for artistic food presentation.

Personalized Nutrition: The most promising applications of 3D printing are in the area od personalized nutrition. 3D food printers can produce dishes tailored to the specific needs of individuals or the health conditions from which they suffer. Foodstuffs can be printed with customized levels of nutrients, such as protein, vitamins, or fibre, suitable for children, athletes, or elderly people. These are so customized that they can fill deficiencies or support a health-related diet.

Sustainable Food Production: 3D printing can be taken as a fight against food waste since it uses raw materials like vegetable peels or food powders sourced from surpluses or unused food items for making new food items. Additionally, 3D printers can produce food on demand thereby reducing the need for mass production. The technology might also be developed to convert plant-based or insect-based proteins into alternatives to meat, especially when produced on a sustainable basis.

Space and Military Applications: NASA and its partners are currently studying the possibility of 3D printing food on-site, where other mechanisms to make food is impossible, such as on a space mission. It could provide fresh printed foods tailored to the astronauts' specific nutrient needs, so it was exempted from carrying several kilogrammes of food. Similarly, research in 3D printing is undertaken about the potential of military use providing soldiers with freshly prepared nutritionally balanced meals in out-of-the-way or high-stress situations.

13.5.2 Benefits of 3D Printing in Food Production

3D printing offers several advantages for food production, including:

Customization and Personalization: One of the greatest thrills of 3D printing is that it can print food to fit one's preference or needs. 3D printing allows for personalization that is hard to achieve with traditional methods. It can create dishes with specific nutrient levels or design edible products in unique shapes.

Sustainability: Sustainable food production systems can be supported through 3D printing because alternative ingredients can be used for the reduction of food waste. For example, food printers could be used to create plant-based ingredients or food by-products, thereby minimizing the amount of resource-intensive meat production required.

Reduced Food Waste: The 3D printers will print foods on demand. This will result in less waste such that products are not overproduced. Manufacturers of foods and restaurants only produce what is needed; hence, they have the lowest surplus and the disposal of unsold products in the market.

Efficiency and Speed: 3D printing accelerates the speed at which such manufacturing is done for some food products. This is true for most complex designs, as it will take a very long time to be produced by hand. Again, the 3D printer automates steps, sometimes in food production, making it much more efficient when used in the kitchen or factory.

Consistency: The concept behind 3D printing is to produce food items with precision and consistency, meaning that any product made is carried out precisely according to specifications. This may be very significant in industrial food production, where uniformity in shape, size, and quality plays such an important role.

13.5.3 Challenges of 3D Printing in Food Production

While 3D printing holds significant promise for the food industry, several challenges must be addressed before it can be widely adopted:

High Initial Costs: The cost of 3D printers and food printing materials is high for small-scale production. As would be expected, with further maturation of technology, prices are likely to decline; however, that remains the case now, which does not favour this widespread adoption.

Limited Ingredient: 3D printers are relatively limited in terms of the number of ingredients that can be used with them. The majority of materials for such printers are provided as powders, pastes, or gels. This shuts the door to the use of a wider variety of textures and flavours. Therefore, the production of more versatile materials for 3D printer applications is an important task.

Speed and Scalability: Although 3D printing is rapid in the case of small-scale, tailor-made applications, it remains challenging to scale this technology to mass production. Printing food layer by layer compared to traditional production is slow, making it unsuitable for high-volume printing.

Consumer Acceptance: Consumers will need to adjust to 3D-printed food as a fact. They will tend not to embrace printed foods because they instinctively fear the taste, texture, or unhealthy ingredients used in the printed food by consumers. Time, education, and demonstration of the benefits of the technology will be required to win over perceptions.

13.6 SUSTAINABLE FOOD PROCESSING TECHNOLOGIES

Sustainable food processing technologies will look into the environmental effects of food production, ensuring that the food developed is safe, nutritious, and of good quality. Food companies have welcomed these technologies because of an increase in the demand from consumers for products that are not only environmentally conscious but also save resources, reduce emissions, and ensure sustainability in companies in the long term. Companies also achieve compliance with regulations besides streamlining the production line.

13.6.1 High-Pressure Processing (HPP)

This innovation promises people a safe future in sustainable food processing. This term refers to non-thermal food preservation, wherein high pressure is applied to inactivate the pathogens, extend shelf life, and retain nutritional and sensorial food qualities.

In HPP, food products are packed into flexible, waterproof packaging and moved into a water-filled chamber. The chamber applies intense pressure of as much as 600 megapascals, thus inactivating the microorganisms such as bacteria, yeasts, or moulds that destroy food products or potentially cause food-borne illnesses. The process does not require heat; thus, it does not lose its taste, texture, or nutrients, making it critical to fresh, raw, or minimally processed foods. Among the benefits of HPP are the following:

- HPP significantly extends the shelf life of perishable products without the need for chemical preservatives or additives, reducing food waste.
- Since HPP does not rely on heat, it preserves the nutritional value of foods, making it ideal for health-conscious consumers.
- Compared to traditional heat-based preservation methods, HPP uses less energy, reducing the carbon footprint of food processing.

13.6.2 Minimal Processing Techniques

These methods retain the "naturalness" of foods while simultaneously improving safety and shelf life. Minimally processed food has gained popularity because it retains the freshness, flavour, and nutritional value of food products with lower usage of additives and preservatives.

Cold Plasma Processing: One new non-thermal technology that has more recently been available is Cold Plasma Processing, which applies ionized gases to inactivate surface pathogens. This technology can be applied to fresh fruits and vegetables, killing bacteria and extending shelf life without having any effect on food taste, texture, or appearance.

Pulsed Electric Fields (PEF): This technology uses electric fields in short pulses for food products. These pulses disrupt cell membranes, which can inactivate microorganisms and extend shelf life. Liquid foods, such as juices and milk, are good prospects for PEF treatment because the process would preserve freshness and nutritional value by applying heat pasteurization.

Ultrasonic Processing: Ultrasonic processing involves using sound waves to increase the efficiency of extraction of bioactive compounds, decrease microbial contamination, and modify the texture of the final food product. The technology is also very efficient and sustainable, offering a wide range of possible applications for foods, ranging from fruits and vegetables to dairy products.

13.6.3 Sustainable Packaging Solutions

Packaging remains one of the significant factors in retaining perishable and safe features through the minimization of food waste. However, the worst drawback of conventional packaging materials is their use of plastics on an environmental scale. Therefore, sustainable solutions for packaging that save waste increase recyclability, and decrease carbon emissions, are being developed.

Biodegradable and Compostable Packaging: Many companies are exploring biodegradable ingredients from natural stocks, including corn starch, sugarcane, and cellulose. Biodegradables break down in the ecosystem, but put less pressure on landfills and oceans. This package benefit of compostable packaging is necessary for food products with shorter shelf lives and also for those that are sold in single-use containers.

Edible Packaging: Another innovation in sustainable food processing is the use of edible packaging. Edible packaging materials are made from ingredients, such as seaweed, rice, or protein, which are eaten with the packaged food, hence leading to no waste. The use of edible packaging is still in its experimental phase, but it might become one of the largest contributors to preventing plastic waste, especially among fast foods and snack products.

Minimalist and Lightweight Packaging: Designing limited packages to reduce the materials used can simplify recycling. The use of lightweight packages also saves on transport costs and emissions, as less fuel is consumed in the transportation of lighter products.

13.6.4 Energy-Efficient Food Processing

Energy efficiency is a crucial factor for food processing companies seeking to reduce environmental impact. Sustainable processing methods include the optimization of machinery to use less energy in producing food, improving processing techniques that will cut down on waste energy, and the recovery of waste energy itself.

Heat Recovery Systems: In this process he waste heat generated during food processing steps is recovered and reused. This recovered heat is used to prewarm air, water and other fluids used in the production cycle. It minimizes the amount of total energy required for the entire process.

Solar-Powered Processing: In this scheme the company invests in renewable energy sources to minimize fossil fuel consumption. Electrical power from solar panels generates most of the electricity needed to power its machinery and lowers the carbon emissions and cost of operation.

Energy-Efficient Refrigeration: Energy saving through improved cooling efficiency of refrigerated or frozen products is an excellent scheme. Low-global-warming-potential refrigerants and variable-speed compressors are among the factors that enhance the efficiency of refrigeration technology, leading to the consumption of less energy and resulting in a lower environmental footprint.

13.6.5 Waste Reduction in Food Processing

A central aim of sustainable food processing is waste reduction. A reduction in the amount of waste produced at any stage of production will result in lower costs and a reduction in pollution.

By-product Utilization: One of the principal forms of waste reduction is in the use of food by-products. Organizations can use food waste, peels, or cuts to make other products such as animal feed, biofuels, or food additives. For instance, citrus peels might end up in pectin, a natural gelling agent used in the manufacture of jams and jellies.

Zero-Waste Processing: Some companies have embraced the "zero-waste" processing model, in which an organization extracts every single part of the raw material. This has less impact on the environment less and also increases profitability because every little bit of value is extracted from the raw material.

Water Recycling and Reuse: Water is the most essential resource in the food processing industry. At present, responsible businesses are finding ways to recycle and reuse water within their operations. Treating and reusing wastewater could be the solution to a company's decreased overall water usage, lowering operational costs, and even reducing the on local water sources.

13.7 BIOTECHNOLOGY IN FOOD PRODUCTION

Biotechnology in food production uses biological processes, organisms, or systems to develop and improve agricultural products and foods. It encompasses mainly genetic improvement, molecular markers, and tissue culture for ensuring efficiency, sustainability, and nutritional value in food production. In short, the greatest benefit of the adoption of biotechnology is that crops and livestock become more productive, robust, and nutritious with fewer chemical inputs, such as fertilizers and pesticides.

13.7.1 Genetically Modified Organisms (GMOs)

The best-known application of biotechnology in foods pertains to the use of genetically modified organisms (GMOs). In this process plants or animals are altered through changes in genetic content so that they acquire specific characteristics, such as resistance to pests, tolerance to herbicides, or nutritional enhancement. The advantages of GMOs include:

Increased Crop Yields: These crops are designed to be grown faster, produce more fruits or grain, and are resistant to pests and diseases, thereby ensuring increased yields and uninterrupted food production.

Reduced Pesticide Use: These GMO pest-resistant crops necessitate less use of chemical pesticides. This is seen as environmentally-friendly because fewer chemicals will run off to waterways. It will therefore reduce the impacts on non-target organisms, like pollinators.

Improved Nutrition: GMOs can be designed to have higher levels of essential nutrients, helping to address micronutrient deficiencies in regions where diets may lack variety. For example, genetically engineered Golden Rice contains precursors to vitamin A to counter vitamin A deficiency in the developing world.

Climate Adaptation: GMOs present a significant value in terms of climatic adaptation as they survive extreme weather effects brought by climatic change, such as drought and floods.

13.7.2 Marker-Assisted Breeding

Marker-assisted breeding is another biotechnology tool that has revolutionized crop and livestock improvement. For example, molecular markers are short DNA segments specific to disease resistance and drought tolerance. Isolation and tracing markers enable scientists and breeders to choose the desired traits much more efficiently in plants or animals without carrying out extensive field trials. It accelerates cycle breeding and makes it possible to increase the selection of valuable traits with a higher degree of accuracy. The benefits of market-assisted breeding include:

High Precision: Marker-assisted breeding allows the choice of targeted traits that may enhance the likelihood of a successful crop improvement programme with some desirable traits.

Faster Breeding Cycles: Traditional breeding methods can take years to develop new crop varieties. Marker-assisted breeding shortens this process by identifying and selecting favorable traits earlier in the breeding cycle.

Non-GMO: Since marker-assisted breeding employs nature's genetic variation and does not introduce any form of modification, it's also less anathema in countries where using GMOs is restricted.

13.7.3 Biofortification

Biofortification is a biotechnology-based process aimed at making crops nutritionally enriched. Biofortification may either be genetically modified or traditionally bred or produced through both approaches. Specialists fortify the bioavailability of crops that absorb and retain such nutrients as iron, zinc, and vitamin A. Nutrient-enriched crops can then be consumed as part of an ordinary diet to solve nutrient deficiencies. Examples of biofortified crops include:

Golden Rice: It has been genetically engineered to produce beta-carotene, the precursor of vitamin A. This has removed the deficiency problem due to vitamin A, in areas where rice is a staple food.

Iron-Enriched Beans: Previously fortified with iron in parts of Africa and Latin America, the beans boost the iron levels of its consumers, mostly ravaged by iron deficiency anaemia, a growing health crisis among developing countries.

Zinc Enrichment in Wheat: Zinc-enriched wheat varieties have already been developed to improve the nutritional quality of diets, where wheat is the major staple in most developing countries.

Biofortification appears to be an attractive strategy to combat malnutrition, largely in regions where low-value staple foods constitute a large proportion of diets. Better-fortified staple crops should decrease the malnutrition burden related to micronutrient inadequacies implicated with serious health consequences, including stunted cognitive development and reduced immune strength.

13.7.4 Sustainable Pest and Disease Control

Biotechnology also assists in developing sustainable pest and disease control methods that provide less dependence on chemical pesticides and herbicides. Once more, scientists through biotechnology have developed crops with built-in resistance to common pests and diseases, thereby avoiding bad agrochemicals.

Bt Crops: Bt crops are genetically engineered to produce a toxin from the bacterium Bacillus thuringiensis (Bt), which is toxic to specific pestiferous insects but non-toxic to humans and other animals. Bt crops such as cotton and corn have made chemical insecticide use substantially less necessary, thereby gaining environmental and farmer benefits.

Disease-Resistant Varieties: Another application of biotechnology is in the development of disease-resistant crops. Such crops can protect mankind from devastating plant diseases. It includes varieties of bananas resistant to Panama disease, a fungal infection that threatens banana plantations worldwide, which scientists are trying to develop using genetic modification with marker-assisted breeding.

13.8 ARTIFICIAL INTELLIGENCE AND DATA ANALYTICS IN FOOD PRODUCTION

The use of artificial intelligence (AI) changes food production by optimizing the processes, enabling people to make better decisions or improve efficiency in whatever they're doing. More than that, it brings AI-driven tools and techniques to the food industry to predict market trends, optimize the supply chain, and food safety, and even design new food products. It will provide food companies with access to an enormous volume of data being created at each step of the food production farm to the consumer and then churn it into actionable insights driving better business outcomes.

13.8.1 Applications of AI in Food Production

AI applications in food production are diverse, with benefits ranging from improving operational efficiency to ensuring food safety and driving innovation in product development.

Predictive Analytics for Demand Forecasting: AI-powered predictive analytics systems examine past sales histories, the behaviour of consumers, weather patterns, and market trends in order to predict future demands. With improved accuracy of demand forecasting, food manufacturers can regulate their inventory levels to avoid unnecessary waste and make proper planning of production orders. For example, an AI system would decide how many kilogrammes of fresh produce are needed, given the expected output due to weather prediction and the buying pattern of the consumers, so that companies meet the demands without over-production or under-supply.

Supply Chain Optimization: AI transforms supply chain management through increased transparency and efficiency. Indeed, with AI-powered algorithms, companies can optimize their logistics by figuring out the best transportation routes, managing inventory levels, and identifying possible occurrences of failure. It reduces transportation costs and assures timely delivery, while a minimum of food is not spoiled due to the delays brought about by the supply chain. AI makes possible better-intermeshed supply chain collaboration since it links data from different suppliers, manufacturers, and retailers.

Food Safety and Quality Control: AI is increasingly used to trace food safety and quality at each step of production. Machine learning algorithms can scan sensor data, camera images, and data streams from monitoring devices to detect contaminants, assess defects, and prevent potential failures to meet safety requirements. For instance, vision systems that rely on AI-based vision systems check food products on the production line for signs of bruising or discoloring, thus, only quality products are passed on to consumers.

Automation in Food Processing: At the core of automation in food processing activities is AI, covering the spectrum from sorting and grading to packaging. By fitting sensor and machine learning devices into an AI robot, tasks such as these can be performed with very high precision, thus displacing human labour with much less human error. This can be applied for repeated, high-volume tasks, such as, for example, sorting fruits and vegetables according to their size and quality, or packaging products rapidly, which may prove useful.

Smart Farming and Precision Agriculture: AI can be applied to smart farming to grow crop yields and resource use efficiency. Using AI-based tools, the health of the soil, weather conditions, and crops' growth can be scanned and analyzed to determine the best times to plant, the irrigation system, and the application of pesticides. AI-enabled drones and robots continue to monitor crops to detect pests or diseases and apply the treatment precisely where needed as they optimize water and chemical application.

Food Product Development: The advent of AI has the potential to revolutionize food product innovation purely because companies can now analyze consumer preferences, nutritional data, or even flavour profiles for novel, innovative products. Machine learning algorithms can mine huge volumes of data relating to ingredients, recipes, and consumer feedback for new formulations, or even modifications to existing ones. Artificial intelligence can go so far as assisting companies with healthier food designs by pinpointing precise ingredient substitutions that would not compromise taste but increase reduction in sugar, salt, or even fat.

13.8.2 Data Analytics in Food Production

Data analytics refers to the collection, processing, and analysis of data to derive insights or make better-informed decisions. In the food production context, it tends to enhance supply chain management and improve operating efficiency while attempting to optimize such processes intended for the production of foods. Increases in data from sensors, smart devices, and cloud-based platforms enable food companies to gain a clear view of operations and relate such opportunities for improvement.

Big Data in Agriculture: Big Data is being utilized by farmers and agro-businesses to assist them in making decisions regarding available resources and market strategies for crop management. It is used to keep track of the health of the soil and crop growth to estimate the effect of patterns in weather on crop yield. All this can help the farmer make better decisions in improving the productivity and sustainability of the crops, such as irrigation schedule adjustments or selection of optimal planting dates.

Optimizing Food Production: Data analytics could help an automobile manufacturing company optimize its production lines, decrease downtime, and increase efficiency. Real-time sensors in the machines collect data about their performance. Thus, it can predict equipment failures and perform scheduled maintenance to reduce downtime and significantly increase productivity and profitability. Data analytics could also find bottlenecks in the production process and bring in suggestions for improvement to make the process leaner.

Improving Customer Experience: Data analytics helps food companies understand consumer preferences and, as such, tailor products for growth demand. The data includes sales, social media postings, and reviews that help tell trends and customer feedback on product development, marketing strategies, and pricing. For example, using data analytics, a company can know the favorite flavours or ingredients of a customer segment that can help prepare them with a targeted marketing campaign or rolling out new variants.

Enhancing Customer Experience: The two main areas of data analytics for food companies include the reduction of waste as well as sustainability. The companies make them sustainable by analyzing production efficiency, the logistics in supply chains, and resources used in their production process. For example, through analytics, they can trace all the stages of food production with waste and strategize how to reuse the by-products or reduce losses from spoilage.

13.9 NANOTECHNOLOGY IN FOOD PROCESSING

Nanotechnology uses nanomaterials or nanoparticles with new physical, chemical, and biological properties compared to their bigger counterparts. With them in the food industry, one can establish a new type of unique food product and offer the possibility of developing more effective food processing techniques and food packaging solutions. The most significant applications of nanotechnology include improved nutritional bioavailability for excellent food safety by antimicrobial packaging.

13.9.1 Applications of Nanotechnology in Food Production and Processing

Nanotechnology offers several promising applications in food processing and production, contributing to enhanced food safety, quality, and sustainability.

Nanoencapsulation for Enhanced Nutritional Content: Nanoencapsulation is the encapsulation of nutrients, vitamins, or bioactive compounds in nanostructures for stabilization, thereby enhancing bioavailability and also controlled delivery in living organisms. The delivery of nutrients is controlled through nanoencapsulation technology so that the gastrointestinal system can absorb them more efficiently. By nanoencapsulation, it is possible to fortify foods with required nutrients like vitamins, minerals, or antioxidants without their corresponding taste, texture, and colour being affected. For example, nanoencapsulation could enhance the intake of fat-soluble vitamins such as vitamin D or omega-3 fatty acids, neither of which is absorbed by an organism in a magnitude suitable for the organism's needs in their natural forms. In this regard, food manufacturers can prepare fortified food commodities that may be nutritionally more beneficial.

Nanofiltration for Water and Waste Management: Nanofiltration is a membrane filtration process in which nanoporous membranes filter contaminants from water or other liquids. It can be used in the food industry as a method for water purification, removing unwanted substances traced to food products, and efficient waste management. This technology is fundamentally important to ensure that liquid food products like juices, dairy, and high-quality beverages, are free from bacteria, viruses, and other harmful particles while retaining their essential nutrients and flavours.

Nanosensors for Food Safety and Quality Control: Nanosensors are extremely sensitive instruments that can pick the smallest environmental changes. That makes nanosensors' application well-fitted to observe food safety and quality at every moment of the production/distribution cycle. Generally speaking, they are well-suited to spot pathogens, spoilage, and contaminants in food items so that unsafe or compromised items do not reach the consumer. For instance, doped nanosensors can be embedded into the packaging or coated onto food surfaces for monitoring in real-time points where temperature, moisture, pH, or counts of inhibitory pathogens are required. It improves food safety since it provides an early warning; thus, the chances of contaminated products are reduced to a minimum and there is less of the chance of them going off during transport and storage.

Nanoparticles for Food Preservation: This process introduces antimicrobial nanoparticles into food coatings or packaging materials with the help of nanotechnology for the better preservation of food nanoparticles in food preservation. Such an antimicrobial nanoparticle typically contains silver, copper, or zinc oxide. The latter possesses antimicrobial activity because of its antibacterial properties, thereby inhibiting the growth of harmful bacteria and fungi. It increases the shelf life of perishable food. This technology keeps food fresher. As a result, it helps combat environmental impacts, such as reducing food waste and the period of the perishable's marketability. Additionally, there is scope for using antimicrobial nanoparticles in packing fresh produce, meats, or dairy products that would delay spoilage and decrease spoilage chemicals.

13.9.2 Nanotechnology in Food Packaging

One of the most promising areas of nanotechnology in the food industry sector is in the area of packaging. The application of nanotechnology opens new perspectives in the development of "intelligent," sustainable and functional packaging that would protect food products and ensure safety, extending "shelf life."

Active Packaging: These interact with the food or its environment for the extension of freshness and safety. Nanotechnology allows the creation of films as a result of absorption of moisture, the controlled release of antimicrobial agents, and the control of gas exchange for the inhibition of deterioration. For instance, nanoparticles that are covered in film absorb the excess oxygen; this controls the oxidation reaction rate which might lead to the deterioration of food products.

Smart Packaging: Nanosensor-based tracking of the condition of food products gives real-time information about the freshness, quality, or safety of contents, thereby making another name for smart packaging systems. Such systems are capable of detecting the changes that could be due to temperature, humidity, or microbial conditions and warn consumers or retailers about the risk of a product getting spoiled. For example, intelligent packaging can contain colour-changing indicators that might raise a warning if a product is (or is not) fresh or experienced unsafe conditions during transport or storage.

Barrier Packaging: Barrier packaging further improves the barrier properties of packaging to keep food away from external environments, moisture, and gases. Generally, nanoparticles may be incorporated into packaging films to increase strength and durability, and is thereby an efficient barrier to oxygen, light, and other factors that cause food spoilage. Thus, it ensures the high quality and freshness of the food and reduces dependence on artificial preservatives.

Biodegradable and Sustainable Nanopackaging: Biodegradable and sustainable materials in nanopackaging are of interest in the present scenario. Nanotechnology is focused on biodegradable and compostable packaging materials that break down naturally in the environment. Usually, biopolymers, such as starch or cellulose, are added to reinforce nanoparticles for strength and functionality. Biodegradable nanomaterials help companies move away from single-use plastics to the adoption of more sustainable solutions for packaging over time.

13.9.3 Benefits of Nanotechnology in Food Processing

Nanotechnology offers several key benefits for food production and processing, including:

Enhanced Food Safety: Nanosensors and antimicrobial nanoparticles improve food safety by detecting contaminants and curbing the development of harmful microorganisms. This limits the chances of food-borne diseases, thereby making it possible for the consumer to receive high-quality, safe products.

Extended Shelf Life: Nanotechnology prevents spoilage; hence, it enhances food preservation by keeping the product fresh for extended periods. It reduces wastage and gives fresher products to consumers.

Improved Nutritional Content: Nanoencapsulation enhances the bioavailability of most nutrients; therefore, food manufacturers can fortify most products with vitamins and minerals without necessarily trading off on taste or texture.

Sustainable Packaging: Nanotechnology supports the development of eco-friendly packaging materials that are biodegradable, compostable, or recyclable. This innovation helps to reduce plastic waste and minimize the environmental impact of food packaging.

13.9.4 Challenges of Nanotechnology in Food

While nanotechnology offers significant benefits, it also presents challenges and potential risks that must be carefully considered.

Regulatory and Safety Concerns: One of the major concerns about the application of nanotechnology in food production is the lack of clear regulations regarding the use of nanoparticles in food products and their use in packaging. There is much debate, and the potential health effects from the ingestion of these nanoparticles are not known at this time. Regulatory agencies, such as the U.S. Food and Drug Administration (FDA) and the European Food Safety Authority (EFSA), are working to establish guidelines and standards for the application of nanotechnology in food.

Consumer Acceptance: Just like any other emerging technology, consumer acceptance of nanotechnology-based food products is limited by the risk perception and the lack of familiarity with the new technology. "Consumer education and public awareness of the benefits and risks" would further add to the building of confidence among consumers.

Cost: The development and implementation of nanotechnology into food processing is expensive, at least in the case of small and medium-sized enterprises. Of course, the longer this technology remains under development, the lower its costs will be, but some producers may avoid entry into the process due to very high initial investments.

13.10 CHAPTER SUMMARY

Technology is playing an increasingly critical role in food production and processing, helping to address some of the most pressing challenges faced by the present-day industry. It is rapidly changing at every stage of the food production process. AI and other data analytics technologies are also optimizing supply chains as developments such as 3D printing, smart farming, and food safety systems also help corporations produce more food with fewer negative inputs and reduced ramifications for the environment.

Implementation of green practices, such as high-pressure processing, and efficient energy use will ensure that food production meets the increasing demands of the world without hurting the environment. There have also been other innovations, including nanoencapsulation and smart packaging, along with other increases in the quality and safety of food, nutritional value, shelf life, and reductions in food waste.

While many benefits accompany much of this progress, these are matched by several problems, ranging from high costs to technical expertise and regulatory hurdles. To enjoy the benefits of most of these developments, the food industry will have to invest in research and development, upskill its workforce, and engage with the regulatory authorities to ensure that new technologies are acknowledged to be safe, effective, and accessible to most.

To sum up, technology will spearhead evolving food production systems into a more sustainable, efficient, and safe food supply chain. In embracing these innovations towards the development of a future where food is produced responsibly and in sustainable ways, the industry will have to satisfy the needs of both the consumer and the planet.

CASE STUDY

Automation in Food Manufacturing at Nestlé

Nestlé is the largest food and beverage company in the world, holding an impressive, if not enviable, position in terms of technological innovation. Of course, with ever-rising global demands for its vast lines of products, the company has sought ways to realize greater efficiency in production, better quality in its products, and lower operational costs through the adoption of automation and robotics. The case study that elaborates on how Nestlé has managed to translate and integrate automation into its manufacturing plant by weighing out the scale hurdles as well as flexibility.

Automation in Nestlé's Manufacturing Process

Nestlé uses automation. The high volumes of production developed by this approach are marked by quality alongside food safety standards. Its plants, as such, cover over 400 factories all over the world which manufacture everything from baby food and instant coffee to bottled water and confectionery products. To meet increased demands, the company has, as a result, adhered to the kind of system that is automated for tasks relating to packaging, sorting, quality control, and even warehousing.

Packaging Automation: Nestlé makes use of high-power robotics which packs its products rapidly and with absolute accuracy. The system can be so designed to handle all the major packaging formats: bottles, cartons pouches, and cans. It can therefore accommodate a plethora of consumer preferences. High-speed robotic arms using vision systems sort product lines and pack them with near-absolute accuracy. This almost eliminates the possibility of human error.

Sorting and Quality Control: Automation has had a positive impact on Nestlé's quality control systems. Advanced vision systems scan for defects such as improper sealing or labelling and the presence of objects inside the products. Automated systems can pick out defective goods in real time and reject them prior to letting high-quality products reach the consumers' hands. This creates a method lesser product recall and waste.

Robotics in Warehousing: Using Automated Guided Vehicles (AGVs) and robotic systems enables the easy movement of goods while warehousing. Nestlé employs these robots during the movement of ready products from production lines to points of storage. This reduces human involvement and thereby increases efficiency. Optimizing areas of space and distribution means that Nestlé products will reach the market sooner.

The Benefits of Automation at Nestlé

The adoption of automation has provided Nestlé with several significant benefits, including:

Increased Efficiency: Through the automation of repetitive processes such as packaging and sorting, Nestlé has increased the velocity of production, which means it produces more goods in less time. This efficiency becomes very critical in product lines where demand outstrips supply quickly, particularly with regard to coffee and bottled water products.

Consistency and Quality: With automation, the same product specifications are guaranteed with each manufacturer. This aspect reduces high variability and ensures quality is consistent. Faults are detected at an early stage by automated quality control systems; hence, defects will not reach consumers and destroy the reputation of the company as a quality goods producer.

Cost Savings: The cost of introducing automation technology is high, yet Nestlé has realized long-term cost savings, mostly in the aspect of labour, wastage, and increased production efficiency. Automation has allowed the company to reduce its reliance on manual labour for repetitive tasks, enabling workers to focus on more value-added activities.

Worker Safety: Automation also produces a safer workplace environment by preventing exposure to as many arduous or hazardous jobs as possible. For example, production-floor tasks involving heavy lifting are undertaken by robotic systems, thus reducing injuries.

Challenges Faced by Nestlé in Automation

While automation has brought numerous benefits to Nestlé, the company faces several challenges in implementing and scaling automation across its global operations.

High Initial Costs: The installation and maintenance of automated systems are expensive. Thus, substantial investments are required, especially when installed in developing markets where labour is cheap. The company must determine if automation benefits the company more than the prohibitively high up-front investment in factories.

Flexibility of Product Lines: Automation also suffers inflexibility problems in the product lines. Since Nestlé produces various products in many markets, each may require different types of packaging and regulation requirements. Though automation is highly effective in almost all products, especially high-volume products, it's is not particularly flexible in changing between different product lines or introducing new ones. The company should develop ways to balance the need for efficiency with the flexibility to meet changing consumer preferences and market demands.

Technological Integration: Integrating new automation technologies into existing equipment at old factory sites can be a problem. Therefore, there is a need for the firm to ensure that its automated systems smoothly integrate with its legacy equipment as it trains and equips the workforce to operate and maintain these new technologies.

Workforce Adaptation: As automation systems perform workforce functions, they should adapt to new roles requiring high technical skills, such as running and maintaining automated systems. On this occasion, the company is upskilled in an attempt to try and ensure their employees work harmoniously with the new technologies.

By investing in automation, Nestlé has enhanced efficiency in production by improving the quality output while reducing costs. Implementing automation to core processes such as packaging, sorting, and warehousing capabilities enabled the company to increase significantly the ability to meet global demand levels with high safety and consistency levels. However, it still needs to address the flexibility of technology integration and the adoption of its workforce as it expands its automation of the varied product lines and markets. With the rapid improvement of these advanced automation technologies, Nestlé is uniquely positioned to continue leading players in the food sector by embracing innovation with operations that are efficient, safe, and responsive to market situations.

Discussion Questions

1. What are the primary benefits that Nestlé has gained from integrating automation into its manufacturing processes? How do these benefits contribute to the company's competitive advantage in the global food industry?
2. What are the key challenges Nestlé faces in scaling automation across its diverse product lines and global operations? How can the company address these challenges while maintaining flexibility and efficiency?

3. How does automation impact the workforce at Nestlé, and what strategies can the company implement to upskill employees and ensure a smooth transition as new technologies are introduced?
4. In what ways can automation contribute to sustainability efforts in food production? How might Nestlé leverage automation to reduce waste, energy consumption, and its overall environmental footprint?
5. How can Nestlé balance the high initial costs of automation with the long-term savings and operational benefits it provides? In which regions or product lines should Nestlé prioritize its automation investments?

BIBLIOGRAPHY

Statista. (2022, May 20). *Food robotics: global market unit volume 2020–2030.* https://www-statista-com.eu1.proxy.openathens.net/statistics/1290022/food-robotics-global-market-unit-volume/

Statista. (2023, October 12). *Smart farming market size worldwide from 2021 & 2027, by type.* https://www-statista-com.eu1.proxy.openathens.net/statistics/957271/market-value-smart-agriculture-worldwide-by-type/

Statista. (2024, August 14). *3D printing industry - worldwide market size 2020–2026.*. https://www-statista-com.eu1.proxy.openathens.net/statistics/315386/global-market-for-3d-printers/

CHAPTER 14

Ethical and Social Issues in the Food Industry

LEARNING OBJECTIVES

By the end of this chapter, you should be able to:

- Understand the ethical considerations in food production.
- Evaluate the ethical implications of labour practices in the supply chain.
- Assess the environmental challenges associated with food production.
- Assess food safety and public health concerns.
- Discuss the ethical challenges of marketing practices and the role of corporate responsibility in branding.
- Understand consumer rights and protection.
- Explore cultural and social responsibility in food marketing.
- Understand the regulatory framework governing food marketing.

14.1 INTRODUCTION

The food industry plays a pivotal role in our daily lives, shaping not only our dietary habits but also our health, environment, and societal structures. As the global population continues to grow and industrialize, the ethical and social dimensions of food production, distribution, and consumption have become increasingly significant. This chapter aims to explore the myriad ethical and social issues that permeate the food industry, highlighting the importance of these considerations in fostering a sustainable and just food system.

Ethics in the food industry encompasses a wide array of practices and principles that ensure the welfare of all stakeholders involved, from farm to table. This includes the treatment of animals, the use of agricultural inputs, labour conditions, and the environmental footprint of food production. Ethical food practices are essential not only for safeguarding public health but also for maintaining consumer trust and corporate integrity.

DOI: 10.1201/9781003473305-14

Social issues in the food industry often intersect with ethical concerns, particularly in relation to labour practices, environmental sustainability, and consumer rights. Workers in the food supply chain, from agricultural labourers to factory employees, frequently face challenges such as poor working conditions, low wages, and lack of job security. Moreover, the environmental impact of food production, including pollution, deforestation, and greenhouse gas emissions, poses significant threats to ecosystems and human communities alike.

Food safety and public health are also central to the ethical discourse in the food industry. Ensuring that food products are safe, nutritious, and accurately labelled is paramount to protecting consumer health and fostering informed choices. Transparency in food labelling, honest marketing practices, and swift response to food recalls are all critical components of a responsible food industry.

Furthermore, the marketing and advertising strategies employed by food companies can raise ethical questions, particularly when they involve misleading claims or target vulnerable populations. The industry's responsibility extends beyond profit-making to include the promotion of healthy lifestyles and the prevention of diet-related diseases.

In addressing these complex issues, this chapter will delve into the various ethical and social challenges faced by the food industry, examining both the problems and the progress being made. Through case studies and real-world examples, we will highlight the efforts of companies and organizations that are leading the way in ethical food practices. By understanding these issues in depth, we can work towards a more equitable, sustainable, and ethical food system that benefits all members of society.

14.2 ETHICAL CONSIDERATIONS IN FOOD PRODUCTION

The production of food is fraught with ethical considerations that influence the entire supply chain, from the way crops are grown to how animals are treated. Ethical food production practices are essential for creating a sustainable and just food system that respects the rights and well-being of all living beings involved. This section explores key areas where ethical considerations come into play, including sustainable agriculture, animal welfare, and the use of chemicals and genetically modified organisms (GMOs).

14.2.1 Sustainable Agriculture Practices

Sustainable agriculture is an approach to farming that aims to meet the needs of the present without compromising the ability of future generations to satisfy their own demands. It emphasizes practices that maintain the health of the environment, support the well-being of farmworkers, and ensure economic viability for farmers.

- **Soil Health and Fertility**: Maintaining soil health is fundamental to sustainable agriculture. Practices such as crop rotation, cover cropping, reduced tillage, and organic fertilization help preserve soil structure, enhance biodiversity, and prevent erosion. These methods reduce dependency on chemical fertilizers and pesticides, promoting a healthier ecosystem.

- **Water Conservation**: Efficient water use is critical in agriculture, especially in regions facing water scarcity. Techniques like drip irrigation, rainwater harvesting, and drought-resistant crop varieties help conserve water and reduce the strain on local water resources.
- **Biodiversity**: Promoting biodiversity on farms through polycultures, agroforestry, and integrated pest management creates a resilient agricultural system. Biodiversity enhances pest control, improves soil fertility, and contributes to overall ecosystem health.

14.2.2 Animal Welfare and Ethical Farming

The ethical treatment of animals in food production is a major concern for consumers and advocacy groups alike. Ethical farming practices prioritize the welfare of animals, ensuring they are treated humanely throughout their lives.

- **Humane Housing and Handling**: Ethical animal farming involves providing adequate space, proper shelter, and opportunities for animals to engage in natural behaviours. This includes avoiding overcrowded conditions, ensuring access to clean water and nutritious food, and handling animals with care in order to minimize stress and suffering.
- **Avoidance of Inhumane Practices**: Practices such as debeaking, tail docking, and confinement in gestation crates are increasingly scrutinized for their cruelty. Ethical farms strive to eliminate these practices and adopt more humane alternatives.
- **Transparent Labelling**: Transparency in labelling, such as certifications for free-range, organic, and pasture-raised products, helps consumers make informed choices about the animal welfare standards behind the food they purchase.

14.2.3 Use of Chemicals and GMOs: Ethical Implications

The use of chemicals and genetically modified organisms (GMOs) in agriculture raises significant ethical questions regarding their impact on health, the environment, and social equity.

- **Chemical Pesticides and Fertilizers**: The widespread use of synthetic pesticides and fertilizers has led to concerns about environmental degradation, human health risks, and the sustainability of agricultural practices. Ethical considerations include the potential for groundwater contamination, harm to beneficial insects and wildlife, and the development of pesticide-resistant pests.
- **Genetically Modified Organisms (GMOs)**: GMOs are created through the manipulation of an organism's genetic material to achieve desired traits, such as resistance to pests or improved nutritional content. Ethical debates surrounding GMOs focus on issues of food safety, environmental impact, and the rights of farmers. Critics argue that GMOs can lead to reduced biodiversity, increased use of chemical inputs, and corporate control over seeds.

- **Alternatives and Innovations**: Ethical food production advocates for the development and adoption of sustainable and less harmful alternatives to traditional chemicals and GMOs. This includes organic farming practices, biological pest control, and advancements in agricultural biotechnology that prioritize environmental and social welfare.

By addressing these ethical considerations, the food industry can move towards more responsible and sustainable production practices. This not only benefits the environment and society but also helps build consumer trust and ensures the long-term viability of the food supply chain.

14.3 LABOUR PRACTICES AND WORKING CONDITIONS

The ethical and social dimensions of labour practices and working conditions in the food industry are critical to ensuring a fair and just system for all workers involved. From farm labourers to factory employees, the conditions under which food industry workers operate have significant implications for their health, well-being, and economic security. This section examines key issues such as fair wages, workers' rights, child labour, and health and safety standards.

14.3.1 Fair Wages and Workers' Rights

Ensuring that workers in the food industry receive fair wages and are treated with respect and dignity is a fundamental ethical concern. Many workers in this sector, especially those involved in agricultural labour, face challenges such as low pay, lack of benefits, and job insecurity.

- **Living Wages**: A living wage is essential for workers to meet their basic needs, including food, housing, healthcare, and education. Ethical labour practices advocate for wages that reflect the true cost of living, allowing workers and their families to lead decent lives.
- **Workers' Rights and Unionization**: The right to organize and join unions is a critical component of worker empowerment. Unions can negotiate for better wages, benefits, and working conditions, providing a collective voice for workers. Ethical employers respect these rights and engage in fair negotiations with worker representatives.
- **Contractual Security**: Many workers in the food industry are employed on a seasonal or a temporary basis, leading to job insecurity and a lack of benefits. Providing stable, long-term employment opportunities is key to improving the economic stability of these workers.

14.3.2 Child Labour and Exploitation in the Food Supply Chain

Child labour remains a persistent issue in the food supply chain, particularly in developing countries where economic pressures and lack of education opportunities force children into work at an early age.

- **Eliminating Child Labour**: Ethical food production requires stringent measures to prevent child labour, including adherence to international labour standards and regular audits of supply chains. Companies must ensure that suppliers do not engage in exploitative practices and that children are given opportunities for education and development instead of work.
- **Supporting Families**: Addressing the root causes of child labour involves supporting the economic stability of families. Initiatives such as providing fair wages to adult workers and community development programmes can help reduce the reliance on child labour.

14.3.3 Health and Safety Standards for Workers

Ensuring safe and healthy working conditions is paramount in the food industry, where workers are often exposed to various hazards, including physical injuries, chemical exposures, and ergonomic risks.

- **Occupational Health and Safety (OHS)**: Implementing comprehensive OHS programmes is essential in protecting workers from workplace hazards. This includes proper training, the provision of protective equipment, and regular monitoring of health and safety practices.
- **Ergonomic Improvements**: Many tasks in the food industry involve repetitive motions and heavy lifting, which can lead to musculoskeletal disorders. Ergonomic interventions, such as adjustable workstations and mechanical aids, can help reduce these risks and improve worker comfort.
- **Chemical Safety**: Workers in agriculture and food processing are often exposed to pesticides, fertilizers, and other chemicals. Ensuring proper handling, storage, and disposal of these substances, as well as providing adequate protective gear, is crucial in minimizing health risks.
- **Mental Health and Well-Being**: The mental health of workers is an often-overlooked aspect of workplace safety. High-stress environments, long working hours, and job insecurity can contribute to mental health issues. Employers should offer support programs, counselling services, and promote a healthy work–life balance.

By addressing labour practices and working conditions, the food industry can create a more ethical and equitable environment for its workforce. Ensuring fair wages, safeguarding workers' rights, eliminating child labour, and prioritizing health and safety not only improve the lives of workers but also enhance productivity and foster a positive corporate image.

14.4 ENVIRONMENTAL IMPACT

The environmental impact of the food industry is profound and far-reaching, affecting ecosystems, biodiversity, and the planet's climate. Ethical considerations in food production must address the ways in which agricultural practices, food processing, and distribution contribute to environmental degradation and explore sustainable alternatives. This section delves into key issues such as pollution and waste management, deforestation and land use, and the industry's carbon footprint.

14.4.1 Pollution and Waste Management

Pollution and waste generated by the food industry pose significant threats to the environment. These include water pollution from agricultural runoff, air pollution from greenhouse gas emissions, and the accumulation of food waste in landfills.

- **Water Pollution**: Runoff from farms often carries pesticides, fertilizers, and animal waste into rivers, lakes, and oceans, leading to water pollution and the disruption of aquatic ecosystems. Implementing sustainable farming practices such as buffer strips, cover crops, and integrated pest management can help reduce agricultural runoff and protect water quality.
- **Air Pollution**: The food industry is a significant source of greenhouse gas emissions, particularly methane from livestock and carbon dioxide from food transportation and processing. Reducing emissions through practices such as improved manure management, the adoption of renewable energy sources, and optimizing transportation logistics can mitigate the industry's impact on air quality and climate change.
- **Food Waste**: Approximately one-third of all food produced globally is wasted, contributing to unnecessary resource consumption and greenhouse gas emissions from decomposing organic matter in landfills. Initiatives to reduce food waste include improving supply chain efficiency, promoting consumer awareness, and developing food recovery programmes to redistribute surplus food to those in need.

14.4.2 Deforestation and Land Use

Deforestation and land conversion for agriculture are major drivers of habitat loss, biodiversity decline, and climate change. The expansion of agricultural land often comes at the expense of forests and other natural ecosystems.

- **Deforestation for Agriculture**: Clearing forests to make way for crop production and livestock grazing leads to significant carbon emissions, loss of biodiversity, and disruption of indigenous communities. Sustainable land management practices, such as agroforestry, reforestation, and conservation agriculture, can help balance the need for food production with environmental preservation.
- **Sustainable Land Use**: Ethical land use involves practices that maintain soil health, protect natural habitats, and promote biodiversity. Techniques such as crop rotation, permaculture, and organic farming contribute to sustainable land management and reduce the environmental impact of agriculture.
- **Supporting Indigenous Rights**: Many deforestation activities infringe upon the rights and lands of indigenous communities. Respecting indigenous land rights and incorporating traditional ecological knowledge into land management practices are crucial for ethical and sustainable agriculture.

14.4.3 Climate Change and the Carbon Footprint of Food Production

The food industry is both a contributor to and a victim of climate change. Agricultural activities, food processing, and transportation generate substantial greenhouse gas emissions, while climate change impacts food production through extreme weather events, changing precipitation patterns, and rising temperatures.

- **Greenhouse Gas Emissions**: Livestock production, particularly cattle, is a major source of methane, a potent greenhouse gas. Reducing meat consumption, improving feed efficiency, and adopting methane capture technologies can help lower emissions. Additionally, reducing food miles through local sourcing and optimizing supply chain logistics can decrease carbon dioxide emissions.
- **Climate-Smart Agriculture**: Implementing climate-smart agricultural practices, such as conservation tillage, agroforestry, and climate-resilient crop varieties, helps mitigate the impact of climate change on food production while reducing the industry's carbon footprint. These practices enhance the resilience of farming systems to climate variability and extreme weather events.
- **Renewable Energy and Resource Efficiency**: Transitioning to renewable energy sources, such as solar and wind power, and improving resource efficiency in food processing and distribution can each significantly reduce the environmental impact of the food industry. Investments in energy-efficient technologies and infrastructure are key to achieving sustainability goals.

By addressing the environmental impact of food production, the industry can contribute to global efforts to combat climate change, preserve biodiversity, and protect natural resources. Sustainable practices not only benefit the environment but also ensure the long-term viability of food production systems, supporting the well-being of future generations.

14.5 FOOD SAFETY AND PUBLIC HEALTH

Ensuring food safety and protecting public health are fundamental ethical obligations of the food industry. From farm to table, every step in the food supply chain must adhere to strict standards to prevent contamination, ensure quality, and provide consumers with accurate information. This section examines the importance of food safety standards, transparency in food labelling, and the ethical handling of food recalls and contamination.

14.5.1 Ensuring Quality and Safety Standards

The food industry must comply with rigorous quality and safety standards to protect consumers from food-borne illnesses and other health hazards. These standards cover various aspects of food production, processing, and distribution.

- **Hygiene and Sanitation**: Maintaining high levels of hygiene and sanitation in food processing facilities is crucial to prevent contamination. This includes proper cleaning and maintenance of equipment, regular inspections, and training employees in food safety practices.
- **Hazard Analysis and Critical Control Points (HACCP)**: Implementing HACCP systems helps identify and manage potential hazards in food production. By monitoring critical control points, businesses can ensure that food safety risks are effectively controlled and mitigated.
- **Microbial Testing and Monitoring**: Regular testing for microbial contaminants, such as bacteria, viruses, and fungi, is essential in ensuring food safety. This involves sampling raw materials, finished products, and processing environments to detect and address potential contamination.

14.5.2 Transparency in Food Labelling and Information

Transparency in food labelling is vital for empowering consumers to make informed choices about the foods they purchase and consume. Accurate and honest labelling helps build trust and ensures that consumers have access to important information about the food they eat.

- **Nutritional Information**: Providing clear and comprehensive nutritional information on food labels allows consumers to understand the nutritional content and make healthier choices. This includes listing ingredients, calorie content, and macronutrient breakdowns.
- **Allergen Labelling**: Clearly identifying potential allergens in food products is critical for protecting individuals with food allergies. Labels should highlight common allergens, such as peanuts, tree nuts, dairy, and gluten, to prevent adverse reactions.
- **Country of Origin and Source Transparency**: Indicating the country of origin and sourcing information on labels helps consumers make informed decisions based on ethical and environmental considerations. This includes details about farming practices, fair trade certifications, and sourcing from sustainable fisheries.

14.5.3 Handling Food Recalls and Contamination

Despite best efforts, food contamination can occur, necessitating prompt and ethical handling of food recalls to protect public health and maintain consumer trust.

- **Traceability Systems**: Implementing robust traceability systems enables quick identification and isolation of contaminated products. This involves tracking food products through each stage of the supply chain, from production to retail, to ensure rapid response in case of a recall.
- **Effective Communication**: Transparent and timely communication with consumers, retailers, and regulatory authorities is essential during a food recall. Providing clear instructions on how to identify and return affected products, as well as offering information on potential health risks, helps manage the situation effectively.
- **Corrective Actions and Preventive Measures**: After a recall, it is important to conduct thorough investigations to determine the cause of contamination and implement corrective actions. This includes reviewing and improving food safety protocols, enhancing employee training, and investing in better technology and infrastructure to prevent future incidents.

By prioritizing food safety and public health, the food industry can uphold its ethical responsibilities and foster a culture of trust and accountability. Ensuring high standards of quality and safety, promoting transparency in labelling, and effectively managing food recalls are crucial steps in protecting consumers and maintaining the integrity of the food supply chain.

14.6 MARKETING AND ADVERTISING ETHICS

Marketing and advertising play a crucial role in the food industry, influencing consumer choices and shaping public perceptions about food products. However, these practices must be conducted ethically to avoid misleading consumers and to promote healthy, informed decision-making. This section explores the ethical considerations surrounding marketing and advertising in the food industry, including the prevention of misleading claims, responsible advertising to vulnerable populations, and the importance of ethical branding and corporate responsibility.

14.6.1 Misleading Claims and Deceptive Practices

One of the most significant ethical issues in food marketing is the use of misleading claims and deceptive practices that can confuse or misinform consumers.

- **False Health Claims**: Exaggerating the health benefits of food products or making unsupported claims can lead consumers to make poor dietary choices. Ethical marketing involves providing accurate, evidence-based information about the nutritional benefits and health impacts of food products.
- **Hidden Ingredients**: Failing to disclose certain ingredients, especially those that are artificial or potentially harmful, undermines consumer trust. Transparent ingredient lists and clear labelling are essential for honest marketing.
- **Misleading Packaging**: Packaging that creates false impressions about the nature, quality, or quantity of a product is unethical. For example, packaging that uses images or words suggesting that a product is healthier or more natural than it actually is can mislead consumers. Ethical packaging should accurately reflect the product inside.

14.6.2 Targeting Vulnerable Populations

Certain marketing practices target vulnerable populations, such as children and the elderly, who may be more susceptible to persuasive advertising.

- **Advertising to Children**: Children are particularly impressionable and may not have the critical thinking skills to evaluate advertising claims. Ethical advertising to children should avoid promoting unhealthy food products and should not exploit their naivety. Regulations and guidelines that restrict the marketing of high-sugar, high-fat, and high-salt foods to children are essential.
- **Elderly Consumers**: The elderly may also be more vulnerable to misleading claims, particularly those related to health and wellness. Marketing that targets this demographic should be sensitive to their needs and avoid making exaggerated claims about the benefits of certain foods.

14.6.3 Ethical Branding and Corporate Responsibility

Ethical branding and corporate responsibility go beyond avoiding deceptive practices; they involve actively promoting values such as sustainability, fairness, and transparency.

- **Corporate Social Responsibility (CSR)**: Companies should engage in CSR initiatives that contribute positively to society and the environment. This can include supporting local communities, investing in sustainable agriculture, reducing environmental impact, and ensuring fair labour practices.

- **Sustainable and Ethical Sourcing**: Brands that prioritize ethical sourcing of ingredients demonstrate a commitment to responsible business practices. This includes sourcing from suppliers who follow sustainable practices, uphold labour rights, and minimize environmental impact.
- **Transparency and Accountability**: Brands that are transparent about their sourcing, production processes, and business practices build consumer trust. This involves clear communication about where ingredients come from, how products are made, and what companies are doing to address ethical issues in their supply chains.
- **Engaging with Consumers**: Ethical marketing involves engaging with consumers in meaningful ways, providing education about healthy eating, and encouraging sustainable consumption habits. This can include initiatives like informative advertising campaigns, public health partnerships, and community outreach programmes.

By adhering to ethical marketing and advertising practices, food companies can contribute to a healthier, more informed consumer base and a more just and sustainable food system. Ethical considerations in marketing not only protect consumers but also enhance the reputation and trustworthiness of food brands, leading to long-term success and positive societal impact.

14.7 CONSUMER RIGHTS AND PROTECTION

Consumer rights and protection are fundamental components of an ethical food industry. Ensuring that consumers have access to safe, nutritious, and honestly marketed food is crucial for maintaining trust and promoting public health. This section examines key areas of consumer rights and protection, including access to healthy food, information disclosure and transparency, and addressing food deserts and inequality.

14.7.1 Access to Healthy and Nutritious Food

Access to healthy and nutritious food is a basic human right. The food industry has a responsibility to ensure that all consumers, regardless of their social and economic status, have the ability to obtain and afford healthy food options.

- **Affordability and Availability**: Making healthy food affordable and widely available is essential for promoting public health. This can be achieved through policies and practices that support local food systems, subsidize healthy foods, and reduce the cost of nutritious options.
- **Education and Awareness**: Educating consumers about the benefits of healthy eating and how to make nutritious choices is critical. Public health campaigns, nutritional labelling, and community programmes can help raise awareness and empower consumers to make informed decisions.

14.7.2 Information Disclosure and Transparency

Transparency in the food industry is key to protecting consumer rights. Providing clear and accurate information about food products enables consumers to make informed choices and builds trust in the industry.

- **Ingredient Transparency**: Full disclosure of all ingredients, including additives and preservatives, is essential for consumer trust. Labels should be clear and easy to understand, avoiding technical jargon that can confuse consumers.
- **Nutritional Information**: Providing detailed nutritional information, including calorie counts, macronutrient content, and the presence of vitamins and minerals, helps consumers make health-conscious decisions. This information should be prominently displayed on packaging.
- **Labelling of GMOs and Allergens**: Transparency about the presence of GMOs and potential allergens in food products is crucial for consumer safety. Clear labelling allows consumers to avoid ingredients they are concerned about or allergic to.

14.7.3 Addressing Food Deserts and Inequality

Food deserts, areas with limited access to affordable and nutritious food, are a significant issue that affects many communities, particularly in low-income and rural areas. Addressing these inequalities is a crucial aspect of consumer protection.

- **Improving Access in Food Deserts**: Initiatives to improve access to healthy food in food deserts can include opening grocery stores in underserved areas, supporting farmers' markets, and providing mobile food markets. Policies that incentivize the establishment of food retailers in these areas can also be effective.
- **Support for Local Food Systems**: Strengthening local food systems by supporting small farmers, local food co-ops, and community gardens can enhance access to fresh, nutritious food. These initiatives also contribute to local economies and promote sustainable agriculture.
- **Equitable Food Policies**: Policies that address food inequality should focus on ensuring that all communities have access to healthy food options. This can include subsidies for healthy foods, nutrition assistance programmes, and efforts to reduce food prices in underserved areas.

14.8 PROTECTING CONSUMER HEALTH AND SAFETY

Ensuring that food products are safe and do not pose health risks is a fundamental aspect of consumer protection. This involves strict adherence to food safety standards and prompt action in case of contamination.

- **Food Safety Standards**: Adherence to rigorous food safety standards, including regular inspections and testing, helps prevent contamination and ensures that food products are safe for consumption. Companies must comply with local, national, and international food safety regulations.
- **Handling Food Recalls**: In the event of a food recall, it is essential to act swiftly and transparently to protect consumers. Clear communication about the recall, including information on affected products and steps for returning or disposing of them, helps minimize health risks.
- **Consumer Feedback and Complaints**: Providing channels for consumers to give feedback and report issues with food products is important for continuous improvement and accountability. Companies should take consumer complaints seriously and address them promptly.

By prioritizing consumer rights and protection, the food industry can ensure that consumers have access to safe, nutritious, and honestly marketed food. Addressing issues of affordability, transparency, inequality, and safety not only benefits consumers but also fosters a more ethical and trustworthy food system.

14.9 CULTURAL AND SOCIAL RESPONSIBILITY

Cultural and social responsibility in the food industry involves recognizing and respecting the diverse food practices and traditions of various communities, engaging with these communities meaningfully, and implementing Corporate Social Responsibility (CSR) initiatives that support and enhance their well-being. This section explores how food companies can act ethically and responsibly by focusing on these areas.

14.9.1 Respecting Cultural Food Practices and Traditions

The food industry operates within a diverse global landscape, where cultural food practices and traditions hold significant importance. Respecting and honouring these practices is vital for ethical business conduct and fostering positive relationships with different communities.

- **Cultural Sensitivity in Product Development**: When developing new products, companies should consider the cultural preferences and dietary restrictions of different communities. This includes understanding traditional diets, religious dietary laws (e.g., kosher, halal), and regional food customs. Offering culturally appropriate food options shows respect and inclusivity.
- **Labelling and Packaging**: Clear and respectful labelling that highlights how products align with cultural and traditional practices can enhance consumer trust. For instance, indicating whether a product is suitable for vegetarian, vegan, kosher, or halal diets helps consumers make informed choices that align with their cultural values.
- **Celebrating Food Heritage**: Marketing campaigns and community events that celebrate and promote traditional food practices can strengthen cultural identity and pride. Partnering with local chefs, food historians, and cultural organizations to showcase traditional recipes and cooking methods is an effective way to honour and preserve culinary heritage.

14.9.2 Community Engagement and Support

Engaging with and supporting local communities is a cornerstone of social responsibility. Food companies can positively impact communities by participating in local initiatives, supporting economic development, and addressing specific community needs.

- **Local Sourcing and Employment**: Sourcing ingredients from local farmers and producers supports regional economies and reduces the environmental impact of transportation. Additionally, hiring local employees and providing fair wages and development opportunities helps strengthen community ties and boosts economic stability.

- **Health and Nutrition Education**: Providing education on health, nutrition, and sustainable food practices benefits communities by promoting healthier lifestyles. Companies can partner with schools, community centres, and health organizations to offer workshops, resources, and support on topics such as balanced diets, food safety, and sustainable cooking.
- **Community-Based Programmes**: Supporting community-based programmes, such as food banks, meal programmes, and community gardens, addresses food insecurity and enhances local food systems. Volunteering, donations, and partnerships with non-profit organizations help ensure that vulnerable populations have access to nutritious food.

14.9.3 Corporate Social Responsibility (CSR) Initiatives

Corporate social responsibility (CSR) initiatives in the food industry demonstrate a company's commitment to ethical practices, sustainability, and social well-being. These initiatives should align with broader societal goals and contribute positively to communities and the environment.

- **Sustainability and Environmental Stewardship**: Implementing sustainable practices throughout the supply chain, such as reducing waste, conserving water, and minimizing carbon emissions, demonstrates environmental responsibility. Companies can also support reforestation projects, protect biodiversity, and promote regenerative agriculture.
- **Ethical Sourcing and Fair Trade**: Committing to ethical sourcing and fair trade practices ensures that suppliers adhere to fair labour standards and environmental protection. Supporting fair trade certification for products like coffee, tea, and chocolate ensures that producers receive fair compensation and work in safe conditions.
- **Health and Wellness Programmes**: Developing and promoting products that support health and wellness, such as low-sugar, low-sodium, and fortified foods, addresses public health concerns. Companies can also advocate for policies that improve food safety and nutrition standards.
- **Philanthropic Activities**: Engaging in philanthropic activities, such as donating to charitable organizations, funding scholarships, and supporting disaster relief efforts, reflects a company's commitment to social responsibility. These activities should align with the company's values and address pressing social issues.

By respecting cultural food practices and traditions, engaging with and supporting local communities, and implementing comprehensive CSR initiatives, food companies can fulfil their cultural and social responsibilities. These actions not only enhance the well-being of communities but also build a positive corporate image and foster long-term sustainability and success in the food industry.

14.10 REGULATORY FRAMEWORK AND COMPLIANCE

Navigating the regulatory framework and ensuring compliance with food safety standards are critical aspects of maintaining ethical practices in the food industry. This section explores the international and national food safety standards that govern food practices, the roles of government and non-governmental organizations in overseeing compliance, and the enforcement mechanisms and penalties for ethical violations.

14.10.1 International and National Food Safety Standards

Food safety standards are essential for protecting consumer health and ensuring that food products meet high-quality benchmarks. These standards are established at both international and national levels and provide guidelines for safe food production, processing, and distribution.

- **International Food Safety Standards**: Various international organizations set food safety standards that influence global practices. The Codex Alimentarius Commission, established by the Food and Agriculture Organization (FAO) and the World Health Organization (WHO), develops international food standards, guidelines, and codes of practice. These standards address food safety, quality, and labelling requirements, providing a framework for international trade and consumer protection.
- **National Food Safety Regulations**: Each country has its own regulatory body responsible for enforcing food safety standards. For example, in the United States, the Food and Drug Administration (FDA) and the United States Department of Agriculture (USDA) oversee food safety regulations. In the European Union, the European Food Safety Authority (EFSA) provides scientific advice and risk assessments. National regulations often align with international standards but may include additional requirements specific to local conditions and practices.
- **Harmonization and Compliance**: Harmonizing national standards with international guidelines helps facilitate global trade and ensure consistent safety practices. Companies operating in multiple countries must navigate different regulatory environments, making compliance with both international and national standards essential for market access and consumer trust.

14.10.2 The Role of Government and Non-Governmental Organizations

Government agencies and non-governmental organizations (NGOs) play crucial roles in establishing, monitoring, and enforcing food safety standards. Their efforts help ensure that the food industry adheres to ethical practices and maintains high standards of safety and quality.

- **Government Agencies**: Government agencies are responsible for developing and enforcing food safety regulations, conducting inspections, and overseeing compliance. They set standards for food labelling, ingredient safety, and manufacturing practices. Agencies also respond to food safety incidents, manage recalls, and provide guidance to the food industry.
- **Non-Governmental Organizations (NGOs)**: NGOs often work alongside government agencies to promote food safety and ethical practices. They may conduct research, advocate for policy changes, and provide education and training to industry stakeholders. NGOs also play a role in raising awareness about ethical issues, such as fair trade and sustainable sourcing.

- **Collaboration and Advocacy**: Collaboration between government agencies, NGOs, and industry stakeholders is vital for addressing complex food safety and ethical issues. Joint initiatives and advocacy efforts can drive improvements in food safety standards, enhance transparency, and promote ethical practices across the supply chain.

14.10.3 Enforcement and Penalties for Ethical Violations

Effective enforcement and penalties are necessary to ensure that food industry stakeholders adhere to established standards and practices. Regulatory frameworks include mechanisms for monitoring compliance and addressing violations.

- **Inspection and Monitoring**: Regular inspections and monitoring by regulatory agencies help ensure that food businesses comply with safety and ethical standards. Inspections may focus on production facilities, distribution centres, and retail outlets to assess adherence to regulations and identify potential issues.
- **Penalties for Non-Compliance**: Penalties for ethical violations vary depending on the severity of the infringement and the regulatory framework in place. Common penalties include fines, product recalls, suspension of business licenses, and legal action. Severe violations, such as those resulting in harm to consumers, can lead to more stringent penalties and criminal charges.
- **Corrective Actions and Compliance Programmes**: In addition to penalties, companies may be required to implement corrective actions to address violations and prevent future issues. Compliance programmes, including employee training, internal audits, and improved quality control measures, help ensure that businesses adhere to ethical standards and regulatory requirements.
- **Consumer Protection Measures**: Protecting consumers is a key focus of enforcement efforts. Regulatory agencies may provide resources and support for consumers to report food safety concerns, access information about recalls, and seek redress for issues related to unethical practices.

By adhering to international and national food safety standards, engaging with government and non-governmental organizations, and implementing effective enforcement and compliance measures, the food industry can uphold ethical practices and ensure the safety and quality of food products. This regulatory framework supports consumer trust, facilitates global trade, and contributes to the overall integrity of the food supply chain.

14.11 CHAPTER SUMMARY

The ethical landscape of the food industry is complex and multifaceted, encompassing a wide range of issues from sustainability and transparency to cultural respect and consumer protection. As the industry evolves, the importance of adhering to ethical practices and complying with regulatory standards becomes increasingly crucial for maintaining trust, ensuring safety, and promoting positive societal impact.

This chapter has explored key areas of ethical responsibility, including:

- **Sustainability and Environmental Stewardship**: The need for food companies to adopt sustainable practices, reduce environmental impact, and support ethical sourcing is essential in addressing global challenges such as climate change and resource depletion.
- **Transparency and Labelling**: Clear and accurate labelling, along with transparent practices, empowers consumers to make informed choices and fosters trust in food products and brands.
- **Cultural Sensitivity and Social Responsibility**: Respecting cultural food practices, engaging with local communities, and supporting initiatives that address social and economic inequalities are vital for building positive relationships and contributing to community well-being.
- **Marketing Ethics**: Ethical marketing practices, including avoiding misleading claims and promoting products responsibly, are crucial for maintaining consumer trust and promoting healthy choices.
- **Consumer Rights and Protection**: Ensuring that consumers have access to safe, nutritious, and honestly marketed food is a fundamental aspect of ethical business conduct.
- **Corporate Social Responsibility (CSR)**: CSR initiatives that focus on sustainability, community support, and ethical business practices contribute to a company's positive impact on society and the environment.
- **Regulatory Framework and Compliance**: Adhering to international and national food safety standards, engaging with regulatory bodies and NGOs, and implementing effective enforcement measures are critical for ensuring compliance and addressing ethical violations.

As the food industry faces ongoing challenges and opportunities, the commitment to ethical practices will play a significant role in shaping its future. Companies that prioritize ethical responsibility, embrace innovation, and actively engage with stakeholders will not only enhance their reputation but also contribute to a more sustainable, fair, and transparent food system.

At last, it can be said that the path to a more ethical food industry requires continuous effort, collaboration, and dedication. By upholding ethical principles and striving for excellence in all aspects of food production and marketing, companies can lead by example and make a meaningful difference in the lives of consumers and communities worldwide.

CASE STUDY

Ethical and Social Responsibility at Green Harvest Foods

Green Harvest Foods is a global food company specializing in organic and sustainable products. The company has gained recognition for its commitment to environmental sustainability and social responsibility. Recently, Green Harvest Foods has been involved in several initiatives aimed at addressing ethical issues in the food industry, including improving transparency in labelling, respecting cultural food practices, and engaging with local communities.

Company Background

The goal of Green Harvest Foods, which was established in 2010, is to provide organic, healthful food alternatives while maintaining a sustainable supply chain. The firm supports fair trade principles and makes community investments by sourcing its raw materials from small farmers. These days, Green Harvest distributes a variety of goods across Europe and North America, including organic dairy, fruits, and vegetables.

Scenario

Green Harvest Foods launched a new line of plant-based protein products designed to appeal to health-conscious consumers and those seeking sustainable food options. The company advertised the products as being "100% organic," "ethically sourced," and "culturally sensitive." However, several concerns have emerged:

Labelling Accuracy**:** Some consumers have questioned the accuracy of the product labels, particularly regarding the "100% organic" claim. There have been reports that the products may contain ingredients that do not meet organic certification standards.

Cultural Sensitivity**:** The plant-based protein products are marketed as suitable for various cultural diets. However, some cultural groups have expressed concerns that the products do not fully respect traditional dietary practices or adequately address religious dietary laws.

Community Engagement**:** Green Harvest Foods has also been involved in community engagement programmes, such as supporting local farmers and funding educational initiatives. While these programmes have been praised, there are concerns about the effectiveness and transparency of these initiatives.

Discussion Questions

1. What steps should Green Harvest Foods take to address concerns about the accuracy of their product labelling?
2. How can the company ensure that its "100% organic" claim is supported by evidence and compliant with certification standards?
3. What role does transparency play in maintaining consumer trust, and how can Green Harvest Foods improve transparency in its labelling practices?

BIBLIOGRAPHY

Costa, R., & Pittia, P. (Eds.). (2018). *Food Ethics Education*. Cham: Springer. https://doi.org/10.1007/978-3-319-64738-8

Index

Pages in *italics* refer to figures.

C

E

F

K

L

M

S

T

U

V

W